Applied Bioinformatics and Biostatistics in Cancer Research

Series editors: Jeanne Kowalski, Steven Piantadosi

For further volumes:
http://www.springer.com/series/7616

Wan Tang • Xin Tu
Editors

Modern Clinical Trial Analysis

Springer

Editors
Wan Tang
Department of Biostatistics
and Computational Biology
School of Medicine and Dentistry
University of Rochester
Elmwood Ave. 601
Rochester, New York, USA

Xin Tu
Department of Biostatistics
and Computational Biology
School of Medicine and Dentistry
University of Rochester
Elmwood Ave. 601
Rochester, New York, USA

ISBN 978-1-4899-8778-5 ISBN 978-1-4614-4322-3 (eBook)
DOI 10.1007/978-1-4614-4322-3
Springer New York Heidelberg Dordrecht London

Printed on acid-free paper

Springer is part of Springer Science+Business Media (www.springer.com)

Preface

With advances in medical sciences, cancer research no longer focuses primarily on effective treatments, but rather a host of other issues such as tools for early diagnosis, cost of treatment and long-term care, and quality of life. Thus, in addition to traditional survival analyses for comparing treatment differences, modern clinical trials in cancer research are also designed to address these new emerging issues. This volume covers classic as well as cutting-edge topics on the analysis of clinical trial data in biomedical and psychosocial research and discusses each topic in an expository and user-friendly fashion. The intent of the book is to provide an overview of the primary statistical and data analytic issues associated with each of the selected topics, followed by a discussion of approaches for tackling such issues and available software packages for carrying out the analyses.

Some of the topics covered are quite standard, such as survival data analysis and longitudinal data. Although in-depth discussions of such classic topics can be found in various topic-specific texts, our coverage highlights their specific and important roles in clinical trials. Further, by presenting the topics in a self-contained fashion, the materials become more assessable by researchers in other disciplines, particularly clinicians and medical investigators who seek a "crash course" to understand the "nuts and bolts" of clinical trials.

With new medical discoveries, advances in technology and treatment delivery, rises in health-care cost, and emphasis on quality of life and patient-centered care, clinical trials have become increasingly complex in design to address all these concerns. For example, in most randomized controlled trials, patients are randomized to a particular treatment at baseline, regardless of whether the treatment is optimized for the patient. Although it is a necessary step to study the efficacy of a treatment, this traditional approach does not fully reflect how patients are treated in practice, when multiple treatments are generally used either sequentially or in combination to effectively treat a patient. In fact, dynamically adjusting treatment in accordance with the patient response to the previously assigned treatment is the only viable option for treating many mental disorders. Some of the chapters are devoted to addressing such cutting-edge issues to reflect advances in statistical methodology for clinical trials.

As a series to provide an overview of the core concepts of clinical trials and a guide to statistical methods for analyzing data from such studies, the chapters are organized in an order following the logical considerations of the issues arising from the design to the execution of clinical trials. Thus, we start with two chapters focusing on the classic topics of survival and longitudinal data analysis. Treatment evaluations in clinical trials generally center on two types of outcome. If the study involves patients with terminal illnesses such as advanced cancers, the duration from a certain time point such as the initiation or termination of the treatment is often of primary interest. Survival analysis models are uniquely suited to comparing survival times between different treatment groups.

In most clinical trials, treatment differences are evaluated by comparing changes in an outcome of interest over time such as tumor volume in cancer studies between different treatment groups. Longitudinal models are applied to facilitate comparisons of such temporal changes. These models extend classic methods for cross-sectional data to address the within-subject correlations in the repeated assessments of the individual and missing data due primarily to premature dropout by study subjects. Chapter 2 discusses these distinctive features of longitudinal data and associated models.

Valid inference not only replies on correct statistical models, but on quality and fidelity of outcomes of clinical trials as well. Although clinical trials typically use measures with established fidelity and reliability, it is important to have some level of understanding of the inner workings of the process to develop and validate such measures, especially for outcomes derived from measures of latent constructs such as quality of life. Chapter 3 provides an overview of such measurement error issues and methods to address them.

Upon settling down on the measures of treatment effect, the next step is to decide on the length of the study. In particular, we may want to know whether we can expedite the trial as soon as evidence of treatment efficacy emerges, especially for treatments with adverse reactions and side effects and studies with serious deterioration of health and fatality outcomes. Chapter 4 discusses dynamic decision rules to stop a trial as soon as there is indication of treatment difference.

As noted earlier, the standard protocol in clinical trials is to randomize patients at baseline, which, although a necessary step to study the efficacy of a single treatment, does not reflect real clinical practice. Also, with new discoveries on genetic linkage in disease predisposition and treatment response and recent emphasis on patient-centered outcome research, this standard treatment protocol does not meet the needs of the new patient-specific treatment and care model. Chapter 5 focuses on this new person-centered treatment approach by dynamically adjusting treatment in accordance with the patient response to the previously assigned treatment. The dynamic treatment regime, which continuously adjusts treatment type and dosage, is particularly effective for effectiveness research, because of the diverse range of patients' conditions and disease progression in such studies.

As the cost for research in developing and delivering new treatment becomes increasingly high, health care has become quite expensive, especially in the USA. In recent years, more and more clinical trials have included a cost–effectiveness component to also examine the cost for the new intervention. The cost–effectiveness analysis allows one to see if the added benefit is worth the increased cost and how to maximize such benefit-to-cost margins for a population of interest. Chapter 6 is devoted to addressing these issues.

In most clinical trials, we are interested in establishing the superiority of the new intervention over existing or conventional treatment. But, in some cases, we may be interested in equivalence between a new and conventional treatment, such as in replacing costly name brand medications with less expensive generic alternatives, and a simplified instrument for diagnosis of disease. More importantly, such equivalence tests are employed in early drug development to assess the potential of drug induced, prolonged duration of ventricular depolarization and subsequent repolarization, or QT interval, as the duration is derived from the interval of ECG tracing from the beginning of Q wave to the end of T wave. For some drugs, significant prolongation of the absolute QT interval has been associated with the precipitation of a potentially fatal cardiac arrhythmia and can degenerate into ventricular fibrillation, leading to sudden cardiac death. To ensure drug safety, thorough QT (TQT) trials are recommended by FDA drug regulatory requirements to assess the treatment response of a new drug and ensure that it does not induce prolonged QT intervals. Thus, the primary object of the TQT is to demonstrate equivalence, rather than superiority as in most clinical trials.

In Chap. 7, we first discuss the fundamental issues arising from the paradigm shift from superiority to equivalence and methods for addressing them under this alternative inference paradigm for equivalence. We then turn our attention to the design and analysis of TQT trials by applying the models for equivalence in Chap. 8.

Randomized controlled clinical trials rely on randomization, the hallmark of modern clinical research, to deliver valid conclusions regarding treatment differences. In some studies, it may not be possible to conduct such a trial. For example, it is clearly unethical to contemplate a randomized controlled trial to study the effect of smoking on lung cancer by randomizing subjects to a smoking group. In some other studies, decision to treat may also depend on the health condition of a subject, in which case treatment assignment is no longer random and treatment differences cannot be evaluated as in randomized trials. Chapter 9 discusses issues in assessing treatment difference in such nonrandomized trials and methods for addressing them to enable valid inference.

We conclude this series with a chapter discussing the opportunities and challenges that lie ahead in developing on person-centered treatment regimens. The advances in cancer biology and the genetics of cancer have rapidly provided us with a better fundamental understanding of cancer. These new developments require a new generation of clinical trials that modernize the processes and methods used to examine the safety and efficacy of novel, gene-based therapies without sacrificing high standards. Chapter 10 revisits some of the basic components of clinical trial design within the context of timely areas of vaccine trials, cancer stem cell trials, and trials of epigenetic targeted therapies.

This book is intended for biostatisticians with special interest in cancer research or medical researchers with some background in biostatistics such as a working knowledge of clinical trial designs and regression analysis. Since the authors for all the chapters are experienced in modern clinical trial data analysis and are at the forefront of their respective areas, this book should enable them to quickly apply these methods to their own studies, especially considering the fact that most chapters contain illustrative real study data and associated software.

We would like to express our appreciation to all who have contributed to this book. We are also thankful to editors Fiona Sarne and Rachel Warren for their patience and continuing support, despite multiple delays on the project on our part.

Contents

Chapter 1
Survival Analysis

Changyong Feng, Yu Han, Pan Wu, and Xin Tu

1.1 Introduction

Survival analysis concerns the time from a well-defined origin to some end event, such as the time from surgery to death of a cancer patient, the time from wedding to divorce, time from layoff to finding a new job, and time between the first and second suicide attempts. Although originated in and driven much by research on lifetime, or survival of an object such as light bulbs and other electric devices in the early days, modern applications of survival analysis include many non-survival events such as the aforementioned examples. Thus, survival analysis may be more appropriately called the *time-to-event analysis*. However, in this chapter we continue to use the classic term *survival analysis* in our discussion of this statistical model and its applications.

Like all other statistical methods, survival analysis deals with random events such as death, divorce, and layoff. Thus, the time to event is a nonnegative random variable and usually right skewed because most events tend to occur in close proximity to each other after being observed for a while. In survival analysis we are primarily interested in: (i) the distribution of survival time, and (ii) the influence on such a distribution by some explanatory variables such as age and gender. In some applications, subjects may be clustered because of certain common features such as genetic traits and shared environmental factors. Clustered survival time data also arise from analysis involving multiple events from the same individual such as repeated suicide events. Note that although estimates of model parameters, or point estimates, obtained from the standard likelihood or partial likelihood-based

C. Feng (✉) • Y. Han • P. Wu • X. Tu
Department of Biostatistics and Computational Biology, University of Rochester, Rochester, NY 14642, USA
e-mail: feng@bst.rochester.edu

W. Tang and X. Tu (eds.), *Modern Clinical Trial Analysis*, Applied Bioinformatics and Biostatistics in Cancer Research, DOI 10.1007/978-1-4614-4322-3_1,

inference are still valid for clustered survival data, estimates of standard errors and associated confidence intervals are generally not requiring additional considerations to address the correlated outcomes (see Sect. 1.6).

Note that the events of interest are not necessarily negative in nature such as the aforementioned events. For example, in a weight-loss program, it may be of interest to study the length of time for an overweight person to reduce weight to a desired level. The survival methodology may even be applicable to outcomes not involving time, but share the properties of time such as space.

The most significant property of survival data is the incomplete observations, with the most prevailing case being the *right censoring*, where the survival time is only known to be beyond the observation period. Many other mechanisms can also cause incomplete observations. For example, truncation arises quite often in epidemiologic studies, when only a portion of the study population is available. For example, in the early years of the AIDS epidemic, interest is centered on estimating the latency distribution between HIV infection and AIDS onset. Data from Centers for Disease Control and Prevention (CDC) and other local (state health departments) surveillance systems are used for this purpose. Since the time of HIV infection is usually unknown due to the lack of screening for HIV during this period, only those of the infected individuals who came down with AIDS symptoms are captured by the surveillance system. Because of the long duration of the latency period (mean is about 10 years), and the relatively short time span covered by the surveillance database, the AIDS subjects in the surveillance systems during the early years represent a sample from the subgroup of the patients' population whose latency times fall within the surveillance time frame. In this chapter we focus on right-censored survival data.

Although methods for survival analysis have been developed and used in actuarial sciences long before their applications in modern medical sciences, the first mathematically rigorous approach for estimating the distribution of survival time or *survival function* did not appear in the literature until the publication of Kaplan and Meier (1958) who formalized the *product-limit estimator*, also known as the *Kaplan–Meier estimator*. Although the development of *Nelson–Aalen estimator* (Nelson 1969, 1972; Aalen 1978) of the *cumulative hazard function* is pretty independent of the Kaplan–Meier estimator of the survival function, the *product integral* (Gill and Johansen 1990) makes a natural connection between the survival function and the cumulative hazard function. The study of the asymptotic properties of these two estimators can be unified under the product-integral and the functional δ-method (Andersen et al. 1993).

Because of censoring, popular nonparametric methods such as the Kolmogorov–Smirnov test, the Wilcoxon rank-sum test, and the Kruskal–Wallis test are not applicable directly. The *log-rank test* (Mantel 1966; Peto and Peto 1972) is the most popular method in survival analysis to compare the distributions in different groups due to its efficiency under the proportional hazards. Another seminal work in the development of modern survival data analysis is the *proportional hazards model* introduced by Cox (1972). Instead of modeling the mean response of the dependent variable as in traditional regression analysis, the Cox regression models the hazard function of the survival time without specifying the baseline hazard using an analytic function, a ground-breaking approach that not only pushed survival analysis to a new level, but also laid the foundation for modern *semiparametric models* in statistics.

As survival data present some unique features and issues, understanding and addressing these distinctive features of survival data are crucial for modeling such data. In Sect. 1.2, we describe the unique features of survival data. In Sect. 1.3, we discuss models for survival data assuming a homogeneous sample. In Sect. 1.4, we discuss regression analysis to accommodate covariates.

In the development above, the survival times from different individuals are assumed to be independent. Some times we may have observations of clustered data (e.g., survival times of family members) or times of multiple events of the same individuals. The frailty model (Vaupel et al. 1979; Oakes 1989; Hougaard 2000) plays a central role in the multivariate survival data.

1.2 Unique Features of Survival Data and Models

The most notable phenomenon in survival analysis is *censoring*. In order to observe the occurrence of the event of interest, the time frame of the observation must be sufficiently long to contain the time when the event, or failure, occurs. However, it is generally not possible to have such an extended time frame to observe the events for all the subjects, due primarily to logistics and cost considerations. For example, many clinical trial studies last two to five years because of considerations of logistics and cost constraints, advances in knowledge and availability of new medication and treatments. If the event does not occur when the study is terminated, then the survival time is not observed or censored. Censoring may also occur if a subject withdraws from the study before the event occurs for treatment, logistics, and related reasons.

The occurrence of event of interest is often called *failure* in the nomenclature of survival analysis. This is because the issue of censoring initially arises from the analysis of life-testing data to determine the life expectancy of certain objects such as light bulbs. Because the life of such objects is typically longer than the observation time as determined by logistics and cost considerations, even under unusual and extreme conditions to accelerate the failure of the object, not all objects will fail at the end of life testing, yielding (right) censored failure times.

1.2.1 *Hazard Rate Function*

Let T (≥ 0) be the survival time of interest, and $F(t) = \Pr(T \leq t)$ denote the (cumulative) distribution function of T for $t \geq 0$. Since primary interest in survival analysis centers on the likelihood of surviving beyond a point in time t, it is more convenient to use the *survival function* $S(t) = 1 - F(t) = \Pr(T > t)$ to describe the distribution of T.

The survival function can only be used to describe the cumulative survival experience. In practice, we may be more interested in the instantaneous failure rate at each time point. Given the individual known to be alive immediately before t, the probability that event will occur within the infinitesimal interval $[t, t + \mathrm{d}t)$ is $\Pr\{t \le T < t + \mathrm{d}t \mid T \ge t\}$.

If F (and hence S) is absolutely continuous, the probability density function $f = F'$ exists. In this case, the *hazard rate function* λ is defined as

$$\lambda(t) = \lim_{\Delta t \to 0} \frac{1}{\Delta t} \Pr\{T \in [t, t + \Delta t) \mid T \ge t\} = \frac{f(t)}{S(t)}.$$

Note that $\lambda(t)$ is the instantaneous failure rate given that the individual is still alive at t and $\lambda(t)\mathrm{d}t = \Pr\{T \in [t, t + \mathrm{d}t) \mid T \ge t\}$. The function $\Lambda(t) = \int_0^t \lambda(u)\mathrm{d}u$ is called the *cumulative (or integrated) hazard function* and it's easy to see that $S(t) = \exp(-\Lambda(t))$. If F is not absolutely continuous, the hazard rate function cannot be defined. However, we can still define the cumulative hazard function Λ by

$$\Lambda(t) = -\int_0^t \frac{\mathrm{d}S(u)}{S(u-)}, \tag{1.1}$$

which is the theoretical basis of Nelson–Aalen estimator. The survival function is the product-integral of the cumulative hazard function, i.e.

$$S(t) = \prod_{[0,t]} (1 - \mathrm{d}\Lambda(u)). \tag{1.2}$$

We will not go into the details of product-integral. Please refer to Gill and Johansen (1990) for more details. Equation (1.2) is the most general duality between the survival function and cumulative hazard function. Specially, if Λ is a step function, then

$$S(t) = \prod_{u \le t} [1 - \Delta\Lambda(u)], \tag{1.3}$$

where $\Delta\Lambda(u) = \Lambda(u) - \Lambda(u-)$, the jump of Λ at time u. Equation (1.3) is the theoretical connection between the Nelson–Aalen estimator and Kaplan–Meier estimator.

1.2.2 Independent Censoring

Let C denote the censoring time, distributed according to G. For censored survival time we only know that $T > C$. Let $\Delta = \mathbf{1}\{T \le C\}$ be the failure indicator, which is 1 if the failure has been observed and 0 otherwise. The observable data can be represented by: $(\tilde{T} = \min(T, C),\ \Delta)$. One convention in survival analysis is that if both censoring and failures occur at the some time t we assume that the failure precedes censoring, with the censoring time denoted by $t+$ (Kalbfleisch and Prentice 2002, pp 108–109).

If there is no censoring, the empirical distribution function

$$\hat{F}(t) = \frac{\#\text{ of failure times} \leq t}{n}$$

is a uniformly consistent estimator of the true distribution function F by Glivenko–Cantelli theorem (Dudley 1999). This means the distribution function can be identified from the data and reliably estimated. In the presence of censored observations, such an estimator cannot be constructed. An alternative is to apply this estimator to the subgroup with observed failures. However, since the censoring mechanism may interact with the event of failure, introducing confounding, the latter ad hoc approach generally yields biased results. However, in survival analysis, the censoring reduces the information available to estimate the distribution of T. Can we still identify the distribution from the survival data, at least within some period of time? To solve this problem, we first discuss the underlying model structure.

There are three different models we need to deal with in survival data:

(i) A model for F without censoring, i.e., all occurrence of the event of interest are observed
(ii) A joint model for the (hypothetical) situation where all occurrences of the event of interest as well as the censoring processes are observed
(iii) A model for the situation with censoring, i.e., for the occurrences of the event actually observed

These three levels of model offer us different information. If there is no censoring, we only need information from model (i) to make statistical inference. All parameters of interest are defined according to model (i). Model (iii) is what we can observe and can be used in the actual statistical inference. The observations of censoring process in model (iii) may change our vision of failures in the future.

The *independent censoring* is the right condition we need to identify F from censored survival data. The right censoring is said to be *independent* if individuals who are still at risk $\tilde{T} \geq t$ have the same risk of having failures in the infinitesimal time interval $[t, t + \mathrm{d}t)$ as those that would have applied without censoring. In probability language, the independent censoring is defined as

$$\Pr\{T \in [t, t + \mathrm{d}t) \mid T \geq t\} = \Pr\{\tilde{T} \in [t, t + \mathrm{d}t),\ \Delta = 1 \mid \tilde{T} \geq t\}. \qquad (1.4)$$

Note that the independence censoring does not mean the censoring time C and the survival time T are independent although (1.4) is always true if they are independent.

The concept of independent censoring is defined based on model (ii). It means that the observations of censoring processes will not offer more information about the instantaneous failure. In a clinical trial, if a patient is censored because the patient's condition is so bad and cannot stay in the current treatment any more, the censoring is no independent as the patients still in the trial do not represent the patients without censoring.

There are several equivalent ways to characterize the independent censoring (Fleming and Harrington 1991, Theorems 1.3.1 and 1.3.2; Gill 1980, Theorem 3.1.1). Aalen 1978, p 57; ABGK). For a rigorous definition of independent censoring of counting processes, please refer to Andersen et al. (1993).

We also say that we have a *random censorship model* (RCM) if the censoring is independent. Note that in RCM model, the survival time and the censoring time are not assumed to be independent. See the comment of Fleming and Harrington (1991) after their Theorem 1.3.1.

The censoring is a special case of missing data. Little and Rubin (2002) defined very general missing data patterns. The independent censoring is the same as sequential *missingness at random* in longitudinal data analysis (Hogan et al. 2004).

There is another concept which is called *noninformative censoring* in survival analysis literature. Many people regard independent censoring and noninformative censoring as the same (e.g., Lu and Tsiatis 2008). In fact, they are totally different concepts. In noninformative censoring, the distribution of the censoring times is assumed to not contain the parameter(s) in the distribution of survival time. Andersen et al. (1993) give an example of independent but informative censoring.

1.3 Estimation of Survival Distribution

As seen in (1.2), failure times are commonly characterized by the survival and hazard functions. As a result, methods are available to estimate these functions for homogeneous data under independent censoring.

The *Nelson–Aalen estimator* is used to estimate cumulative hazard function in survival data (Nelson 1969, 1972; Aalen 1978). There are many ways to derive this estimator. In this section, we drive it from the definition of independent censoring.

1.3.1 The Nelson–Aalen Estimator of Cumulative Hazard Function

Consider a study with n individuals. The observed data are given by

$$\tilde{T}_i = \min(T_i, C_i),\ \Delta_i = \mathbf{1}\{T_i \le C_i\},\ \ i = 1, \ldots, n.$$

For each ith subject, let $N_i(t) = \mathbf{1}\{\tilde{T}_i \le t,\ \Delta_i = 1\}$, $Y_i = \mathbf{1}\{\tilde{T}_i \ge t\}$. Hence $N_i(t)$ defines an indicator for observed failure for individual i for at time t, while $Y_i(t)$ indicates if the individual is still under observation, or at risk, immediately before time t. These variables N_i and Y_i are step functions of time and are known as the counting and at-risk processes in survival literature, respectively. Thus, the sum $Y.(t) = \sum_{i=1}^{n} Y_i(t)$ and $\mathrm{d}N.(t) = \sum_{i=1}^{n} \mathrm{d}N_i(t)$ represent the number of individuals still under observations immediately at t and the number of failures occurring within the infinitesimal interval $[t, t + \mathrm{d}t)$.

From the definition of independent censoring (1.4) we have

$$\mathrm{d}\Lambda(t) = \Pr\{\tilde{T} \in [t, t+\mathrm{d}t),\ \Delta = 1 \mid \tilde{T} \geq t\} = \frac{\Pr\{\tilde{T} \in [t, t+\mathrm{d}t),\ \Delta = 1\}}{\Pr\{\tilde{T} \geq t\}}.$$

Note that $\Pr\{\tilde{T} \in [t, t+\mathrm{d}t), \Delta = 1\}$ and $\Pr\{\tilde{T} \geq t\}$ can be estimated by $\mathrm{d}N.(t)/n$ and $Y.(t)/n$, respectively. Hence $\mathrm{d}\Lambda(t)$ can be estimated by $(\mathrm{d}N.(t)/n)/(Y.(t)/n) = \mathrm{d}N.(t)/Y.(t)$. Taking the possibility that $Y.(u) = 0$ for large u, let $J(u) = \mathbf{1}\{Y.(u) > 0\}$. The Nelson–Aalen estimator of the cumulative hazard function is

$$\hat{\Lambda}(t) = \int_0^t \frac{J(u)}{Y.(u)} \cdot \mathrm{d}N.(u) = \sum_{u \leq t} \frac{J(u)}{Y.(u)} \cdot \Delta N.(u). \tag{1.5}$$

In this section we discuss both the small and large sample properties of the Nelson–Aalen estimator.

Using properties of martingales we can prove that

$$\mathrm{E}\hat{\Lambda}(t) = \int_0^t \Pr\{Y.(u) > 0\}\mathrm{d}\Lambda(u) < \Lambda(t),$$

which means that the Nelson–Aalen estimator is an (under)-biased estimator of the cumulative hazard function. However, when sample size is large enough, the probability $\Pr\{Y.(u) = 0\}$ will be very small. The bias will not be a concern in large sample cases.

There are two widely used consistent estimators of the variance of $\hat{\Lambda}(t)$, which are

$$\hat{\sigma}(t) = \int_0^t \frac{J(u)}{Y.(u)^2} \cdot \mathrm{d}N.(u) = \sum_{u \leq t} \frac{J(u)}{Y.(u)^2} \cdot \Delta N.(u),$$

and

$$\tilde{\sigma}(t) = \int_0^t \frac{J(u)[Y.(u) - \Delta N.(u)]}{Y.(u)^3} \cdot \mathrm{d}N.(u) = \sum_{u \leq t} \frac{J(u)}{Y.(u)} \cdot \frac{\Delta N.(u)}{Y.(u)} \cdot \left(1 - \frac{\Delta N.(u)}{Y.(u)}\right),$$

which has a very intuitive interpretation from binomial distribution at each time point. Klein (1988, 1991) studied the properties of these two estimators by simulations and shows that neither one is uniformly better than the other. However, $\tilde{\sigma}(t)$ can be used to handle tied observations in practice due to the grouping or rounding in observations.

The pointwise confidence interval of Λ can be easily obtained by Wald test or by δ-method from appropriate transformation of $\hat{\Lambda}$.

In the large sample case we can prove under some regularity conditions that the Nelson–Aalen estimator is a uniformly consistent estimator. Further $\sqrt{n}(\hat{\Lambda}(t) - \Lambda(t))$ converges weakly to a mean-zero Gaussian process. This fact can be used to construct confidence bands for the estimator.

1.3.2 The Kaplan–Meier Estimator of Survival Function

The Kaplan–Meier estimator (Kaplan and Meier 1958) is the first mathematically rigorous estimator of the survival function based from censored survival data. This estimator was derived as a nonparametric maximum likelihood estimator (NPMLE). Kalbfleisch and Prentice (2002) also provided a more straightforward derivation for this estimator. In this section we give an alternative derivation of the Kaplan–Meier estimator from the relationship between the cumulative hazard function and survival function in (1.2).

Since the Nelson–Aalen estimator $\hat{\Lambda}$ is a step function, substituting this estimator in place of Λ in (1.3) immediately yields the following estimator of the survival function:

$$\hat{S}(t) = \prod_{u \le t}(1 - \Delta\hat{\Lambda}(u)) = \prod_{u \le t}\left\{1 - \frac{\Delta N(u)}{Y(u)}\right\}. \tag{1.6}$$

The above is known as the *Kaplan–Meier estimator*, which as noted earlier was first derived by Kaplan and Meier (1958) based on the nonparametric maximum likelihood. This estimator is also called the *product-limit estimator*.

The Kaplan–Meier estimator is a nonnegative, but biased estimator of S. However, akin to the Nelson–Aalen estimator, the bias diminishes to zero as the sample size n increases. Using the *functional δ-method* (van der Vaart 1998), we can readily derive an estimator of the variance of $\hat{S}(t)$,

$$\widehat{\mathrm{Var}}(\hat{S}(t)) = \left[\hat{S}(t)\right]^2 \int_0^t \frac{\mathrm{d}N(u)}{Y(u)[Y(u) - \Delta N(u)]}. \tag{1.7}$$

The above is also known as the *Greenwood formula* in honor of Greenwood, who first derived this using a more elementary approach (Greenwood 1926). See Oakes (2001, p 102) for an interesting comment on this formula. If there is no censoring, the Kaplan–Meier estimator in (1.6) reduces $1 - \hat{F}(t)$ and the Greenwood formula reduces to the formula from binary distribution. The pointwise confidence interval of S can be easily obtained by Wald test.

Like the Nelson–Aalen estimator, the Kaplan–Meier estimator is also uniformly consistent on any closed interval. Confidence bands in the whole range of observation can also be derived by applying the functional δ-method (Hall 1980).

1.4 Comparison of Two Survival Distributions

In most clinical trials and observation studies, we often need to compare survival distributions from two treatment groups. Let S_k $(k = 1, 2)$ be the survival functions of these two groups. Our interest in whether there is any difference between the two groups can be formulated as testing the null hypothesis $H_0 : S_1(t) = S_2(t)$ for all $t \ge 0$. In other words, the two treatment groups have identical survival distributions under the null hypothesis. Thus rejection of H_0 at a level of significance such as type I error $\alpha = 0.05$ implies a significance difference between the two survival distributions.

If the distributions are modeled by some mathematical functions such as the exponential distribution, popular statistics such as the likelihood ratio test can be used to test the null, in which case, H_0 is essentially a statement of whether the parameters of the assumed distributions are equal. If no such model is assumed for $S_k(t)$, it becomes much more difficult to test the null involving such *nonparametric* models, since most methods for comparing two *nonparametric* distributions only apply when there is no censoring, such as the Kolmogorov–Smirnov test and the Wilcoxon rank sum test.

The log-rank test was proposed by Mantel (1966) based on the Mantel and Haenszel (1959) statistic for stratified contingency tables. Peto and Peto (1972) gave the name of "log-rank test." This test is the most popular nonparametric method in survival analysis, since it is the most efficient nonparametric test under the proportional hazards assumption.

In the Cox proportional hazards model (discussed in Sect. 1.5), if the treatment group indicator (0-1 variable) is the only covariate in the model, the score test from the Cox partial likelihood is exactly the log-rank test. This also explains the efficiency of the log-rank test.

Let $N_{k\cdot}(t)$ and $Y_{k\cdot}(t)$ be the number of observed failures up to time t and number of individuals still at risk at time $t-$ in group k, $k = 1, 2$. Let Λ_k be the cumulative hazard function in group k. The null hypothesis is equivalent to $\Lambda_1 = \Lambda_2$ due to the 1–1 relation between the cumulative hazard function and the survival function. The log-rank test is defined as

$$\mathrm{LR} = \left(\frac{n_1 + n_2}{n_1 n_2}\right)^{1/2} \int_0^{\infty} \frac{Y_{1\cdot}(t) Y_{2\cdot}(t)}{Y_{1\cdot}(t) + Y_{2\cdot}(t)} \left\{ \frac{\mathrm{d}N_{1\cdot}(t)}{Y_{1\cdot}(t)} - \frac{\mathrm{d}N_{2\cdot}(t)}{Y_{2\cdot}(t)} \right\}. \tag{1.8}$$

Under the null hypothesis, $\mathrm{E}[\mathrm{LR}] = 0$. Its asymptotic variance can be consistently estimated by

$$\hat{\sigma}^2 = \left(\frac{n_1 + n_2}{n_1 n_2}\right) \int_0^{\infty} \left\{ \frac{Y_{1\cdot}(t) Y_{2\cdot}(t)}{Y_{1\cdot}(t) + Y_{2\cdot}(t)} \right\}^2 \cdot \left(1 - \frac{\Delta N_{1\cdot}(t) + \Delta N_{2\cdot}(t)}{Y_{1\cdot}(t) + Y_{2\cdot}(t)} \right) \frac{\mathrm{d}(N_{1\cdot}(t) + N_{2\cdot}(t))}{Y_{1\cdot}(t) + Y_{2\cdot}(t)}.$$

Under the null hypothesis, $\mathrm{LR}/\hat{\sigma} \xrightarrow{\mathscr{D}} N(0, 1)$, providing the basis for inference about H_0. If there are $K(\geq 2)$ treatment groups in the study, the log-rank test has asymptotic chi-squared distribution with $k - 1$ degrees of freedom. See Gill (1980), Fleming and Harrington (1991), and Andersen et al. (1993) for more details.

1.5 Cox Regression Models for Survival Data

In the preceding sections, we have assumed that patients within each group are homogeneous, and used the survival function to describe survivability for a single population or investigate differential risks of failure between two groups. However, in practice, variability in survival times can often be explained by a host of covariates, and we are interested in identifying such explanatory variables and characterizing their effects on the distribution of survival time.

Regression is a well-known modeling paradigm for investigating such relationships between a response, survival time within the current context, and a set of covariates. Specifically, if Y is some response of interest and X is a vector of covariate, a linear regression relating X to Y is given by

$$\mathrm{E}[g(Y)|X] = X'\beta, \tag{1.9}$$

where $g(\cdot)$ is some (monotone) transformation. The essence of regression is to model the conditional mean of $g(Y)$ given X. However, in he survival analysis, since the survival time is usually censored, it is difficult to model the mean of T directly as in the above traditional setup. An alternative formulation is then used to bypass this difficulty, which we discuss next.

1.5.1 Proportional Hazards Model

The Cox proportional hazards model is the most popular regression approach for survival data. Instead of modeling the survival mean response as in the classic regression, this approach models the (conditional) hazard function, which not only addresses the aforementioned technical difficulty, but also leads to a more meaningful and informative model and its parameter interpretation.

Let $Z(t)$ be a vector of explanatory variables consisting of both time-invariant and time-dependent covariates. Note that since the survival distribution is a process of time, it is often of interest to include covariates that also change over time. The conditional hazard function of the survival time T given the covariates $Z(t)$ is assumed to be of the form

$$\lambda(t \mid Z) = \lambda_0(t)\exp(Z(t)'\beta), \tag{1.10}$$

where $\lambda_0(\cdot)$ is the baseline hazard function (when $\beta = 0$). Recently, Feng et al. (2012) have developed a unified definition of conditional expectation to interpret the conditional part in (1.10) as there is no clear definition for this kind of conditional expectation in the literature.

If $Z(t)$ consists a single time-invariant covariate Z, then for two values z_1 and z_2 of Z, it follows from (1.10) that

$$\frac{\lambda(t \mid z_1)}{\lambda(t \mid z_2)} = \exp\left((z_1 - z_2)'\beta\right), \tag{1.11}$$

a constant independent of time t. In other words, under the model in (1.10), the two hazard functions $\lambda(t \mid z_1)$ and $\lambda(t \mid z_2)$ are proportional for all $t > 0$, hence giving rise to the name of the Cox proportional hazards model.

Note that if $Z(t)$ is a time-dependent covariate, the ratio of hazards in (1.11) generally depends on t. Despite this fact, (1.10) is still referred to as the Cox proportional hazards model. Note also that the constant ratio of hazard functions in (1.11) clearly imposes a strong assumption on the way the covariates exert their effects on the survival time, which must be carefully checked when applying this model to read study data.

1.5.2 Inference for Proportional Hazards Model

The method of maximum likelihood is widely used for inference about model parameters. The maximum likelihood estimator, or MLE, is not only easy to compute, but also is the most efficient (i.e., most powerful in testing hypotheses concerning parameters). However, this popular approach is mostly applied to parametric regression models, i.e., those that assume a mathematical distribution such as the normal distribution for the response Y in (1.9). Within the current context, if $\lambda_0(\cdot)$ is modeled using a parametric distribution such as the exponential, a popular choice for survival time, then $\lambda_0(\cdot)$ becomes a function of a vector of parameters γ, and both β and γ can be jointly estimated by maximum likelihood. However, in practice, it is generally difficult to specify this baseline hazard function that will adequately fit the data at hand. Thus in most applications, the model in (1.10) is applied with the baseline hazard function left completely unspecified. Indeed, this is also the primary reason for the popularity of the Cox regression, since finding an appropriate mathematical distribution for $\lambda_0(\cdot)$ in a given study is not only quite labor intensive, but more importantly may not yield a satisfactory solution because of the limited number of mathematical distributions.

With $\lambda_0(\cdot)$ unspecified, the model in (1.10) is called a semiparametric model in the sense that it is neither fully parametric nor nonparametric. As noted earlier, such models are more applicable than their parametric counterparts to real study data. However, inference is more difficult for the former, since in particular the conventional maximum likelihood does not apply because the baseline hazard $\lambda_0(\cdot)$ is not determined by a vector of parameters, but rather than an unspecified curve of infinite dimension.

The Cox partial likelihood circumvents this problem by focusing some component of the likelihood that does not involve the baseline hazard $\lambda_0(\cdot)$, but has sufficient information to estimate β, the parameters of primary interest. To illuminate the idea, let $N(t)$ denote the number of failures in the time interval $[0, t]$ and $\mathscr{F}_t$ be the collective information of failures and censoring, or observed history, up to, but not including, time t. The likelihood of the observed data can be expressed as

$$\Pr\{\text{Data}\} = \underbrace{\prod_t \{\Pr\{dN.(t) \mid \mathscr{F}_{t-}\}\}^{dN.(t)}}_{\text{First factor}} \cdot \underbrace{\prod_t \prod_i \{\Pr\{dN_i(t) \mid dN.(t), \mathscr{F}_{t-}\}\}^{dN_i(t)}}_{\text{Second factor}}$$

$$\cdot \underbrace{\prod_t \Pr\{\text{other events in } [t, dt) \mid dN.(t), \mathscr{F}_{t-}\}}_{\text{Third factor}}.$$

There are three factors in the likelihood at time point t:

(i) Whether or not a failure occurs at time t
(ii) If there is a failure at t, to which individual it occurs
(iii) Given (i) and (ii), one looks at the possible censoring or changes in covariates The Cox partial likelihood only keeps part (ii) in the decomposition of the full likelihood. For more discussion of the construction of the partial likelihood in the counting processes, please refer to Jacod (1975). The theory of partial likelihood was discussed thoroughly in Wong (1986).

Given that there is a failure at t, i.e., $\Delta N.(t) = 1$, the probability that the failure occurs to individual i is

$$\Pr\{\Delta N_i(t)=1 \mid \Delta N.(t)=1, \mathscr{F}_{t-}\} = \frac{Y_i(t)\lambda(t|Z_i)}{\sum_{l=1}^n Y_l(t)\lambda(t|Z_l)} = \frac{Y_i(t)\exp(Z_i(t)'\beta)}{\sum_{l=1}^n Y_l(t)\exp(Z_l(t)'\beta)}.$$

Hence, the contribution of individual i to the partial likelihood during the follow-up is

$$\prod_{t\geq 0} \left\{ \frac{Y_i(t)\exp(Z_i(t)'\beta)}{\sum_{l=1}^n Y_l(t)\exp(Z_l(t)'\beta)} \right\}^{\Delta N_i(t)} = \prod_{t\geq 0} \left\{ \frac{\exp(Z_i(t)'\beta)}{\sum_{l=1}^n Y_l(t)\exp(Z_l(t)'\beta)} \right\}^{\Delta N_i(t)}.$$

The Cox partial likelihood is the product of contributions from all individuals,

$$\text{PL}(\beta) = \prod_{i=1}^n \prod_{t\geq 0} \left\{ \frac{\exp(Z_i(t)'\beta)}{\sum_{l=1}^n Y_l(t)\exp(Z_l(t)'\beta)} \right\}^{\Delta N_i(t)}. \tag{1.12}$$

Since the partial likelihood is constructed based on part of the observed data, it is not the full data likelihood. Indeed, the partial likelihood in (1.12) is not even in the form of a product of independent terms as in the traditional likelihood function. Nonetheless, it turns out that we can treat it as a regular likelihood function and obtain the MLE of β and its variance estimator for valid inference about β (Andersen and Gill, 1982). For example, just like MLEs, the partial likelihood estimator of β is asymptotically normal, and popular tests such as the Wald and score statistics can be used for inference about hypotheses concerning β. Also, treatment indicators are the only covariates in the Cox proportional hazard model; the partial-likelihood-based score test reduces to the log-rank test.

Although the regression coefficient β is of major interest, we may also want to estimate the baseline (cumulative) hazard function sometimes. Several estimators of the cumulative hazard function have been proposed, among which the Breslow (1972) is the most popular one:

$$\hat{\Lambda}_0(t) = \int_0^t \frac{\sum_{i=1}^n \mathrm{d}N_i(u)}{\sum_{i=1}^n Y_i(u)\exp(Z_i(u)'\hat{\beta})},$$

where $\hat{\beta}$ is the partial likelihood estimator of β from the Cox regression. Efron (1977) proposed another estimator which can deal with possible more than 1 failure at a time point. Both estimators are available in many statistical software packages such as SAS. In addition to these two popular estimators, we may also use smoothing methods to provide more continuous-looking estimators of $\Lambda_0(t)$ to capture subtle changes in the baseline hazards (Andersen et al. 1993, Chap. 4).

Just like the probability density function, there are many methods that can be used to estimate the baseline hazard function. The most popular one is the kernel smoothing method. Andersen et al. (1993), Chap. 4, give kernel smoothing method for the hazard function under the homogeneous sample. The kernel smoothing and functional δ-method can be combined to study the weak convergence of the baseline hazard function.

1.6 Analysis of Clustered Survival Data

So far we have only considered the case with independent observations. In other words, for a sample of n subjects, the survival (or censoring) time T_i (C_i) are independent across the n subjects. In many applications, the survival times are correlated because of certain common features such as genetic traits or shared environmental factors, known as *clusters*. For example, the lifetimes of twins share the same genetic makeup. In multicenter clinical trials, the patients within the same center are correlated as well due to similar quality of care received by these patients. To analyze such correlated or clustered survival outcomes, we must take the correlation into account.

1.6.1 Frailty Model

The frailty model is widely used to explicitly model the correlation structure from clustered survival data. Consider a study with s clusters and n_i individuals within each ith cluster ($1 \leq i \leq s$). Let T_{ij} denote the survival time for the jth subject within each ith cluster ($1 \leq j \leq n_i$, $1 \leq i \leq s$). We assume a common latent condition of frail due to genetic traits and/or shared environmental effects, W_i, and given this shared risk for failure, the hazard of T_{ij} for each jth subject within the ith cluster is modeled according to the proportional hazard assumption:

$$\lambda(t \mid W_i, X_{ij}) = W_i \lambda_0(t) \exp(X_{ij}' \beta), \quad 1 \leq j \leq n_i, 1 \leq i \leq s. \tag{1.13}$$

In the *shared frailty* model above, the W_i 's are assumed to be independent and identically distributed positive random variables. This approach uses random effects to account for correlations across the subjects within the cluster, and as such is analogous to generalized linear mixed-effects models (GLMM) for longitudinal data (Fitzmaurice et al. 2004). If W_i is a constant, (1.13) reduces to the standard proportional hazard model.

In theory, any positive random variable can serve as the frailty in (1.13). However, due to mathematical tractability, the log-normal, gamma, and positive stable frailties are frequently used in data analysis. As noted in Elbers and Ridder (1982), the frailty distribution can be identified in (1.13) if W_i has a finite mean value such as in the case of the Gamma and log-normal frailty model. Although the conditional hazards given the frailty are proportional, the marginal hazards after integrating out the frailty W_i in (1.13) are generally not, except when W_i has a positive stable distribution (Hougaard 2000).

Since W_i are unobservable, the Cox partial likelihood cannot be used for inference about the model parameters (β plus the parameters for the distribution of W_i). One popular approach is to regard the frailty as a missing covariate and use the expectation-maximization (EM) algorithm (e.g., Dempster et al. 1977) to fit the Cox partial likelihood (Klein, 1991; Nielsen et al. 1992; Petersen et al. 1996). The frailty model is available from major software packages such as SAS and R. For example, one may use the procedure PHREG in SAS 9.3 to fit the shared frailty model in (1.13) with log-normal frailty.

It is necessary to distinguish between the shared frailty model and correlated frailty model. In the correlated gamma frailty model, the frailty of the jth individual in the ith cluster is of the form $W_{ij} = W_{i0} + W_{ij*}$, where W_{i0} and W_{ij*} are independent nonnegative random variables. Here W_{i0} is the shared part common to all individuals in the cluster and W_{ij*} is the part specific to the $i$the individuals. The distribution of the correlated frailty model cannot be identified from the data even if $E(W_{ij}) < \infty$. The asymptotic results for the shared gamma frailty model were discussed thoroughly in Murphy (1994, 1995) and the asymptotic results for the correlated was given in Parner (1998).

1.6.2 Marginal Models

Another widely used method for correlated survival outcomes is the marginal model. This approach does not explicitly model the within-cluster correlations, but rather using the sandwich variance estimates to implicitly account for correlated outcomes, akin to the generalized estimating equations (GEE) for modeling longitudinal data (Liang and Zeger 1986; Kowalski and Tu 2007; Tang et al. 2012).

Specifically, the hazard function for each individual is assumed to follow the standard Cox regression model in (1.10). Inference proceeds using a *working independence partial likelihood*, which in appearance has the same form as (1.12), but differs in how estimates are obtained. Specifically, the marginal hazard for the ith individual in the jth cluster is assumed to be

$$\lambda(t \mid X_i) = \lambda_0(t)\exp(\gamma' X_{ij}). \tag{1.14}$$

Note that in general, the marginal hazard function obtained from the shared frailty model in (1.13) does not have the form of (1.14). Therefore the regression coefficients of these two models generally have totally different interpretations. This difference is similar to the difference between subject-specific (SS) mixed model and the marginal model in generalized estimation equation (see, e.g., Zhang et al. 2011, 2012).

Given the marginal hazard (1.14), a working independence partial likelihood function can be constructed:

$$\mathrm{PL}(\beta) = \prod_{i=1}^{n}\prod_{j=1}^{n_i}\prod_{t\ge 0}\left\{\frac{\exp(X_{ij}(t)'\beta)}{\sum_{l=1}^{n}\sum_{m=1}^{n_i} Y_{lm}(t)\exp(X_{lm}(t)'\beta)}\right\}^{\Delta N_{ij}(t)}. \tag{1.15}$$

The estimator of $\hat{\beta}$ is obtained by solving for β the "score" function of (1.15): equation can be obtained from

$$s(\hat{\beta}) = 0, \quad s(\beta) = \frac{\partial}{\partial\beta}\log(\mathrm{PL}(\beta)).$$

Like most results from estimating equations, the estimator $\hat{\beta}$ is asymptotically normal (Cai and Prentice 1995), and the asymptotic variance is estimated by a sandwich variance estimator:

$$\widehat{\mathrm{Cov}}(\hat{\beta}) = \left[\frac{1}{K}\frac{\partial^2}{\partial\beta\partial\beta'}\log(\mathrm{PL}(\hat{\beta}))\right]^{-1}\left[\frac{1}{K}s(\hat{\beta})\cdot s(\hat{\beta})^{\top}\right]\left[\frac{1}{K}\frac{\partial^2}{\partial\beta\partial\beta'}\log(\mathrm{PL}(\hat{\beta}))\right]^{-1},$$

where $K = \sum_i n_i$.

1.7 Examples

In this section, we illustrate the considerations in the preceding sections with two real study data. The first is standard survival analysis, while the second involves clustered outcomes.

1.7.1 PBC Study

This data set is from the Mayo Clinic trial in primary biliary cirrhosis (PBC) of the liver conducted between 1974 and 1984. A total of 424 PBC patients, referred to Mayo Clinic during that ten-year interval, met eligibility criteria for the randomized placebo-controlled trial of the drug D-penicillamine. Of the referred patients, 312 participated in the randomized trial, with quite largely complete data, while the remaining 112 patients only consented to have basic measurements recorded. Six of the latter cases were lost to follow-up shortly after diagnosis. For our illustration, we only analyzed data from the 312 study participants. The data contains following variables and can be found in the Appendix of Fleming and Harrington (1991). It can also be downloaded from the webpage: http://mayoresearch.mayo.edu/mayo/research/biostat/upload/therneau_upload/pbc.dat.

- Case number
- Number of days between registration and the earlier of death, transplantation, or study analysis time in July, 1986
- Status: 0 = alive, 1 = liver transplant, 2 = dead
- Drug: 1= D-penicillamine, 2=placebo
- Age in days
- Sex: 0=male, 1=female
- Presence of ascites: 0=no 1=yes
- Presence of hepatomegaly 0=no 1=yes
- Presence of spiders 0=no 1=yes
- Presence of edema 0=no edema and no diuretic therapy for edema; .5 = edema present without diuretics, or edema resolved by diuretics; 1 = edema despite diuretic therapy
- Serum bilirubin in mg/dl
- Serum cholesterol in mg/dl
- Albumin in gm/dl
- Urine copper in ug/day
- Alkaline phosphatase in U/liter
- SGOT in U/ml
- Triglicerides in mg/dl
- Platelets per cubic ml/1000
- Prothrombin time in seconds
- Histologic stage of disease

We first estimated the cumulative hazard and survival functions for each treatment group, followed by the log-rank test to compare the survival times between the two groups. We then controlled for some important demographic and baseline disease severity variables. We performed the analysis in SAS version 9.3.

The SAS codes below were used to fit the data to obtain the Kaplan–Meier estimator of survival function, the Nelson–Aalen estimator of cumulative hazard function, and the log-rank test for comparing the two treatment groups (drug: 1=D-penicillamine, 2=placebo).

```
OPTIONS LS=80 PS=60 PAGENO=1 NOFMTERR;

LIBNAME SAVE 'Z:\RESEARCH\PAPER\BOOK_PUBLICATION';
```

```
DATA SAVE.PBC;
    INFILE 'Z:\RESEARCH\PAPER\BOOK_PUBLICATION\PBC.
      TXT';
    INPUT ID FUTIME STATUS DRUG AGE SEX ASCITES
      HEPATO SPIDERS EDEMA
    BILI CHOL ALBUMIN COPPER ALK_PHOS SGOT TRIG
      PLATELET PROTIME STAGE;
    AGE = AGE/365.25;
RUN;

DATA PBC1;
    SET SAVE.PBC;
    IF ID <=312;
    DRUG = 2 - DRUG;
RUN;

ODS GRAPHICS ON;
PROC LIFETEST DATA=PBC1 NELSON OUTSURV=A PLOTS=S
  (NOCENSOR ATRISK TEST);
    TIME FUTIME*STATUS(0 1);
    STRATA DRUG;
PROC PRINT DATA=A;
RUN;
ODS GRAPHICS OFF;
```

Here is a subset of results from the SAS output of Kaplan–Meier estimator of survival functions and Nelson–Aalen estimator of cumulative hazard function for the two groups.

Stratum 1: DRUG = 1
Survival Function and Cumulative Hazard Rate

FUTIME	Product-Limit Survival: Survival	Product-Limit Survival: Failure	Survival Standard Error	Nelson-Aalen Cum Haz: Cumulative Hazard	Nelson-Aalen Cum Haz: Standard Error	Number Failed	Number Left
0.00	1.0000	0	0	0	.	0	158
41.00	0.9937	0.00633	0.00631	0.00633	0.00633	1	157
71.00	0.9873	0.0127	0.00889	0.0127	0.00898	2	156
131.00	0.9810	0.0190	0.0109	0.0191	0.0110	3	155
140.00	0.9747	0.0253	0.0125	0.0256	0.0128	4	154
179.00	0.9684	0.0316	0.0139	0.0321	0.0143	5	153
198.00	0.9620	0.0380	0.0152	0.0386	0.0158	6	152
223.00	0.9557	0.0443	0.0164	0.0452	0.0171	7	151
334.00	0.9494	0.0506	0.0174	0.0518	0.0183	8	150
348.00	0.9430	0.0570	0.0184	0.0585	0.0195	9	149
388.00	0.9367	0.0633	0.0194	0.0652	0.0206	10	148
400.00	0.9304	0.0696	0.0202	0.0719	0.0217	11	147

......

```
                      Stratum 2: DRUG = 2
            Survival Function and Cumulative Hazard Rate

             Product-Limit            Nelson-Aalen
               Survival                  Cum Haz
                            Standard Cumulative Standard  Number   Number
FUTIME   Survival Failure   Error      Hazard     Error    Failed    Left

  0.00    1.0000        0        0          0          .       0      154
 51.00    0.9935  0.00649  0.00647    0.00649    0.00649       1      153
 77.00    0.9870   0.0130  0.00912     0.0130    0.00921       2      152
110.00    0.9805   0.0195   0.0111     0.0196     0.0113       3      151
130.00    0.9740   0.0260   0.0128     0.0262     0.0131       4      150
186.00    0.9675   0.0325   0.0143     0.0329     0.0147       5      149
191.00    0.9610   0.0390   0.0156     0.0396     0.0162       6      148
207.00    0.9545   0.0455   0.0168     0.0464     0.0175       7      147
216.00    0.9481   0.0519   0.0179     0.0532     0.0188       8      146
264.00         .        .        .          .          .       9      145
264.00    0.9351   0.0649   0.0199     0.0669     0.0211      10      144
304.00    0.9286   0.0714   0.0208     0.0738     0.0223      11      143
321.00    0.9221   0.0779   0.0216     0.0808     0.0233      12      142
326.00    0.9156   0.0844   0.0224     0.0878     0.0244      13      141
460.00    0.9091   0.0909   0.0232     0.0949     0.0254      14      140
......
```

The SAS procedure PROC LIFETEST also provides results of the log-rank test. The test statistic was $1.7811/\sqrt{31.1917} = 0.3189$, with a p-value $= 0.7498$, indicating no significant difference between the two treatment groups.

```
Testing Homogeneity of Survival Curves for FUTIME
  over Strata

               Rank Statistics

     DRUG         Log-Rank    Wilcoxon
     1              1.7811     -52.000
     2             -1.7811      52.000

  Covariance Matrix for the Log-Rank Statistics

     DRUG                1             2
     1             31.1917      -31.1917
     2            -31.1917       31.1917

  Covariance Matrix for the Wilcoxon Statistics

     DRUG                1             2
     1             1522246      -1522246
     2            -1522246       1522246
```

Test of Equality over Strata

Test	Chi-Square	DF	Pr > Chi-Square
Log-Rank	0.1017	1	0.7498
Wilcoxon	0.0018	1	0.9664
-2Log(LR)	0.0634	1	0.8013

Finally the Cox proportional hazards model was fit using the following codes to control for the effects of age, sex, log(Albumin), log(Bilirubin), and edema when comparing the survival times between the two groups:

```
ODS HTML;
PROC PHREG DATA=PBC1;
    MODEL FUTIME*STATUS(0 1)= AGE SEX DRUG ALBUMIN
      BILI EDEMA;
RUN;
ODS HTML CLOSE;
PROC FREQ DATA=PBC1; TABLES DRUG; RUN;
```

The SAS output from the above codes is shown as follows. The results indicate that all included covariates had a significant effect on the survival time (p-values < 0.05). The p-value of the treatment is 0.94, again showing no significant treatment effect.

Fleming and Harrington (1991, p13) considered a slightly expanded model by adding another covariate, log(Prothrombin Time), but the treatment effect remained insignificant (p-value = 0.73).

The PHREG Procedure

Analysis of Maximum Likelihood Estimates

Parameter	DF	Parameter Estimate	Standard Error	Chi-Square	Pr > ChiSq	Hazard Ratio
AGE	1	0.03166	0.00941	11.3318	0.0008	1.032
SEX	1	-0.57843	0.24772	5.4522	0.0195	0.561
DRUG	1	0.01348	0.18563	0.0053	0.9421	1.014
ALBUMIN	1	-1.22542	0.23993	26.0857	<.0001	0.294
BILI	1	0.12536	0.01485	71.2409	<.0001	1.134
EDEMA	1	1.02150	0.31708	10.3786	0.0013	2.777

1.7.2 Litter-Clustered Survival Data

The data for this example first appeared in Mantel et al. (1977). The study is a litter-matched tumorigenesis experiment involving 50 litters of rats and 3 rats within each litter. One of the 3 rats received was in the treatment and the remaining two was in the control group. For genetic reasons, the times to tumor appearance are in principle correlated for the rats within the same litter, but independent across the different litters. Censoring occurs if the rat dies before tumor appears. The main interest is the treatment effect.

The EM algorithm was used to fit the semiparametric positive stable frailty model. The data set and a SAS macro written used to fit the data can be downloaded from the webpage of Dr. J. P. Klein (http://www.mcw.edu/biostatistics/Faculty/Faculty/JohnPKleinPhD.htm).

Note that in the positive stable frailty model, the Laplace transform of the frailty W is $p(s) = \exp(-s^{\alpha})$, $0 < \alpha \leq 1$. If the conditional hazard function given W is of the form

$$\lambda(t \mid W, Z) = W\lambda_0(t)\exp(Z\beta), \tag{1.16}$$

then the marginal hazard function is

$$\lambda(t \mid Z) = \alpha\Lambda_0^{\alpha-1}(t)\lambda_0(t)\exp(Z\gamma), \tag{1.17}$$

where $\Lambda_0(t) = \int_0^t \lambda_0(u)\mathrm{d}u$ and

$$\gamma = \alpha\beta. \tag{1.18}$$

If there is no litter effect, i.e., rats within the litter are all independent, then $\alpha = 1$ and the marginal hazard in (1.17) reduces to the standard proportional hazard function.

Shown below is the output from the SAS macro.

```
~~~~~~~~~~~~~~~~~~~~~~~~~~~~~~~~~~~~~~~~~~~~~~~~~~~~~~~~~~~~
                 Summary  of  Final  Results
~~~~~~~~~~~~~~~~~~~~~~~~~~~~~~~~~~~~~~~~~~~~~~~~~~~~~~~~~~~~

        /******************************************/
        /*        Usual Cox Independence Model    */
        /******************************************/

   Log partial likelihood = -181.8874
   Log full likelihood = -208.3643

              Analysis of Maximum Likelihood Estimates

          Variable      DF  Estimate    Stderr     p_Wald         RR
          DRUG           1   0.89747   0.31741    0.00469    2.45340

        /******************************************/
        /*       Positive Stable Frailty Model    */
        /******************************************/

   Dependence Parameter: ALPHA = 0.92837
   Kendall's TAU = 0.07163
   Log full likelihood =  -208.026

        Likelihood Ratio Test of Independence Model (H0: THETA=1)
                     DF     Chi-square      p-value
                      1        0.67654      0.41078
              Analysis of Maximum Likelihood Estimates
```

```
Effect         DF   Estimate     Stderr     p_Wald  RR_within  RR_between
Depend Parm     1    0.92837    0.09676    0.45911          .           .
DRUG            1    0.92922    0.32596    0.00436    2.53253     2.36945
```

From the output we can see that the estimator of parameter in the positive stable distribution was $\widehat{\alpha} = 0.9284$ with p-value = 0.41 (from testing the null hypothesis $H : \alpha = 1$). This means that we cannot reject the null hypothesis that the survival times within the same litter are independent. The treatment effect is very significant (p-value = 0.004).

It is interesting to compare the results from the shared frailty model with those from the marginal Cox working independence model. Remember that under the positive stable frailty assumption with the conditional proportional hazards in (1.16), the marginal hazards are still proportional (see equation (1.17). However, the regression coefficients are attenuated (see (1.18)). In the working independence model, the estimate of γ for the effect of treatment was $\widehat{\gamma} = 0.8975$, which was quite similar to the product 0.8627 of the estimated $\widehat{\alpha} = 0.9284$ and $\widehat{\beta} = 0.9292$ in the frailty model.

1.8 Discussion and Further Reading

In this chapter we introduced the four building blocks of modern survival analysis: the Nelson–Aalen estimator of the cumulative hazard function, the Kaplan–Meier estimator of survival functions, the log-rank test and the Cox proportional hazards model for independent survival data. We focused on the interpretation of the assumptions underlying the independent censoring and the Cox partial likelihood function. We also extended the discussion to introduce the correlated and shared frailty models for correlated survival times due to considerations of multiple events from the same individual such as repeated suicide events and hospitalizations or single events from multiple subjects sharing common features such as genetics traits.

Due to space considerations, we skipped some topics such as competing risk analysis (Fine and Gray 1999) and multivariate survival analysis (Hougaard 2000). A nice brief summary of the development of survival analysis from 1901 to 2001 is given by Oakes (2001). For the past three decades, the martingale theory has been the major driving for developing new models and inference procedures (Aalen et al. 2009). As survival analysis finds new applications, survival models have become increasingly complex, leading to new theoretical challenges beyond the scope of the martingale theory. The next generation empirical process theory has been used over the past decade to lay a new foundation to address such complex issues. There are several good texts on this topic, such as Pollard (1984, 1990), van der Vaart and Wellner (1996), and Kosorok (2008). The last book is particularly accessible for biostatisticians.

References

Aalen OO (1978) Nonparametric estimation of partial transition probabilities in multiple decrement models. Ann Stat 6:534–545

Aalen OO, Andersen PK, Borgan Ø, Gill RD, Keiding N (2009) History of applications of martingales in survival analysis. Electron J Hist Probab Stat 5: 1–28.

Aalen OO, Borgan Ø, Gjessing HK (2008) Survival and event history analysis. Springer, New York

Andersen PK, Borgan Ø, Gill RD, Keiding N (1993) Statistical models based on counting processes. Springer, New York

Andersen PK, Gill RD (1982) Cox's regression model for counting processes: a large sample study. Ann Stat 10:1100–1120

Breslow NE (1972) Discussion of Professor Cox's paper. J Roy Stat Soc B 34:216–217

Cai J, Prentice RL (1995) Estimating equations for hazard ratio parameters based on correlated failure time data. Biometrika 82:151–164

Cox DR (1972) Regression models and life tables (with discussion). J Roy Stat Soc B 34:187–220

Dempster AP, Laird NM, Rubin DB (1977) Maximum likelihood from incomplete data via the EM algorithm. J Roy Stat Soc B 39:1–38

Dudley RM (1999) Uniform central limit theorems. Cambridge University Press, Cambridge

Efron B (1977) The efficiency of Cox's likelihood function for censored data. J Am Stat Assoc 72:557–565

Elbers C, Ridder G (1982) True and spurious duration dependence: the identifiability of the proportional hazard model. Rev Econ Stud 49:403–409

Feng C, Wang H, Tu XM (2012) A unified definition of conditional expectation and its applications in survival analysis. Adv Appl Stat 27:81–95

Fine JP, Gray RJ (1999) A proportional hazards model for the subdistribution of a competing risk. J Am Stat Assoc 94:496–509

Fitzmaurice GM, Laird NM, Ware JH (2004) Applied longitudinal analysis. Wiley, Hoboken, NJ

Fleming TR, Harrington DP (1991) Counting processes and survival analysis. Wiley, Chichester, UK

Gill R (1980) Censoring and stochastic integrals. Tract 124, Amsterdam Mathematical Centre, Amsterdam

Gill RD, Johansen S (1990) A survey of product-integration with a view towards applications in survival analysis. Ann Stat 18:1501–1555

Greenwood M (1926) The natural duration of cancer. Reports of Public Health and Related Subjects, vol 33. HMSO, London

Hall WJ, Wellner JA (1980) Confidence bands for a survival curve from censored data. Biometrika 67:133–143

Hogan JW, Roy J, Korkontzelou C (2004) Handling drop-out in longitudinal studies. Stat Med 23:1455–1497

Hougaard P (2000) Analysis of multivariate survival data. Springer, New York

Jacod J (1975) Multivariate point processes: predictable projection, Radon-Nikodym derivatives, representation of martingales. Zeitschriftfür Wahrschein-lichkeitstheorie und verwandte Gebiete 31:235–253

Kalbfleisch JD, Prentice RL (2002) The statistical analysis of failure time data, 2nd edn, Wiley, New York

Kaplan EL, Meier P (1958) Nonparametric estimator from incomplete observations. J Am Stat Assoc 53:457–481

Klein JP (1988) Small sample properties of censored data estimators of the cumulative hazard rate, survivor function, and estimators of their variance. Research Report 88/7. Statistical Uunit, University of Copenhagen.

Klein JP (1991) Small sample moments of some estimators of the variance of the Kaplan-Meier and Nelson-Aalen estimators. Scandinavian J Stat 18:333–340

Kosorok MR (2008) Introduction to empirical processes and semiparametric inference. Springer, New York
Kowalski J, Tu XM (2007) Modern Applied U Statistics. Wiley, New York
Liang KY, Zeger S (1986) Longitudinal data analysis using generalized linear models. Biometrika 73:13–22
Little RJA, Rubin DB (2002) Statistical analysis with missing data, 2nd edn. Wiley, New York
Lu X, Tsiatis AA (2008) Improving the efficiency of the logrank test using auxiliary covariates. Biometrika 95:679–694
Mantel N (1966) Evaluation of survival data and two new rank order statistics arising from its consideration. Canc Chemother Rep 50:163–170
Mantel N, Haenszel W (1959) Statistical aspects of the analysis of data from the retrospective analysis of disease. J Natl Canc Inst 22(4):719–748
Mantel N, Bohidar NR, Ciminera JL (1977) Mantel-Haenszel analyses of litter-matched time-to-response data, with modifications for recovery of interlitter information. Canc Res 37:3863–3868
Murphy SA (1994) Consistency in a proportional hazards model incorporating a random effect. Ann Stat 22:712–731
Murphy SA (1995) Asymptotic theory for the frailty model. Ann Stat 23:182–198
Nelson W (1969) Hazard plotting for incomplete failure data J Qual Tech 1:27–52
Nelson W (1972) Theory and applications of hazard plotting for censored failure data. Technometrics 14:945–965
Nielsen GG, Gill RD, Andersen PK, Sørensen TIA (1992) A counting process approach to maximum likelihood estimation in frailty models. Scandinavian J Stat 19:25–44
Oakes D (1989) Bivariate survival models induced by frailties. J Amer Stat Assoc 84:487–493
Oakes D (2001) Biometrika centenary: survival analysis. Biometrika 88:99–142
Parner E (1998) Asymptotic theory for the correlated gamma-frailty model. Ann Stat 26:183–214
Peto R (1959) Rank tests of maximal power against Lehmann-type alternatives. Biometrika 59:472–475
Petersen JH, Andersen PK, Gill RD (1996) Variance components models for survival data. Statistica Neerlandica 50:193–211
Peto R, Peto J (1972) Asymptotically efficient rank invariant test procedures (with discussion). J Roy Stat Soc A 135:185–206
Pollard D (1984) Convergence of stochastic processes. Springer, New York
Pollard D (1990) Empirical processes: theory and applications. NSF-CBMS regional conference series in probability and statistics, vol. 2. Institute of Mathematical Statistics
Tang W, He H, Tu XM (2012) Applied categorical data analysis. Chapman & Hall/CRC, New York
van der Vaart AW (1998) Asymptotic statistics. Cambridge University Press, Cambridge
van der Vaart AW, Wellner J (1996) Weak convergence and empirical processes. Springer, New York
Vaupel JW, Manton KG, Stallard E (1979) The impact of heterogeneity in individual frailty on the dynamics of mortality. Demography 16:439–454
Wong WH (1986) Theory of partial likelihood. Ann Stat 14:88–123
Zhang H, Xia Y, Chen R, Lu N, Tang W, Tu XM (2011) On modeling longitudinal binomial responses—implications from two dueling paradigms. Appl Stat 38:2373–2390
Zhang H, Yu Q, Feng C, Gunzler D, Wu P, Tu XM (2012) A new look at the difference between GEE and GLMM when modeling longitudinal count responses. J Appl Stat 39:2067–2079

Chapter 2
Longitudinal Data Analysis

Wan Tang, Naiji Lu, Rui Chen, and Hui Zhang

2.1 Introduction

Longitudinal studies are quite common in modern clinical trials and cohort studies. Unlike cross-sectional designs, where observations from study subjects are available only at a single time point, individuals in longitudinal or cohort studies are assessed repeatedly over time. By taking advantages of multiple snapshots of a group over time, data from longitudinal studies captures both between-individual differences and within-individual dynamics, affording the opportunity to study more complicated biological, psychological, and behavioral hypotheses than their cross-sectional counterparts.

For example, if we want to test whether exposure to some chemical agent can cause some type of cancer, the between-subject difference observed in cross-sectional data can only provide evidence of an association or correlation between the exposure and disease. The within-individual dynamics in longitudinal data allows for inference of a causal nature for such a relationship.

Longitudinal data presents multiple methodological challenges in study designs and data analyses. The primary problem is the correlation among the repeated responses of the same subject. Classic models for cross-sectional data analysis such as multiple linear and logistic regressions are based on the independence of observations and thus in general do not apply to longitudinal data. For example, in

W. Tang (✉) • N. Lu • R. Chen
Department of Biostatistics and Computational Biology, University of Rochester, Rochester, NY 14642, USA
e-mail: wan_tang@urmc.rochester.edu; naiji_lu@urmc.rochester.edu; rui_chen@urmc.rochester.edu

H. Zhang
Department of Biostatistics, St. Jude Children's Research Hospital, Memphis, TN 38105, USA
e-mail: hui.zhang@stjude.org

W. Tang and X. Tu (eds.), *Modern Clinical Trial Analysis*, Applied Bioinformatics and Biostatistics in Cancer Research, DOI 10.1007/978-1-4614-4322-3_2,

studies to compare quality of life (QOL) between two treatment conditions, patients are repeatedly assessed for their quality of life over time. Since such repeated assessments are unavoidably correlated, special handling of such correlation is required in order to compare the QOL measures between the treatment conditions.

Another issue common to longitudinal studies is missing values. Since patients are followed up for a period of time in longitudinal studies, it is common that some patients will drop out of the studies. This is of particular importance for cancer research because of the high mortality rate and severely deteriorated health condition of the patients. One naive approach would be to simply delete the subjects with any missing value during the study period. However, as it does not utilize all available data, such an approach is not efficient. But a more serious issue is the bias in the estimate arising from "listwise" deletions, yielding misleading findings and even wrong conclusions, since the data is obviously biased towards those who survive at the end of the study.

Note that in traditional cancer studies, the primary interest is often the time to some significant event such as death. In such situations, each patient is followed for a certain period of time to record the occurrence as well as the timing of the event, and the interest is to estimate the distribution of such *survival times*. As the event may not occur for some patients during the study period, the event times are *censored* for these subjects, requiring specialized methods in survival analysis to handle such censored event times. Although longitudinal data also involves time, it is only used to index the temporal assessment of the subject, rather than the primary focus as in survival analysis. Thus, longitudinal models in general bear little resemblance to those in survival analysis methodology.

2.2 Parametric Models

If the data is well behaved, we may be able to model it using mathematical distributions such as the multivariate normal. Such parametric models have been and still are used for modeling longitudinal data, although semi-parametric or distribution-free alternatives have become increasingly popular. We start with the classic normal-based multivariate linear regression model.

2.2.1 Multivariate Linear Regression Model

Consider a longitudinal study with n subjects assessed at m different time points. Let y_{it} denote some continuous response, dependent variable, and $\mathbf{x}_{it}$ a vector of independent variables, also known as covariates or predictors, from the ith subject at assessment time t. Let

$$\mathbf{y}_i = (y_{i1}, \ldots, y_{im})^\top, \quad \mathbf{x}_i = \left(\mathbf{x}_{i1}^\top, \ldots, \mathbf{x}_{im}^\top\right)^\top.$$

Then, we model the linear relation between y_{it} and $\mathbf{x}_{it}$ as follows:

$$y_{it} = \mathbf{x}_{it}^{\top}\beta + \varepsilon_{it}, \quad \text{or} \quad \mathbf{y}_i = \mathbf{x}_i^{\top}\beta + \varepsilon_i,$$
$$\varepsilon_i = (\varepsilon_{i1}, \ldots, \varepsilon_{im})^{\top} \sim \text{i.i.d. } N(0, \Sigma), \tag{2.1}$$

where ε_{it} (ε_i) denotes the model error and i.i.d. stands for independently and identically distributed. Note that if different β_t are used for different times, the model in (2.1) is known as the *seemingly unrelated regression.*

In many controlled, randomized longitudinal trials, we are interested in treatment difference at a posttreatment or a follow-up time ($1 < t \leq m$). In this case, we can apply the Analysis of Variance (ANOVA) model for comparing different treatment conditions or the Analysis of Covariance (ANCOVA) if we want to control for covariates at pre-treatment or baseline ($t = 1$). These methods yield valid inference in the absence of missing data. However, when there is missing data, both generally gives rise to biased estimates. We address this issue in Sect. 2.4.

Although multivariate normality-based models are widely used in biomedical and psychosocial research, many studies have indicated problems with estimates derived from such models because of the strong distribution assumption imposed. For example, most instruments used for QOL measures are based on item scores, which are intrinsically discrete. If a variable has a relatively large range such as the total score of the popular QOL instrument SF-36, the conventional approach by treating such a variable as a continuous outcome is sensible. However, because these variables are inherently discrete, normal-based parametric models are fundamentally flawed and, in some cases, such distribution assumptions may be severely violated (Lu et al. 2009). Thus, whenever possible, distribution-free alternatives should also be considered.

2.2.2 *Linear Mixed-Effects Model*

The two major limitations of the classic multivariate linear model are (1) its limited ability to deal with missing data, and (2) its requirement of common assessment times for all subjects. The mixed-effects (or latent variable) modeling approach and the distribution-free models provide an effective solution to both issues. We start with the linear mixed-effects model (LMM) for continuous responses, which is a direct extension of the classic multivariate linear models.

The LMM is a general class of models widely used to model the linear relationship between a (continuous) response and a set of independent variables within a longitudinal data setting. LMM addresses the correlated responses by modeling the between-subject variability using random effects, or latent variables, rather than directly correlating the responses as in the classic multivariate linear model. As a result, this approach enables one to address the difficulty in modeling correlated responses arising from varying assessment times as in some longitudinal cohort studies.

As data clustering arises in research studies employing longitudinal study designs and multi-level sampling strategies (e.g., sampling subjects from classes nested within schools) across a wide range of disciplines, various applications of LMM are found under different guises such as random coefficient models, random regression, hierarchical linear models (HLM), latent variable models, mixed models, and multilevel linear models (Goldstein 1987; Bryk et al. 1996; Laird and Ware 1982; Strenio et al. 1983).

Consider a longitudinal study with n subjects. Assume first a set of fixed assessment times for all subjects, $t = 1, 2, \ldots, m$. If the mean of y_{it} is a linear function of time t, then the classic linear model has the form:

$$\begin{aligned} & y_{it} = \beta_0 + \beta_1 t + \varepsilon_{it}, \quad \varepsilon_i = (\varepsilon_{i1}, \ldots, \varepsilon_{im})^{\top} \sim \text{i.i.d. } N(\mathbf{0}, \Sigma), \\ & 1 \leq i \leq n, \quad 1 \leq t \leq m. \end{aligned} \tag{2.2}$$

In the above, Σ is the variance of $\mathbf{y}_i = (y_{i1}, \ldots, y_{im})^{\top}$, which contains both the between- and within-subject variation. The idea behind the LMM is to break up the two sources of variation by modeling each separately.

For each subject i, let b_{i0} and b_{i1} denote the intercept and slope of the response y_{it} and let $\mathbf{b}_i = (b_{i0}, b_{i1})^{\top}$. Then, for each subject i, we model the within-subject variation as follows:

$$\begin{aligned} & y_{it} \mid \mathbf{b}_i = \beta_0 + \beta_1 t + b_{i0} + b_{i1} t + \varepsilon_{it}, \\ & \varepsilon_{it} \sim \text{i.i.d. } N\left(0, \sigma^2\right), \quad 1 \leq t \leq m. \end{aligned} \tag{2.3}$$

In other words, for the ith individual, the response y_{it} is modeled as a linear function of time with intercept, $\beta_0 + b_{i0}$, and slope, $\beta_1 + b_{i1}$. Thus, by modifying the population mean $\beta = (\beta_0, \beta_1)^{\top}$ using the individual specific $\mathbf{b}_i$ to account for between-subject differences in the linear predictor, the error terms $\varepsilon_{i1}, \varepsilon_{i2}, \ldots, \varepsilon_{im}$, in (2.3) can be assumed to be i.i.d. This is in stark contrast to the assumption of the error term for the classic multivariate linear regression in (2.2), in which the model error ε_{it} are correlated over t to account for individual differences. By letting $\mathbf{b}_i$ vary across the subjects, we obtain a LMM that accounts for both between- and within-subject variations. In many applications, $\mathbf{b}_i$ are assumed to follow a multivariate normal $N(\mathbf{0}, D)$.

By combining the two-level specifications, we can express this LMM in a hierarchical form:

$$\begin{aligned} & y_{it} = \beta_0 + t\beta_1 + b_{i0} + b_{i1} t + \varepsilon_{it} = \mathbf{x}_{it}^{\top} \beta + \mathbf{z}_{it}^{\top} \mathbf{b}_i + \varepsilon_{it}, \\ & \mathbf{y}_i = \mathbf{x}_i^{\top} \beta + \mathbf{z}_i^{\top} \mathbf{b}_i + \varepsilon_i, \\ & \varepsilon_{it} \sim \text{i.i.d. } N\left(0, \sigma^2\right), \quad \mathbf{b}_i = N(b_{i0}, b_{i1})^{\top} \sim \text{i.i.d. } N(\mathbf{0}, D), \end{aligned} \tag{2.4}$$

where $\mathbf{x}_{it} = \mathbf{z}_{it} = (1, t)^\top$ and $\mathbf{x}_i = \mathbf{z}_i = \left(\mathbf{x}_{i1}^\top, \mathbf{x}_{i2}^\top, \ldots, \mathbf{x}_{im}^\top\right)^\top$. The linear predictor of the LMM in (2.4) has two parts; the first part $\mathbf{x}_{it}^\top \beta$ describes the change of the population mean over time, while the second $\mathbf{z}_{it}^\top \mathbf{b}_i$ models the deviation of each individual subject from the population mean. Since $\mathbf{b}_i$ is random, $\mathbf{b}_i$ (or $\mathbf{z}_{it}^\top \mathbf{b}_i$) is called the *random effect*. The vector of population-level parameters β (or $\mathbf{x}_{it}^\top \beta$) is called the *fixed effect* and hence the name of the LMM. The hierarchical form and the latent nature of $\mathbf{b}_i$ explain the alternative names of the LMM mentioned earlier.

The covariance between the sth and tth assessment points for the classic (2.2) and linear mixed-effects (2.3) models are given by

$$\operatorname{Cov}(y_{is}, y_{it}) = \sigma_{st}, \quad \operatorname{Cov}(y_{is}, y_{it}) = \mathbf{z}_{is}^\top D \mathbf{z}_{it} + \sigma^2, \quad 1 \le s, t \le m.$$

If assessment times vary across individuals, σ_{ist} will depend on i and become inestimable except in special cases with some particular covariance structures. For example, if the covariance between y_{is} and y_{it} follows the uniform compound symmetry assumption, $\sigma_{st} = \sigma$ and is estimable. However, this issue does not arise for the LMM, since D and σ^2 are well defined regardless of the spacing structure of assessment. Further, the variance parameters for the random effect in LMM have well-defined interpretations; the diagonals of D measure the variability in individual intercepts and slopes among different subjects in the study population. Thus, in addition to β, inference about D is also often of interest to assess the variability of individual intercepts or slopes or both.

Suppose that each subject has a varying number as well as times of assessments. An LMM for this general setting has the following form:

$$\begin{aligned} y_{it_{ij}} &= \mathbf{x}_{it_{ij}}^\top \beta + \mathbf{z}_{it_{ij}}^\top \mathbf{b}_i + \varepsilon_{it_{ij}}, \quad \mathbf{y}_i = \mathbf{x}_i \beta + \mathbf{z}_i \mathbf{b}_i + \varepsilon_i, \quad \mathbf{b}_i \perp \varepsilon_i, \\ \mathbf{b}_i &\sim \text{i.i.d. } N(0, D), \quad \varepsilon_i \sim \text{i.i.d. } N\left(0, \sigma^2 \mathbf{I}_{m_i}\right), \quad 1 \le j \le m_i, \end{aligned} \tag{2.5}$$

where $\mathbf{x}_{it}^\top \beta$ is the fixed and $\mathbf{z}_{it}^\top \mathbf{b}_i$ the random effect, $\perp$ denotes stochastic independence, $\mathbf{I}_m$ denotes the $m \times m$ identity matrix, and $\mathbf{u}_i = (\mathbf{u}_{it_{i1}}^\top, \ldots, \mathbf{u}_{it_{im_i}}^\top)^\top$. For growth-curve analysis (change of y_{it} over time as in longitudinal studies), $\mathbf{z}_{it}$ is often equal to $\mathbf{x}_{it}$. It follows from the assumptions of the LMM that

$$E(\mathbf{y}_i \mid \mathbf{x}_i, \mathbf{z}_i) = \mathbf{x}_i^\top \beta, \quad \operatorname{Var}(\mathbf{y}_i \mid \mathbf{x}_i, \mathbf{z}_i) = \mathbf{z}_i^\top D \mathbf{z}_i + \sigma^2 \mathbf{I}_{m_i}. \tag{2.6}$$

Clustered data also often arises from nested studies. For example, in a multi-center trial, subjects are nested within each center or site, causing clustered responses even when analyzing the data at a single time point. For simplicity, consider modeling treatment differences at a posttreatment assessment time in a multi-site, randomized trial with two treatment conditions. Let y_{ij} denote some response of interest from the jth subject within the ith site and x_{ij} be a binary indicator for the treatment received by the j subject at the ith site. Then, an appropriate LMM is given by:

$$y_{ij} = \beta_0 + x_{ij}\beta_1 + b_i + \varepsilon_{ij}, \quad b_i \sim \text{i.i.d. } N\left(0, \sigma_b^2\right), \quad \varepsilon_{ij} \sim \text{i.i.d. } N\left(0, \sigma^2\right),$$

where b_i denotes the (random) effect of site. We can assess whether there is a significant site effect by testing the null: $H_0 : \sigma_b^2 = 0$. If this null is not rejected, we can simplify the model by dropping b_i.

When the number of sites is small such as two or three, we may want to model potential site difference using a fixed effect. In this case, site difference is interpreted with respect to the particular sites in the study. If there are a large number of randomly selected sites, it is sensible to model site differences using random effects. A significant random effect implies differences not only among the participating sites but also across similar sites not included in the study.

2.2.3 Generalized Linear Mixed-Effects Models

A major limitation of the LMM is that it only applies to continuous response. To model other types of responses such as binary and count data, we must use the generalized linear mixed-effects model (GLMM). We first review the generalized linear model (GLM), the premise underlying GLMM, and then discuss how this class of models is extended to GLMM by adding random effects.

The GLM frames a wide range of seemingly disparate problems of statistical modeling and inference under a unified framework. GLM extends linear regression for a continuous response to models for other types of response such as binary and categorical outcomes. Examples of GLMs include linear regression, logistic regression for binary outcomes, and log-linear regression for count data. We give a brief review of GLM.

The classic multiple linear regression model has the form

$$\begin{aligned} y_i \mid \mathbf{x}_i &\sim \text{i.d.} N\left(\mu_i, \sigma^2\right), \\ \mu_i &= \beta_0 + \beta_1 x_{i1} + \ldots + \beta_p x_{ip} = \beta_0 + \boldsymbol{\beta}^\top \mathbf{x}_i, \end{aligned} \tag{2.7}$$

where i.d. means independently distributed. The response y_i conditional on the covariates $\mathbf{x}_i$ is assumed to have a normal distribution with mean μ_i and common variance σ^2. In addition, μ_i is a linear function of $\mathbf{x}_i$. Since the right side of the model, $\eta_i = \beta_0 + \boldsymbol{\beta}^\top \mathbf{x}_i$, has a range in the real line R, concurring with the range of μ_i on the left side, the linear model is not appropriate for modeling other types of noncontinuous responses. For example, if y_i is binary, the conditional mean of $y_i \mid \mathbf{x}_i$ is

$$\mu_i = E\left(y_i \mid \mathbf{x}_i\right) = \Pr\left(y_i = 1 \mid \mathbf{x}_i\right). \tag{2.8}$$

Since μ_i is a value between 0 and 1, it is not sensible to model μ_i directly as a linear function of $\mathbf{x}_i$ as in (2.7). In addition, the normal distribution assumption does not apply to binary responses.

To generalize the linear model to accommodate other types of response, we must modify (1) the normal distribution assumption; and (2) the relationship between the conditional mean μ_i in (2.8) and the linear predictor η_i in (2.7). GLM addresses both issues by extending (2.7) in the respective directions:

1. *Random component.* This part specifies the conditional distribution of the response y_i given the dependent variables $\mathbf{x}_i$.
2. *Deterministic component.* This part links the conditional mean of y_i given $\mathbf{x}_i$ to the linear predictor $\mathbf{x}_i$ by a one-to-one *link* function g:

$$g(\mu_i) = \eta_i = \mathbf{x}_i^\top \beta, \quad \text{or} \quad \mu_i = h\left(\mathbf{x}_i^\top \beta\right), \quad h = g^{-1}. \tag{2.9}$$

Thus, the linear regression is obtained as a special case if y given $\mathbf{x}$ follows a normal distribution, and $g(\mu)$ is the identity function, $g(\mu_i) = \mu_i$. By varying the distribution function for the random component and the link function $g(\cdot)$ in the deterministic part, we can use GLM to model a variety of response types with different distributions. For example, if a binary response y given $\mathbf{x}$ follows a Bernoulli distribution Bern (μ) with the probability of success given by: $E(y \mid \mathbf{x}) = \pi(\mathbf{x}) = \pi$ and the conditional mean μ is linked to the linear predictor η by the logit function, $\eta = g(\pi) = \log(\pi/(1-\pi))$, we obtain the logistic regression model. For a count response y, GLM yields a Poisson log-linear model, if y given $\mathbf{x}$ follows a Poisson distribution Poisson (μ), with the mean μ linked to the linear predictor by the log function.

To extend GLM to a longitudinal data setting, we simply add random effects, akin to LMM. Consider a study with n subjects and m assessments. For notational brevity, we assume a set of fixed assessment times, $1 \le t \le m$. Let y_{it} denote some response and $\mathbf{x}_{it}$ a vector of covariates from the ith subject at time t ($1 \le i \le n$, $1 \le t \le m$). The principle of extending GLM to a longitudinal data setting is the same as in generalizing the classic univariate linear regression to LMM.

For each subject, we first model the within-subject variability using a GLM:

$$y_{it} \mid \mathbf{x}_{it}, \mathbf{z}_{it}, \mathbf{b}_i \sim \text{i.d. } f(\mu_{it}), \quad g(\mu_{it}) = \mathbf{x}_{it}^\top \beta + \mathbf{z}_{it}^\top \mathbf{b}_i, \tag{2.10}$$

where $g(\cdot)$ is a link function, $\mathbf{z}_{it}$ is a sub-vector of $\mathbf{x}_{it}$, $\mathbf{b}_i$ denotes the random effects, and $f(\mu)$ some probability distribution with mean μ. Note that the model specification in (2.10) is quite similar to the cross-sectional GLM except for the added individual effect $\mathbf{b}_i$. By adding a distribution for $\mathbf{b}_i$ to explain the between-subject variation, we obtain from (2.10) the class of *GLMM*. As in the case of LMM, $\mathbf{b}_i$ is often assumed to follow a multivariate normal $\mathbf{b}_i \sim N(\mathbf{0}, D)$ (a common assumption in most studies), although other types of more complex distributions such as mixtures of normals may also be specified.

To use GLMM for modeling a binary y_{it}, we set $f(\mu_{it}) = \text{Bern}(\mu_{it})$, a Bernoulli with mean μ_{it}. The most popular link for modeling such a response is the logit function. If we model the trajectory of y_{it} over time as a linear function of t with a bivariate normal random effect for the mean and slope, the GLMM becomes

$$y_{it} \sim \text{i.d. BI}(\mu_{it};1), \quad \text{logit}(\mu_{it}) = \mathbf{x}_{it}^\top \beta + \mathbf{z}_{it}^\top \mathbf{b}_i,$$
$$\mathbf{b}_i \sim \text{i.i.d. } N(\mathbf{0}, D), \quad 1 \le i \le n, 1 \le t \le m, \tag{2.11}$$

where $\mathbf{x}_{it}^\top = \mathbf{z}_{it}^\top = (1,t)^\top$ and $\text{logit}(\mu_{it}) = \log(\frac{\mu_{it}}{1-\mu_{it}})$. As in the LMM case, $\beta_0 + \beta t$ describes the change over time for the population average, while the random effect $b_{i0} + b_{i1}t$ accounts for individual deviations.

For a count response y_{it}, we may assuming a Poisson distribution Poisson (μ_{it}) and a log link $\log(\mu_{it})$ to obtain a random-effects based the log-linear model for the trajectory of a count response y_{it} over time:

$$y_{it} \sim \text{i.d. Poisson}(\mu_{it}), \quad \log(\mu_{it}) = \mathbf{x}_{it}^\top \beta + \mathbf{z}_{it}^\top \mathbf{b}_i,$$
$$\mathbf{b}_i \sim \text{i.i.d. } N(\mathbf{0}, D), \quad 1 \le i \le n, 1 \le t \le m. \tag{2.12}$$

The fixed and random effects have the same interpretation as in the GLMM for binary data.

In addition to binary and count response, we can similarly generalize the generalized logit and proportional odds models for multi-level nominal and ordinal responses. Interested readers may consult Kowalski and Tu (2008); Tang et al. (2012) for details.

2.2.4 *Maximum Likelihood Inference*

As the multivariate linear regression in (2.1) is not widely used for longitudinal data analysis, we discuss inference only for the LMM and the GLMM. Readers interested in inference for the classic linear models may consult Seber (1984); Kowalski and Tu (2008). We start with LMM.

Consider inference for the LMM in (2.5). Let $\theta = \left(\beta^\top, \text{vec}^\top(D), \sigma^2\right)^\top$, where $\text{vec}^\top(D)$ denotes the vector operator to convert the symmetric $q \times q$ matrix D into a column $\frac{1}{2}q(q+1) \times 1$ vector consisting of the $\frac{1}{2}q(q+1)$ distinct elements of D. The log-likelihood is given by

$$l_n(\theta) = \sum_{i=1}^{n} \log\left[\int f_{y|x,z,b}(\mathbf{y}_i \mid \mathbf{x}_i, \mathbf{z}_i, \mathbf{b}_i)\, f_b(\mathbf{b}_i)\, d\mathbf{b}_i\right], \tag{2.13}$$

where $f_{y|x,z,b}(\mathbf{y}_i \mid \mathbf{x}_i, \mathbf{z}_i, \mathbf{b}_i)$ denotes the density function of a multivariate normal $N(\mu_i, \Sigma_i)$ with mean $\mu_i = \mathbf{x}_i^\top \beta + \mathbf{z}_i^\top \mathbf{b}_i$ and variance $\Sigma_i = \mathbf{z}_i^\top D \mathbf{z}_i + \sigma^2 \mathbf{I}_{m_i}$, and $f_b(\mathbf{b}_i)$ the density function of the normal random effect $\mathbf{b}_i$. The integral in (2.13) is the result of integrating out the latent $\mathbf{b}_i$.

A major technical problem with inference for mixed-effect models is how to deal with such an integral. Fortunately for the normal–normal based LMM, this integral can be completed in closed form. Since the marginal of $\mathbf{y}_i$ is again normal, $N\left(\mathbf{x}_i^\top \beta, V_i = \left(\mathbf{z}_i^\top D \mathbf{z}_i + \sigma_b^2 \mathbf{I}_{m_i}\right)\right)$, the log-likelihood function is given by

$$l_n(\theta) = -\frac{1}{2}\left[N \ln \sigma^2 + \sum_{i=1}^{n} \log |V_i| + \sigma^2 \sum_{i=1}^{n} \mathbf{e}_i^\top V_i \mathbf{e}_i\right], \tag{2.14}$$

where $N = \sum_{i=1}^{n} m_i$ and $\mathbf{e}_i = \mathbf{y}_i - \mathbf{x}_i^\top \beta$. Note that although the dimension of D is fixed, V_i and $\mathbf{e}_i$ both have a varying dimension from subject to subject. The maximum likelihood estimate (MLE) of θ is obtained by maximizing the log-likelihood $l_n(\theta)$.

Although straightforward in principle, it is actually quite difficult to maximize $l_n(\theta)$ in (2.14), since obtaining the derivatives of $l_n(\theta)$ in closed form is quite a daunting task. Because of the analytic complexity in calculating the derivatives, several algorithms have been proposed. Among them, a popular approach is to use the expectation/maximization (EM) algorithm and its various enhanced versions such as the expectation/maximization either (ECME) algorithm (Dempster et al. 1977; Meng and Van Dyk 1997). These approaches maximize the alternative, expected log-likelihood, allowing one to obtain the MLE in closed form. However, such algorithms are notorious for their slow convergence. In addition, they require additional methods to compute its asymptotic variance, the latter of which is often quite complex and a problem in its own right. Most software packages use the Newton–Raphson algorithm based on the analytic first- and second-order analytic derivatives of $l_n(\theta)$, which has a much faster convergence rate than EM type algorithms (Lindstrom and Bates 1988; Wolfinger and O'connell 1993; Demidenko 2004).

Inference for GLMM with noncontinuous responses such as binary and count outcomes is more difficult, since unlike LMM integration cannot even be completed in closed form for models with the simplest normal random effects. For example, consider the GLMM for binary response in (2.11). Let $\alpha = \text{vec}^\top(D)$ and $\theta = \left(\beta^\top, \alpha^\top\right)^\top$. Since $\mathbf{b}_i$ is latent, the log-likelihood is given by:

$$\begin{aligned} l_n(\theta) &= \sum_{i=1}^{n} \log\left[\int f_{y|x,z,b}(\mathbf{y}_i \mid \mathbf{b}_i, \beta) f_b(\mathbf{b}_i \mid \alpha)\, d\mathbf{b}_i\right] \\ &= \sum_{i=1}^{n} \log\left[\int \prod_{t=1}^{\mathrm{m}} \mu_{it}^{y_{it}} (1-\mu_{it})^{1-y_{it}} f_b(\mathbf{b}_i \mid \alpha)\, d\mathbf{b}_i\right], \end{aligned} \tag{2.15}$$

where $f_{y|x,z,b}(\mathbf{y}_i \mid \mathbf{b}_i, \beta) = \prod_{t=1}^{\mathrm{m}} \mu_{it}^{y_{it}} (1-\mu_{it})^{1-y_{it}}$ is the conditional joint density of m independent binary components of $\mathbf{y}_i = (y_{i1}, \ldots, y_{im})^\top$ given the random effect $\mathbf{b}_i$ with mean $\mu_{it} = \exp(\beta_0 + \beta t + b_{i0} + b_{i1} t)$ and $f_b(\mathbf{b}_i \mid \alpha)$ is the joint density of the binary normal $\mathbf{b}_i = (b_{i0}, b_{i1})^\top$. Even in this simplest setting, $l_n(\theta)$ cannot be expressed in closed form, making it impossible to obtain $\frac{\partial}{\partial\theta} l_n(\theta)$ in closed form.

Different algorithms have been proposed to tackle the computational challenges. Because of the lack of consensus, different approaches have been adopted by and implemented in major statistical software packages such as SAS and R. As a result, it is common to obtain different estimates for the same model, when different packages are used. Understanding some basics of the computational aspects may be helpful in dealing with the discrepancies in real study applications.

The two most popular approaches for estimation in GLMM as implemented in major software packages such as SAS and R are: (1) approximating the log-likelihood function, and (2) approximating the model. Algorithms in the second category are developed using linearization of or Laplace approximation to the model. They employ some types of expansions such as the Taylor series expansion to approximate the model by one based on pseudo-data with fewer nonlinear components. The process of computing the linear approximation must be repeated several times until some criteria indicate lack of further improvement (Pinheiro and Bates 2000; Schabenberger and Gregoire 1996). The fitting methods based on such a linearization process typically involve two levels of iterations. The GLMM is first approximated by a LMM based on the current values of the covariance parameter estimates, and the resulting LMM is then fit, which itself is an iterative process. On convergence, the new parameter estimates are used to update the linearization, resulting in a new LMM. The iterative process terminates when the difference in parameter estimates between successive LMM fits falls within a specified tolerance level.

The advantages of the linearization approach include a relatively simple form of the linearized model that typically can be fit based only on the mean and variance in the linearized form. This approach can fit models with a large number of random effects, crossed random effects, multiple types of subjects, and even correlated y_{it}'s after conditioning on $\mathbf{x}_{it}$, $\mathbf{z}_{it}$, and $\mathbf{b}_i$. However, as it does not maximize the underlying log-likelihood function, this approach produces estimates with unknown asymptotic properties, except in the special LMM case for which this approach does produce MLE (Demidenko 2004). A recent study based on simulation show that this linearization approach generally provides invalid inference (Zhang et al. 2011b). In addition, algorithms implementing this approach can fail at both levels of the double iteration scheme.

In contrast, the integral approximation approach aims to directly approximate the log-likelihood in (2.15) and maximize the approximated function (Davidian and Gallant 1993; Demidenko 2004; Breslow and Clayton 1993). Since the accuracy of estimates and inference is determined by the quality of the approximation to the log-likelihood, various techniques have been proposed to compute the approximation, with the Newton and Gauss–Hermite quadratures being the most popular. The advantage of this alternative approach is to provide an actual objective function for optimization, albeit it is an approximated log-likelihood. Nonetheless, this enables likelihood-based fit statistics such as the likelihood ratio test to be used for inference.

Further, unlike its linearization counterpart whose approximation accuracy and thus the asymptotic properties of the estimates are limited by the type of models fit (e.g., linear, logistic, etc.), the approximation to the log-likelihood under this approach can be improved to any degree by increasing the precision of numerical integration, at least in principle.

Thus, algorithms based on the integral approximation are expected to provide better estimates than those based on linearization. In particular, since they can get arbitrarily close to the MLE as the precision of numerical integration increases, estimates obtained under this approach have the same nice large sample properties as the MLE such as consistency, asymptotic normality and efficiency. However, the quality of approximation seems to vary tremendously across the different packages. For example, as shown by Zhang et al. (2011a) using simulated data, none of the available functions in R provides valid inference. Thus, if this approach is sought, the NLMIXED procedure in SAS should be used unless improvements have been made to the others, as it is the only software to provide correct inference at the moment of this writing among a number of available packages evaluated based on simulated data by Zhang et al. (2011a).

2.2.5 *Composite Hypothesis Testing*

Hypotheses concerning the parameter vector $\theta = (\beta, \alpha)$ for most applications can be expressed in the following form:

$$H_0 : C\theta = \mathbf{b}, \quad \text{vs.} \quad H_a : C\theta \neq \mathbf{b}, \tag{2.16}$$

where $\mathbf{b}$ is a known constant vector, C is some known full rank $k \times p$ matrix with $p\ (\geq k)$ denoting the dimension of θ. If $\mathbf{b} = \mathbf{0}$, H_0 becomes a linear contrast. If θ only consists of parameters for the fixed-effect, i.e., $\theta = \beta$, we can re-express the linear hypothesis in terms of a linear contrast by performing the transformation: $\gamma = \beta - C^\top \left(C^\top C\right)^{-1} \mathbf{b}$. When expressed in the new parameter vector γ, the linear predictor will contain an offset term.

For example, the linear predictor for the GLMM in (2.10) under this transformation becomes

$$\eta_{it} = \mathbf{x}_{it}^\top \beta + \mathbf{z}_{it_i}^\top \mathbf{b}_i = c_{it} + \mathbf{x}_{it}^\top \gamma + \mathbf{z}_{it_i}^\top \mathbf{b}_i,$$

where $c_{it} = \mathbf{x}_{it}^\top \left(C^\top \left(C^\top C\right)^{-1} \mathbf{b}\right)$ is the offset. Except for LMM, where c_{it} can be absorbed into the response by redefining the dependent variable as $z_{it} = y_{it} - c_{it}$, the offset must be specified when fitting GLMM using software packages.

The most popular tests for linear hypothesis are the Wald, score, and likelihood ratio statistics. We briefly review these tests below.

If $\widehat{\theta} \sim N\left(\theta, \frac{1}{n}\Sigma_\theta\right)$, then it follows from the properties of multivariate normal distribution that $K\widehat{\theta} \sim N\left(K\theta, \frac{1}{n}K\Sigma_\theta K^\top\right)$. Thus, under the null (2.16), $K\widehat{\theta} \sim N$ $\left(\boldsymbol{b}, \frac{1}{n}K\Sigma_\theta K^\top\right)$. The Wald statistic,

$$Q_n^2 = n\left(K\widehat{\theta} - \boldsymbol{b}\right)^\top \left(K\Sigma_\beta K^\top\right)^{-1}\left(K\widehat{\theta} - \boldsymbol{b}\right),$$

follows asymptotically a chi-square distribution with l degrees of freedom $\left(\chi_l^2\right)$, where l is the rank of K. Note that because K is full rank, $K\Sigma_\theta K^\top$ is invertible. The likelihood-ratio statistic is defined as

$$LR = 2\log R\left(\widetilde{\theta}\right) = 2\left[\log L\left(\widehat{\theta}\right) - \log L\left(\widetilde{\theta}\right)\right],$$

where $L(\theta)$ denotes the likelihood function, $\widehat{\theta}$ the MLE of θ, and $\widetilde{\theta}$ the MLE of the constrained model under the null hypothesis. The likelihood-ratio statistic also follows asymptotically a chi-square distribution with l degrees of freedom.

Both the Wald and likelihood ratio tests require the existence and estimate of the MLE of θ. A third popular alternative, the score test, only requires the existence and the MLE of θ under the restricted model. In general, we can reparameterize θ through a linear transformation so that θ can be decomposed as $\theta = \left(\theta_1^\top, \theta_2^\top\right)^\top$, with the null in (2.16) expressed as $\theta_2 = \mathbf{c}$ (a constant vector). Let $l_i(\theta)$ be the log-likelihood associated with the ith subject. The score equation is given by

$$\mathbf{w}_n^{(1)}(\theta) = \frac{1}{n}\sum_{i=1}^{n}\frac{\partial l_i(\theta)}{\partial \theta_1} = \mathbf{0}, \quad \mathbf{w}_n^{(2)}(\theta) = \frac{1}{n}\sum_{i=1}^{n}\frac{\partial l_i(\theta)}{\partial \theta_2} = \mathbf{0}.$$

Let $\widetilde{\theta}_1$ denote the MLE estimate of θ_1 under the null obtained by solving $w_n^{(1)}(\theta_1, \mathbf{c}) = \mathbf{0}$. Then the score statistic,

$$T_s\left(\widetilde{\theta}_1, \mathbf{c}\right) = n\left(\mathbf{w}_n^{(2)}\left(\widetilde{\theta}_1, \mathbf{c}\right)\right)^\top \widehat{\Sigma}_2^{-1}\left(\widetilde{\theta}_1, \mathbf{c}\right)\mathbf{w}_n^{(2)}\left(\widetilde{\theta}_1, \mathbf{c}\right),$$

follows a χ_q^2, where $\widehat{\Sigma}_2$ is the asymptotic variance of $w_n^{(2)}$ and q is the dimension of θ_2. Since only θ_1 is estimated based on $\theta_2 = \mathbf{c}$, the score statistic does not require the existence of MLE for the full model.

As the likelihood ratio test only depends on the height, rather than the curvature of the likelihood function, it usually derives more accurate inference than the Wald statistic. Thus, although asymptotically equivalent, it is generally preferred. Note that the likelihood ratio test is only appropriate for parametric models, while the Wald and score tests also apply to distribution-free models to be discussed next.

2.3 Distribution-Free Models

A major problem with using random effects to account for correlated responses is the difficulty in empirically validating the distribution assumption for the random effects because of their latent or unobservable nature. To further exacerbate the problem, GLMM also relies on parametric assumptions for the response (conditional on the fixed- and random-effects) such as normality for inference. If either of these assumptions is violated, estimates will be biased. Unfortunately, such assumptions are often unacknowledged, laying the basis for inconsistent and spurious findings. A popular alternative is to use models that are free of such assumptions, yielding estimates that are robust to a much wide range of data distributions. We start with distribution-free models for cross-sectional data.

2.3.1 Distribution-Free Models for Cross-Sectional Data

In contrast to the parametric GLM in Sect. 2.2.3, distribution-free alternatives only posit a relationship between the response and the set of predictors without the random component that postulates an analytic distribution for the response (conditional on the predictors), thereby providing more robust estimates regardless of the complexity of the data distribution. As noted in Sect. 2.1, outcomes from most instruments such as those for assessing quality of life are prone to violations of distribution assumptions, because they are not even continuous, let alone following the normal distribution. The removal of the random component of GLM, however, entails serious ramifications for inference about model parameters; without a distribution model specified in this part of GLM, it is not possible to use maximum likelihood for parameter estimation. Thus, we need an alternative paradigm for inference about parameters.

By removing the random component in GLM and modeling the mean of response, we obtain from (2.9) the distribution-free GLM:

$$g\left(\mu_i\right) = \mathbf{x}_i^\top \beta, \quad 1 \leq i \leq n, \tag{2.17}$$

where $g(\cdot)$ has the same interpretation as in (2.9). For example, for a binary (count) response, $g(\mu) = \text{logit}(\mu)$ $(g(\mu) = \log \mu)$. Because of the absence of a parametric distribution model, MLEs cannot be computed for distribution-free GLM. Inference is typically based on a set of estimating equations (EE):

$$\mathbf{w}_n(\beta) = \sum_{i=1}^{n} D_i V_i^{-1} S_i = \mathbf{0},$$

$$D_i = \frac{\partial}{\partial \beta}\mu_i, \quad V_i = v\left(\mu_i\right), \quad S_i = y_i - \mu_i, \tag{2.18}$$

In the above, S_i is called the *theoretical residual* (to differentiate it from the *observed residual* with estimated μ_i). The quantity V_i is assumed to be a function of μ_i. With the right selection of V_i, the estimating equations in (2.18) yield the MLE of β for the parametric GLM, when y_i is modeled by the exponential family of distributions (e.g., Kowalski and Tu 2008, Chap. 4).

For example, for a count response y_i following the Poisson, $\text{Var}(y_i \mid \mathbf{x}_i) = \mu_i$. Setting $V_i = \mu_i$ in (2.18) yields the MLE of β, if y_i is modeled by a Poisson log-linear regression. However, the advantage of the estimating equations is that even when y_i does not follow a Poisson the estimate obtained from (2.18) with $V_i = \mu_i$ is still consistent and asymptotically normal.

For example, the Poisson distribution is inappropriate for modeling count data with overdispersion, i.e., $\text{Var}(y_i \mid \mathbf{x}_i) > \mu_i$. Under parametric analysis, we must choose a different model to make allowances for such overdispersions. The negative binomial is a popular choice to address overdispersed count data, as it has the same mean as but a larger variance than the Poisson. However, since the distribution-free model only involves the mean response, the distinction between the Poisson and NB in parametric analysis does not arise for distribution-free models. In fact, such a model yields valid inference regardless of whether y_{it} follows the Poisson, NB, or any other distributions, so long as the relationship between the (conditional) mean and the covariates specified in (2.18) is correct. This feature is particularly important for longitudinal data analysis, since the conditional variance is much more complex and difficult to specify for correlated longitudinal responses.

2.3.2 Distribution-Free Models for Longitudinal Data

Consider a longitudinal study with n subjects and m assessment times, and again for notational brevity, assume a set of fixed assessment times $1 \leq t \leq m$, with y_{it} denoting a response and $\mathbf{x}_{it}$ a set of independent variables of interest from the ith subject at time t, as in the discussion of parametric models in Sect. 2.2. By applying the distribution-free GLM in (2.17) to each time t, we obtain a class of distribution-free regression models within the current context of longitudinal data:

$$E(y_{it} \mid \mathbf{x}_{it}) = \mu_{it}, \quad g(\mu_{it}) = \mathbf{x}_{it}^\top \beta, \quad 1 \leq t \leq m, \quad 1 \leq i \leq n, \qquad (2.19)$$

where β is a $p \times 1$ vector of parameters of interest. Note that at each time t, the distribution-free GLM above is exactly the same as the cross-sectional version in (2.17). For example, if y_{it} is a continuous response, $\mu_{it} = \mathbf{x}_{it}^\top \beta$ models the mean of y_{it} as a function of $\mathbf{x}_{it}$ at each time t, while for a binary y_{it}, $\mu_{it} = E(y_{it})$, which relates to $\mathbf{x}_{it}$ via a link function $g(\mu_{it})$ such as the logit.

Note that y_{it} in (2.19) can be a continuous, binary, or count response. Distribution-free models are also available for multi-level categorical and ordinal responses. See Kowalski and Tu (2008, Chap. 4) for details.

2.3.3 Inference for Distribution-Free Models

Consider the class of distribution-free models in (2.19). At each time t, (2.19) reduces to the distribution-free model for cross-sectional data in (2.17). The generalized estimating equations (GEE) are used to provide inference for the longitudinal GLM in (2.19) by extending the estimating equations in (2.18) for a single time t to multiple times across all assessments. This is achieved by capitalizing on the fact that the use of a wrong correlation matrix has no impact on the consistency of the GEE estimate of β, just as in the univariate case that the misspecification of $V(\cdot)$ does not affect the consistency of the estimating equations estimate. Let

$$\mathbf{y}_i = \left(y_{i1}, \ldots, y_{im}\right)^\top, \quad \mu_i = \left(\mu_{i1}, \ldots, \mu_{im}\right)^\top,$$
$$\mathbf{x}_i = \left(\mathbf{x}_{i1}^\top, \ldots, \mathbf{x}_{im}^\top\right)^\top, \quad S_i = \mathbf{y}_i - \mu_i.$$

In analogy to (2.18), the GEE are defined by

$$\mathbf{w}_n = \sum_{i=1}^{n} G_i(\mathbf{x}_i) S_i = \sum_{i=1}^{n} G_i(\mathbf{x}_i)(\mathbf{y}_i - \mu_i) = \mathbf{0}, \tag{2.20}$$

where $G_i(\mathbf{x}_i)$ is some matrix function of $\mathbf{x}_i$. In most applications, $G_i(\mathbf{x}_i)$ has the form

$$G_i(\mathbf{x}_i) = D_i V_i^{-1}, \quad D_i = \frac{\partial}{\partial \beta}\mu_i, \quad A_i = \operatorname{diag}(v(\mu_{it})),$$
$$V_i = A_i^{\frac{1}{2}} R(\alpha) A_i^{\frac{1}{2}}, \tag{2.21}$$

where $R(\alpha)$ denotes a *working correlation* matrix parameterized by α, diag$(v(\mu_{it}))$ a diagonal matrix with $v(\mu_{it})$ on the tth diagonal. As in the case of cross-sectional data, $v(\mu_{it})$ can be set equal to $\operatorname{Var}(y_{it} \mid \mathbf{x}_{it})$ under some parametric assumptions, such as the mean μ_{it} for a count response based on the Poisson distribution. The phrase "working correlation" is used to emphasize the fact that $R(\alpha)$ is not necessarily the true correlation matrix. For example, we may simply set $R = \mathbf{I}_m$. In this case, the correlated components of $\mathbf{y}_i$ are treated as if they were independent. In addition, there is no parameter associated with this particular *working independence model*. Another popular choice is the uniform compound symmetry correlation matrix, $R(\alpha) = C_m(\rho)$, which assumes a common correlation ρ for any pair of responses y_{is} and y_{it} $(1 \leq s, t \leq m)$. This working correlation matrix involves a single parameter ρ.

Under the specification in (2.21), (2.20) can be expressed as

$$\mathbf{w}_n(\beta) = \sum_{i=1}^{n} D_i V_i^{-1} S_i = \sum_{i=1}^{n} D_i V_i^{-1}(\mathbf{y}_i - \mu_i) = \mathbf{0}. \tag{2.22}$$

The above is identical to the estimating equations in (2.18) for cross-sectional data analysis, except that D_i in (2.21) is a $p \times m$ matrix rather than a $p \times 1$ vector. Although the GEE in (2.22) is a function of both β and α, we express it explicitly as a function of β to emphasize the fact that (2.22) is used to obtain the estimate of the parameter vector of interest β. If α is known as in the case of working independence model, we can obtain the solution of β to (2.22) by the following recursive relation based on the Newton–Raphson algorithm:

$$\beta^{(k+1)} = \beta^{(k)} + \left(\sum_{i=1}^{n} D_i V_i^{-1} D_i \right)^{-1} \mathbf{w}_n \left(\beta^{(k)} \right), \tag{2.23}$$

where $\beta^{(0)}$ denotes some initial value. The GEE estimate $\widehat{\beta}$ is obtained by iterating the above until convergence.

When α is unknown, we must estimate it so that (2.22) can be used to find estimates of β. For example, consider modeling a binary response y_{it} with the logistic model:

$$E\left(y_{it} \mid \mathbf{x}_{it}\right) = \mu_{it}, \quad \log\left(\frac{\mu_{it}}{1-\mu_{it}}\right) = \mathbf{x}_{it}^{\top}\beta, \quad 1 \le i \le n, 1 \le t \le m. \tag{2.24}$$

Since $\text{Var}\left(y_{it} \mid \mathbf{x}_{it}\right) = \mu_{it}\left(1-\mu_{it}\right)$, $A_i = \text{diag}(\mu_{it}\left(1-\mu_{it}\right))$. If α is known, $\widehat{\beta}$ can be obtained from (2.23). Otherwise, an estimate of α is needed to compute β using the Newton–Raphson algorithm. As an example, consider the completely unstructured working correlation matrix, i.e., $R(\alpha) = [\rho_{st}]$. We can estimate ρ_{st} by the Pearson correlation:

$$\widehat{\rho}_{st} = \frac{\sum_{i=1}^{n}\left(\widehat{e}_{is} - \overline{r}_{\cdot s}\right)\left(\widehat{e}_{it} - \overline{\widehat{e}}_{\cdot t}\right)}{\sqrt{\sum_{i=1}^{n}\left(\widehat{e}_{is} - \overline{\widehat{e}}_{\cdot s}\right)^2 \sum_{i=1}^{n}\left(\widehat{e}_{it} - \overline{\widehat{e}}_{\cdot t}\right)^2}}, \quad \widehat{e}_{it} = y_{it} - \mathbf{x}_{it}^{\top}\widehat{\beta},$$

$$\overline{\widehat{e}}_{\cdot t} = \frac{1}{n}\sum_{i=1}^{n}\widehat{e}_{it}, \quad 1 \le i \le n, \quad 1 \le t \le m. \tag{2.25}$$

Note that since the correlation $\rho_{i,st} = \text{Corr}\left(y_{is}, y_{it} \mid \mathbf{x}_{is}, \mathbf{x}_{it}\right)$ for binary response generally varies across subjects, a constant $R(\alpha)$ is hardly ever the true correlation matrix except for some special cases. Further, $\rho_{i,st}$ must also satisfy an additional set of *Frechet bounds* (e.g., Shults et al. 2009):

$$\max\left\{-\left[\frac{\mu_{is}\mu_{it}}{(1-\mu_{is})(1-\mu_{it})}\right]^{1/2}, -\left[\frac{(1-\mu_{is})(1-\mu_{it})}{\mu_{is}\mu_{it}}\right]^{1/2}\right\}$$

$$\leq \rho_{i,st} \leq \min\left\{\left[\frac{\mu_{is}(1-\mu_{it})}{(1-\mu_{is})\mu_{it}}\right]^{1/2}, \left[\frac{(1-\mu_{is})\mu_{it}}{\mu_{is}(1-\mu_{it})}\right]^{1/2}\right\}. \tag{2.26}$$

Because of these constraints, α is typically selected by some ad hoc rules rather than estimated. For example, under the uniform compound symmetry assumption, $R(\rho) = C_m(\rho)$, we may select ρ to satisfy the *Frechet bounds* in (2.26).

While primary interest lies in β, α must be estimated to proceed with the computation of the GEE estimate of β. Although the consistency of the GEE estimate $\widehat{\beta}$ is independent of how α is estimated, judicious choices of the type of estimates of α not only ensure the asymptotic normality but also simplify the asymptotic variance of $\widehat{\beta}$. To this end, we require that $\widehat{\alpha}$ be $\sqrt{n}$-consistent, i.e., $\widehat{\alpha}$ converges to some point α and $\sqrt{n}(\widehat{\alpha} - \alpha)$ is bounded in probability. Most popular estimates of $\widehat{\alpha}$ such as the moment estimate are asymptotically normal and thus are $\sqrt{n}$-consistent. Given such an estimate of $\widehat{\alpha}$, the GEE estimate $\widehat{\beta}$ is consistent and asymptotically normal, with the asymptotic variance given by

$$\Sigma_\beta = B^{-1} E\left(G_i S_i S_i^\top G_i^\top\right) B^{-\top}, \quad B^{-\top} = \left(B^{-1}\right)^\top. \tag{2.27}$$

A consistent estimate of Σ_β is given by

$$\widehat{\Sigma}_\beta = \frac{1}{n}\widehat{B}^{-1}\sum_{i=1}^{n}\left(\widehat{G}_i \widehat{S}_i \widehat{S}_i^\top \widehat{G}_i^\top\right)\widehat{B}^{-\top}. \tag{2.28}$$

where $\widehat{A}$ denotes the estimated A obtained by replacing β and α with their respective estimates.

Note that in some cases $\sqrt{n}$-consistent estimates $\widehat{\alpha}$ and their limits α may give rise to some or all entries of the working correlation matrix that exceed one (Crowder 1995). However, such examples are rare in practice, but even when this occurs, the GEE estimate $\widehat{\beta}$ is still consistent and asymptotically normal, although efficiency may be an issue.

2.4 Missing Values

Missing data occurs in most longitudinal studies, including well-designed and carefully executed clinical trials. In longitudinal studies, subjects may simply quit the study or they may not show up at follow-up visits because of problems with transportation, weather condition, relocation, and so on. We characterize the impact of missing data on model estimates through modeling assumptions. Such assumptions allow one to ignore the multitude of reasons for missing data and focus instead on addressing their impact on inference.

The *missing completely at random* (MCAR) assumption is used to define a class of missing data that does not affect model estimates when completely ignored. For example, missing data resulting from relocation and lack of transportation typically follows this model. In clinical trials, missing data may also be the results of patients' deteriorated or improved health conditions due to treatment and treatment-related complications. For example, if some patients in a study feel that the interventions received have resulted in no change or even deteriorated health conditions, and any further treatment will only worsen their overall physical and mental health, they may simply quit the study. On the other hand, some others may feel that they have completely responded to the treatment and do not see any additional benefit in continuing the treatment, they may also choose to stop participating in the study. In both scenarios, missing data does not follow the MCAR mechanism, since the missingness of these patients' data is related to treatment effects.

The *missing at random* (MAR) assumption, which posits that the occurrence of a missing response at an assessment time depends on the observed data prior to the assessment point, attempts to model such a treatment-dependent missing data mechanism. In the two missing data scenarios above, MAR is a reasonable model, since the missingness is a function of past treatment responses.

Missing data satisfying either the MCAR or MAR model is known as *ignorable missing data*. Thus, the *Nonignorable nonresponse* (NINR) or *missing not at random* (MNAR) mechanism encompasses the remaining class of missing data whose occurrence depends on unobserved data, such as current and/or future responses as in a longitudinal study. This category of missing data is generally quite difficult to model without additional information (from other sources), because of lack of information from the study data.

Note that the term "ignorable missing" may be a misnomer. For parametric models, we can indeed ignore such missing data since MLEs are consistent. However, the validity of inference for parametric models depends on the model assumptions; if there is some serious violation of the assumed distributional models, MLEs will be biased (Lu et al. 2009). On the other hand, distribution-free models such as the distribution-free GLM discussed in Sect. 2.3 provide valid inference without any distribution assumption. But for such robust models, "ignorable missing" data may not be ignored. For example, GEE estimates discussed in Sect. 2.3 are generally biased when missing data follows MAR (Lu et al. 2009). We discuss a new class of estimating equations to provide valid inference under MAR in Sect. 2.4.2.

We focus on MCAR and MAR, which apply to most studies in biomedical and psychosocial research, but will mention some popular approaches for NINR in Discussion. In addition, we only consider missingness in the response, as missingness in the independent variables is much more complex to model.

2.4.1 Inference for Parametric Models

Let y_{it} be the response and $\mathbf{x}_{it}$ a vector of independent variables of interest for the ith subject at time t from a longitudinal study design with n subjects and m assessments. We define a missing, or rather observed, data indicator as follows:

$$r_{it} = \begin{cases} 1 \text{ if } y_{it} \text{ is observed} \\ 0 \text{ if } y_{it} \text{ is missing} \end{cases}, \quad \mathbf{r}_i = (r_{i1}, \ldots, r_{im})^\top. \tag{2.29}$$

We assume no missing data at baseline $t = 1$ so that $r_{i1} = 0$ for all $1 \le i \le n$.

Let $\mathbf{y}_i = (y_{i1}, \ldots, y_{im})^\top$ and $\mathbf{x}_i = (\mathbf{x}_{i1}^\top, \ldots, \mathbf{x}_{im}^\top)^\top$. Let $\mathbf{y}_i^{\mathrm{o}}$ and $\mathbf{y}_i^{\mathrm{m}}$ denote the observed and unobserved responses, respectively. Thus, $\mathbf{y}_i^{\mathrm{o}}$ and $\mathbf{y}_i^{\mathrm{m}}$ form a partition of $\mathbf{y}_i$. Under likelihood-based parametric inference, the joint density, $f(\mathbf{y}_i, \mathbf{r}_i \mid \mathbf{x}_i, \theta)$, can be factored into:

$$f(\mathbf{y}_i, \mathbf{r}_i \mid \mathbf{x}_i) = f(\mathbf{y}_i \mid \mathbf{x}_i)\, f(\mathbf{r}_i \mid \mathbf{y}_i, \mathbf{x}_i). \tag{2.30}$$

Under MAR, the distribution of $\mathbf{r}_i$ only depends on the observed responses, $\mathbf{y}_i^{\mathrm{o}}$, yielding

$$f(\mathbf{r}_i \mid \mathbf{y}_i, \mathbf{x}_i) = f\left(\mathbf{r}_i \mid \mathbf{y}_i^{\mathrm{o}}, \mathbf{y}_i^{\mathrm{m}}, \mathbf{x}_i\right) = f\left(\mathbf{r}_i \mid \mathbf{y}_i^{\mathrm{o}}, \mathbf{x}_i\right). \tag{2.31}$$

It then follows from (2.30) and (2.31) that

$$\begin{aligned} f\left(\mathbf{y}_i^{\mathrm{o}}, \mathbf{r}_i \mid \mathbf{x}_i\right) &= \int f\left(\mathbf{y}_i^{\mathrm{o}}, \mathbf{y}_i^{\mathrm{m}} \mid \mathbf{x}_i\right) f\left(\mathbf{r}_i \mid \mathbf{y}_i^{\mathrm{o}}, \mathbf{x}_i\right) d\mathbf{y}_i^{\mathrm{m}} \\ &= f\left(\mathbf{y}_i^{\mathrm{o}} \mid \mathbf{x}_i, \theta_y\right) f\left(\mathbf{r}_i \mid \mathbf{y}_i^{\mathrm{o}}, \mathbf{x}_i, \theta_{y|r}\right), \end{aligned} \tag{2.32}$$

where θ_y and $\theta_{y|r}$ denote a partition of θ. If θ_y and $\theta_{y|r}$ are disjoint, it follows from (2.32) that the log-likelihood based on the joint observations $\left(\mathbf{y}_i^{\mathrm{o}}, \mathbf{r}_i\right)$ can be expressed as

$$\begin{aligned} l(\theta) &= l_1\left(\theta_y\right) + l_2\left(\theta_{y|r}\right) \\ &= \sum_{i=1}^{n} \log\left(f\left(\mathbf{y}_i^{\mathrm{o}} \mid \mathbf{x}_i, \theta_y\right)\right) + \sum_{i=1}^{n} \log\left(f\left(\mathbf{r}_i \mid \mathbf{y}_i^{\mathrm{o}}, \mathbf{x}_i, \theta_{y|r}\right)\right). \end{aligned} \tag{2.33}$$

Within the context of GLMM, $l_1\left(\theta_y\right)$ above is the likelihood in (2.13) based on the observed data. Thus, we can simply use $l_1\left(\theta_y\right)$ for inference about the regression relationship between $\mathbf{y}_i$ and $\mathbf{x}_i$. In other words, under MAR, missing data can be "ignored,"insofar as this regression rclationship is concerned.

Note that inference based on $f\left(\mathbf{y}_i^{\mathrm{o}} \mid \mathbf{x}_i, \theta_y\right)$ may be incorrect, if θ_y and $\theta_{y|r}$ are not disjoint. In practice, it is difficult to validate this disjoint assumption. However, under MCAR, it follows from (2.31) and (2.32) that $f\left(\mathbf{r}_i \mid \mathbf{y}_i^{\mathrm{o}}, \mathbf{x}_i, \theta_{y|r}\right) = f\left(\mathbf{r}_i \mid \mathbf{x}_i, \theta_{y|r}\right)$, implying that θ_y and $\theta_{y|r}$ are disjoint. Thus, only under MCAR can missing data be truly ignored when making inference about the relationship between $\mathbf{y}_i$ and $\mathbf{x}_i$.

Note also that if there is some serious violation of the assumed distributional models, estimates obtained by maximizing $l_1(\theta_y)$ are generally not consistent. Some robust estimates have been adopted to improve the validity of inference (Goldstein 1995; Rasbash et al. 2009; Raudenbush and Bryk 2002). These estimates are essentially equivalent to the GEE estimates. However, as GEE estimates are consistent only under MCAR, rather the more general MAR, these robust adjustments do not provide valid inference either (Lu et al. 2009). To obtain consistent estimates under MAR and violations of distribution assumptions, we must use a new class of estimating equations.

2.4.2 Inference for Distribution-Free Model

Define the *probability weight* π_{it} and *inverse probability weight* Δ_{it} function as follows:

$$\pi_{it} = \Pr(r_{it} = 1 \mid \mathbf{x}_i, \mathbf{y}_i), \quad \Delta_{it} = \frac{r_{it}}{\pi_{it}}, \quad \Delta_i = \operatorname{diag}_t(\Delta_{it}). \tag{2.34}$$

Given Δ_i, we can use the following weighted generalized estimating equations (WGEE) to estimate β:

$$\mathbf{w}_n(\beta) = \sum_{i=1}^{n} G_i \Delta_i S_i = \sum_{i=1}^{n} D_i V_i^{-1} \Delta_i (\mathbf{y}_i - \mathbf{h}_i) = \mathbf{0}, \tag{2.35}$$

where D_i, V_i, and S_i are defined the same way as those in (2.22). If the probability weight function π_{it} is known, the WGEE above can be readily solved for β using the Newton–Raphson algorithm discussed in Sect. 2.3.3. In some multi-stage studies where only a fraction of subjects move to the next stage, π_{it} is known and the WGEE can be used to provide inference for such studies. However, in most longitudinal studies, π_{it} is unknown and must be estimated. In general, it is quite difficult to model and estimate π_{it}, as it may depend on unobserved current, $\mathbf{x}_{it}$ and y_{it}, or even future, $\mathbf{x}_{is}$ and y_{is} $(t \le s \le m)$, observations. But, if missing data follows either MCAR or MAR, π_{it} can be modeled as discussed below.

Under MCAR, $\mathbf{r}_i$ is independent of $\mathbf{x}_i$ and $\mathbf{y}_i$. It follows from (2.34) that π_{it} is a constant and is readily estimated by the sample moment:

$$\widehat{\pi}_t = \frac{1}{n} \sum_{i=1}^{n} r_{it}, \quad 2 \le t \le m, \quad 1 \le i \le n.$$

Under MAR, π_{it} is not a constant, and the above is not a valid estimate. However, as π_{it} is a function of observed $\mathbf{x}_i$ and $\mathbf{y}_i$, we should be able to model π_{it}. For example, if $m = 2$ as in a pre-post study, π_{i2} is a function of observed $\mathbf{x}_{i1}$ and y_{i1}, $\pi_{i2} = \Pr(r_{i2} = 1 \mid y_{i1}, \mathbf{x}_{i1})$, which may be modeled using logistic regression.

However, as m increases, it becomes more difficult to model π_{it}; unlike the pre-post design discussed earlier, there could be many different patterns (2^{m-1} total) in the observed data. Thus, we generally assume the monotone missing data pattern (MMDP) to facilitate modeling. Under MMDP, a subject with missing y_{it} and $\mathbf{x}_{it}$ implies that all subsequent components, y_{is} and $\mathbf{x}_{is}$ ($t \le s \le m$), are also missing. With the help of MMDP, we can express π_{it} as

$$\pi_{it} = \Pr(r_{it} = 1 \mid \mathbf{y}_{it-}, \mathbf{x}_{it-}), \quad 1 \le i \le n, \quad 2 \le t \le T, \tag{2.36}$$

where $\mathbf{y}_{it-} = \left(y_{i1}, \ldots, y_{i(t-1)}\right)^\top$ and $\mathbf{x}_{it-} = \left(\mathbf{x}_{i1}, \ldots, \mathbf{x}_{i(t-1)}\right)^\top$. Since

$$\pi_{it} = \prod_{s=2}^{t} p_{is}, \quad p_{is} = \Pr(r_{is} = 1 \mid \mathbf{y}_{is-}, \mathbf{x}_{is-}, r_{i(s-1)} = 1). \tag{2.37}$$

we can estimate π_{it} through estimates of one-step transition probabilities p_{is} ($2 \le s \le t$).

We can readily model p_{it} by logistic regression:

$$\operatorname{logit}(p_{it}(\gamma_t)) = \gamma_{0t} + \gamma_{xt}^\top \mathbf{x}_{it-} + \gamma_{yt}^\top \mathbf{y}_{it-}, \quad 2 \le t \le m, \tag{2.38}$$

where $\gamma_t = \left(\gamma_{0t}, \gamma_{xt}^\top, \gamma_{yt}^\top\right)^\top$ denotes the model parameters. We can estimate each γ_t using either maximum likelihood or estimating equations. For example, if using maximum likelihood, the score equations for each γ_t are given by

$$\mathbf{Q}_{it} = \frac{\partial}{\partial \gamma_t} \left\{ r_{i(t-1)} \left[r_{it} \log(p_{it}) + (1 - r_{it}) \log(1 - p_{it}) \right] \right\},$$
$$2 \le t \le m, \quad i = 1, 2, \ldots n.$$

By combining the score equations across all γ_t's, we obtain a set of estimating equations for joint γ_t as

$$\mathbf{Q}_n(\gamma) = \sum_{i=1}^{n} \left(\mathbf{Q}_{i2}^\top, \ldots, \mathbf{Q}_{im}^\top\right)^\top = \mathbf{0},$$

where $\gamma = \left(\gamma_2^\top, \ldots, \gamma_m^\top\right)^\top$.

As in the case of GEE, the WGEE estimate $\widehat{\beta}$ is asymptotically normal if $\widehat{\alpha}$ is $\sqrt{n}$-consistent. However, the asymptotic variance is more complex, as it must reflect the additional variability in the estimated $\widehat{\gamma}$:

$$\Sigma_\beta = B^{-1}E\left(G_i \Delta_i S_i S_i^\top \Delta_i G_i^\top\right)B^{-\top} + B^{-1}\Phi B^{-\top},$$

$$B = E\left(D_i V_i^{-1}\Delta_i D_i^\top\right), \quad C = E\left[\frac{\partial}{\partial\gamma}\left(D_i V_i^{-1}\Delta_i S_i\right)\right]^\top, \quad H = E\left(\frac{\partial}{\partial\gamma}\mathbf{Q}_{ni}\right)^\top,$$

$$\Phi = -CH^{-\top}C^\top - E\left(D_i V_i^{-1}\Delta_i S_i \mathbf{Q}_{ni}^\top H^{-\top}C^\top\right) - \left[E\left(D_i V_i^{-1}\Delta_i S_i \mathbf{Q}_{ni}^\top H^{-\top}C^\top\right)\right]^\top. \tag{2.39}$$

The first term of Σ_β in the above is identical to the asymptotic variance of the GEE estimate in (2.27), while the second term accounts for the additional variability in $\widehat{\gamma}$. A consistent estimate of Σ_β is obtained by substituting consistent estimates of the respective parameters.

Note that if π_{it} is known, the asymptotic variance $\widehat{\beta}$ is given by the first term of Σ_β in (2.39). In some multi-stage studies, selection of subjects for each stage may follow models such as those defined by (2.37) and (2.38), but with known γ. We may use π_{it} based on the designed values of γ when estimating β, in which case the asymptotic variance of the WGEE estimate is again given by the first term of Σ_β. However, one may also estimate γ as in the above and use the estimated version in (2.35), in which case the asymptotic variance of the WGEE estimate should include the second term. The latter approach may be preferred, since it may yield more efficient estimates (Tsiatis 2006).

2.5 Software for Fitting Longitudinal Models

The two popular parametric and distribution-free modeling approaches discussed above have been implemented in many general-purpose commercial statistical packages such as R (R Development Core Team 2011), SAS (SAS Institute 2011), SPSS (SPSS Inc. 2009), and Stata (StataCorp 2011). They are also available in some specialized packages for longitudinal data analysis such as HLM (Raudenbush et al. 2007). For example, in SAS, we can use MIXED, GLMMIX, and NLMIXED for fitting GLMM, but only the latter two can be used for fitting noncontinuous responses such as binary. As inference for distribution-free models is typically based on the GEE or WGEE, such models are commonly referred to simply as GEE (WGEE). The SAS GENMOD procedure fits the distribution-free GEE models. Currently, WGEE is not available as a SAS-supported procedure, and some user-written SAS macros may be used to facilitate such inference. In R, the primary functions (packages) for fitting GLMM are lme4, ZELIG, and glmmML, while the functions gee and geepack are used to obtain GEE estimates.

Note that as the log-likelihood for GLMM involves high-dimensional integration, due to the needs to integrate out the latent random effects $\mathbf{b}_i$, numerical approximations must be used to approximate the likelihood function, except for linear regression Zhang et al. (2011a). As the approximation is quite complex and the accuracy of estimate depends critically on the precision of such approximations, performance varies considerably across these available procedures. For example, when fitting binary responses, none of the currently available functions and packages in R at the moment of this writing time yields correct inference, while for the two SAS procedures, only NLMIXED provides good results (Zhang et al. 2011b). Thus, before using a package for fitting GLMM in a real study, it is important to check for information about its performance to ensure valid inference.

2.6 A Real Data Example

The Medical Outcomes Study 36-Item Short-Form Health Survey (SF-36) is widely used as a significant health outcome indicator (Lubetkin et al. 2003; Wan et al. 2011). It is a multipurpose and self-reported health status measure, yielding an eight-subscale profile of scores (domains) as well as the Physical and Mental Health summary measures. The higher measurement precision, reduced floor and ceiling effects, and superior responsiveness make the SF-36 a popular measure of quality of life in research and clinical studies. As a result, the SF-36 has been translated into many foreign languages, including the simplified Chinese (Mandarin) version of SF-36 (CSF-36), and used in more than 40 different countries (Wan et al. 2011).

A recent study was conducted to evaluate the performance of CSF-36 when used to assess health-related quality of life (HRQOL) for patients with hypertension, coronary heart diseases, chronic gastritis, or peptic ulcer in mainland China. The study population consisted of inpatients with one of these four types of chronic diseases seen at the first affiliated hospital of Kunming Medical University, Kunming province, China. There were no exclusions based on age, clinical stage of disease, and treatment course, but the subjects were required to be able to read, understand, and complete the questionnaires, which led to exclusions based on illiteracy and advanced disease status.

Among the 534 patients who completed the initial survey, 40–50 % of these were randomly selected to take the questionnaire a second time, 1–2 days after hospitalization, to assess the test–retest reliability. Another sub-sample of patients (80–90 %) was selected to complete the questionnaire a third time at discharge (after about 2 weeks of treatment) to evaluate clinical change and responsiveness of the instrument to detecting such changes.

For illustration purposes, we focused on one component of the study that examined the criterion-related validity. As there was no agreed-upon gold standard, the Quality of Life Instruments for Chronic Diseases-General Module (QLICD-GM), developed by Wan et al. (2005, 2007), was used to provide the criterion for validating each of the eight subscales: Physical Function (PF), Role-Physical (RP), Bodily Pain (BP), General Health (GH),Vitality (VT), Social Function (SF), Role-Emotional (RE), and Mental-Health (MH). These subscales were used as predictors in three separate regression models with the QLICD-GM's three domain scores, Physical Function, Psychological Function, and Social Function, serving as the respective responses. To utilize all available data, we modeled each regression

Table 2.1 Estimates, standard errors (S.E), and p-values from linear mixed-effects model for the Quality of Life Study Data

Estimates for eight domains of CSF-36 from linear mixed effects model with response based on each domain score of QLICD-GM

	PF	RP	BP	GH	VT	SF	RE	MH
Physical health from QLICD-GM								
Estimate	0.407	−0.028	0.025	0.153	0.116	−0.025	−0.005	0.063
S.E.	0.019	0.014	0.022	0.031	0.032	0.023	0.014	0.032
P-value	<0.01	0.055	0.263	<0.01	<0.01	0.275	0.711	0.047
Psychological function from QLICD-GM								
Estimate	0.143	−0.043	0.010	0.167	0.006	0.105	0.047	0.296
S.E.	0.019	0.014	0.022	0.031	0.032	0.023	0.014	0.032
P-value	<0.01	<0.01	0.650	<0.01	0.848	<0.01	<0.01	<0.01
Social function from QLICD-GM								
Estimate	0.103	−0.005	−0.001	0.093	0.063	0.028	−0.003	0.201
S.E.	0.016	0.011	0.017	0.025	0.025	0.018	0.010	0.025
P-value	<0.01	0.617	0.959	<0.01	0.012	0.123	0.732	<0.01

using the longitudinal methods discussed above based on the data from all three assessment times.

For each of the three domains of CSF-36, let y_{it} denote the domain score, and x_{ikt} the kth subscale of QLICD-GM from the ith subject at time t, with k denoting the subscales, PF, RP, BP, GH, VT, SF, RE, and MH, in order for $1 \leq k \leq 8$. Let $x_{i0t} = 1$ and $\mathbf{x}_{it} = (x_{i0t}, x_{i1t}, \ldots, x_{i8t})^{\top}$.

2.6.1 LMM

We first fit the LMM (LMM) in (2.4) with $\mathbf{z}_{it} = \mathbf{x}_{it}$. Shown in Table 2.1 are the estimates of parameters from the LMM, along with the standard errors (S.E.) and p-values. Overall, the estimates were larger (in absolute value) for the predictors that correspond to the same and similar domains than to different and non-similar domains of the CSF-36. The estimate for PF was the largest for the LMM with Physical Health as the response. In addition, GH, VT, and MH also significantly predicted this outcome. For the LMM with Psychological Function as the response, MH had the largest beta weight, followed by PF, GH, SF, RE, and RP, which were all related to this outcome. All these domains significantly predicted Psychological Function. The MH domain also had the largest beta weight for the LMM with Social Function as the response. Additionally, PF, GH, and VT were also significantly associated with this outcome.

Table 2.2 Estimates, standard errors (S.E), and p-values from distribution-free model with inference based on GEE WGEE for the Quality of Life Study Data

Estimates for eight domains of CSF-36 from distribution-free model with response based on each domain score of QLICD-GM								
	PF	RP	BP	GH	VT	SF	RE	MH
Physical health from QLICD-GM								
Estimate	0.400	−0.017	0.016	0.137	0.111	0.001	−0.001	0.032
S.E.	0.023	0.017	0.025	0.031	0.036	0.025	0.018	0.035
P-value	<0.01	0.293	0.519	<0.01	<0.01	0.973	0.958	0.370
Psychological function from QLICD-GM								
Estimate	0.112	−0.030	−0.012	0.128	0.033	0.122	0.059	0.322
S.E.	0.023	0.014	0.023	0.033	0.035	0.029	0.015	0.038
P-value	<0.01	0.041	0.598	<0.01	0.357	<0.01	<0.01	<0.01
Social function from QLICD-GM								
Estimate	0.134	0.011	−0.002	0.094	0.032	0.054	−0.008	0.197
S.E.	0.019	0.015	0.024	0.032	0.038	0.022	0.014	0.037
P-value	<0.01	0.441	0.932	<0.01	0.390	0.017	0.551	<0.01

2.6.2 GEE

The estimates from LMM will be valid, if there is no serious violation against the normal–normal assumption for the random effects and error terms. Although various types of residuals may be used to help describe the distribution of the response, it is difficult to formally test the normal–normal assumption. However, it is possible to conduct a test to validate the marginal normality, i.e., the sum of random effects and errors follows a multivariate normal (Tan et al. 2005; Feng et al. 2009). If this marginal normality holds true, the LMM above still yields valid inference, provided that robust variance estimates such as the sandwich variance estimates are used. In fact, inference in this case is equivalent to that based on GEE. Thus, we also applied the distribution-free model in (2.19), with the same explanatory variables $\mathbf{x}_{it}$ using GEE for inference.

Shown in Table 2.3 are the GEE estimates for this distribution-free model. By comparing these with the results for LMM in Table 2.2, it is seen that the order of magnitude of estimate is the same for the first three largest beta weights. However, there were some changes for the order of the remaining estimates. For example, as in LMM, MH, PF, and GH were also the first three largest beta weights for Social Function for the distribution-free model. However, VT, the 4th largest beta weight under LMM, drops down to the 5th place for the distribution-free model.

Table 2.3 Estimates, standard errors (S.E), and p-values from logistic regression for modeling missingness under MAR for each of the three domain scores of QLICD-GM for the Quality of Life Study Data

Estimates of parameters of logistic regression parameters for modeling missingness for each domain score of QLICD-GM at time 2/3			
Predictors	Estimates	S.E.	P-value
Prior physical health	−0.0005/0.0002	0.0091/0.0073	0.99/0.92
Prior psychological function	0.006/0.005	0.009/0.006	0.41/0.34
Prior social function	0.001/0.002	0.012/0.022	0.88/0.75

2.6.3 *Testing MCAR*

As discussed in Sect. 2.4, estimates from GEE may not be consistent under MAR; it is only valid under the more stringent MCAR. In principle, the missing data should follow MCAR in this study, as it is the result of random selection of a subgroup at each of the two follow-up times. But, we can also formally test this assumption to see if the random selection was successful. To this end, we modeled the MAR following the discussion in Sect. 2.4.2. Since we did not consider any covariate in the analysis, the logistic regression in (2.38) reduced to

$$\operatorname{logit}\left(p_{it}\left(\gamma_t\right)\right) = \gamma_{0t} + \gamma_{1t} y_{it^-}, \quad t = 2, 3, \tag{2.40}$$

where $y_{i2^-} = y_{i1}$ and $y_{i3^{-1}} = y_{i2}$.

We applied the logistic regression in (2.40) to each of the three domain scores of QLICD-GM. Shown in Table 2.3 are the estimates of parameters from the logistic model, their standard errors, and corresponding p-values. As neither the response at time 1 nor at 2 significantly predicted the missingness at time 2 and 3, the results confirmed the random selection process, i.e., the missing data follows the MCAR mechanism. Since the missingness in our study followed MCAR, GEE provides consistent estimates, regardless of the data distribution, and inference based on this distribution-free would be preferred over its parametric counterpart LMM.

2.7 Discussion

We discussed longitudinal designs and associated advantages of such models in clinical trial and observational studies. Unlike cross-sectional studies that only indicate associations between outcomes, longitudinal designs provide temporal changes of such associations, thereby offering a framework for investigating the causal mechanisms of such relationships. However, adding the time dimension in longitudinal data creates several methodological issues, which in particular preclude applications of standard statistical models and procedures for cross-sectional data analysis. As a result, longitudinal study data must be analyzed by methods that address the unique data-analytic problems associated with such data.

The two most popular approaches for longitudinal data are the GLMM and the distribution-free GLMs. As inference for the latter is often based on the GEE or WGEE, this class of models is typically referred to simply as GEE (WGEE). GLMM explicitly models the two sources of between- and within-subject variability using random effects, while GEE (WGEE) tackles the correlated responses over time directly by basing inference based on the marginal distribution of the subject's responses. Although it carries no effect on linear models for continuous responses, this conceptual difference does have significant implications for interpretations of estimates when modeling noncontinuous responses such as binary and count data. In the validation study of CSF-36, the estimates are quite similar, as they estimate the same underlying quantities. However, when modeling the binary or count response using GLMM and GEE (WGEE), the parameters for the same explanatory variable in GLMM (fixed-effect) and GEE are generally on a different scale (Diggle et al. 2002; Zeger et al. 1988; Zhang et al. 2011a,b). For example, when using the logit link for model binary responses, the parameter may not have the familiar log odds ratio interpretation for the GLMM (Zhang et al. 2011a). Thus, we must be mindful when applying these approaches to real study data, as such differences are likely the source for discrepant findings.

We focused on MCAR and MAR when addressing missing data, as these are the most likely missing data mechanisms for real studies. Although MAR takes into account the dependence of missing visit on the response, such a dependent relationship relates only the *observed* response to the missing visit. Thus, if the missingness relates to the subject's future response, i.e., following NINR, both GLMM and WGEE will yield biased inference. In cancer studies, the subject may drop out of the study either because of death or intolerance of the medication. Thus, the two causes of missingness may be associated with different responses of disease progression, invalidating the MAR mechanism. A number of approaches have been proposed to address this issue. For example, in *pattern mixture* models, different regression relationships are posited based on the different missing data patterns (Birmingham and Fitzmaurice 2002). The most popular approach, however, is to model the relationship between missingness and future responses through the random effects, $\mathbf{b}_i$ in (2.10). For example, in the cancer study example, the potential differential effects of the two causes of missingness on future responses are accommodated by adding a survival model to relate $\mathbf{b}_i$ as a predictor to the outcome of time of death (Tsiatis and Davidian 2004). If $\mathbf{b}_i$ significantly predicts the time to death, then $f\left(\mathbf{r}_i \mid \mathbf{y}_i^{\mathrm{o}}, \mathbf{y}_i^{\mathrm{m}}, \mathbf{x}_i\right) \neq f\left(\mathbf{r}_i \mid \mathbf{y}_i^{\mathrm{o}}, \mathbf{x}_i\right)$, thereby invalidating the MAR condition in (2.31) (Degruttola and Tu 1994; Hogan and Laird 1997). As the two log-likelihood functions, $l_1\left(\theta_y\right)$ and $l_2\left(\theta_{y|r}\right)$ in (2.33), share a set of parameters for the variance of $\mathbf{b}_i$, this joint modeling approach is also known as the *shared parameter* models (Follmann and Wu 1995).

References

Birmingham J, Fitzmaurice G (2002) A pattern-mixture model for longitudinal binary responses with nonignorable nonresponse. Biometrics 58(4):989–996

Breslow N, Clayton D (1993) Approximate inference in generalized linear mixed models. J Am Stat Assoc, 88:9–25

Bryk A, Raudenbush S, Congdon R (1996) HLM: Hierarchical linear and nonlinear modeling with the HLM/2L and HLM/3L programs. Scientific Software International, Lincolnwood, IL

Crowder M (1995). On the use of a working correlation matrix in using generalised linear models for repeated measures. Biometrika 82(2):407–410

Davidian M, Gallant A (1993) The nonlinear mixed effects model with a smooth random effects density. Biometrika 80(3):475–488

Degruttola V, Tu XM (1994) Modeling progression of cd4-lymphocyte count and its relationship to survival-time. Biometrics 50(4):1003–1014

Demidenko E (2004) Mixed models: theory and applications, vol 518. LibreDigital, Austin, TX

Dempster A, Laird N, Rubin D (1977) Maximum likelihood from incomplete data via the em algorithm. J Roy Stat Soc B 39(1):1–38

Diggle PJ, Heagerty PJ, Liang K-Y, Zeger SL (2002) Analysis of longitudinal data, 2nd edn. Oxford statistical science series, vol 25. Oxford University Press, Oxford

Feng C, Su H, Wang H, Tang W, Yu Q, Tu XM (2009) Testing normality for longitudinal studies with missing data. Far East J Theor Stat 28(2):133–155

Follmann D, Wu M (1995) An approximate generalized linear model with random effects for informative missing data. Biometrics, 51:151–168

Goldstein H (1987) Multilevel models in education and social research. Charles Griffin & Co, London, England; Oxford University Press, New York

Goldstein H (1995) Hierarchical data modeling in the social sciences. J Educ Behav Stat 20(2): 201–204

Hogan J, Laird N (1997) Mixture models for the joint distribution of repeated measures and event times. Stat Med 16(3):239–257

Kowalski J, Tu XM (2008) Modern applied U-statistics. Wiley series in probability and statistics. Wiley-Interscience, Hoboken, NJ

Laird N, Ware J (1982) Random-effects models for longitudinal data. Biometrics, 38:963–974

Lindstrom M, Bates D (1988) Newton-raphson and em algorithms for linear mixed-effects models for repeated-measures data. J Am Stat Assoc, 83:1014–1022

Lu NJ, Tang W, He H, Yu Q, Crits-Christoph P, Zhang H, Tu X (2009) On the impact of parametric assumptions and robust alternatives for longitudinal data analysis. Biom J 51(4):627–643

Lubetkin E, Jia H, Gold M (2003) Use of the SF-36 in low-income Chinese American primary care patients. Medical Care 41(4):447–457

Meng XL, Van Dyk D (1997) The em algorithm—an old folk-song sung to a fast new tune. J Roy Stat Soc B Stat Meth 59(3):511–567

Pinheiro J, Bates D (2000) Mixed-effects models in S and S-PLUS. Springer, Berlin

R Development Core Team (2011) R: A language and environment for statistical computing. R doundation for statistical computing, Vienna, Austria. ISBN 3-900051-07-0

Rasbash J, Steele F, Browne W, Goldstein H (2009) A User's Guide to MLwiN, v2.10. Centre for Multilevel Modelling, University of Bristol

Raudenbush S, Bryk A (2002) Hierarchical linear models: Applications and data analysis methods, vol 1. Sage Publications, Thousand Oaks, CA

Raudenbush S, Bryk A, Congdon R (2007) Hlm for windows (version 6.04)[computer software]. Scientific Software International, Lincolnwood, IL

SAS Institute (2011) Base SAS 9.3 procedures guide. SAS Institute, Cary, NC

Schabenberger O, Gregoire T (1996) Population-averaged and subjectspecific approaches for clustered categorical data. J Stat Comput Simulat 54(1-3):231–253

Seber G (1984) Multivariate observations, vol 41. Wiley Online Library, Hoboken, NJ

Shults J, Sun WG, Tu X, Kim H, Amsterdam J, Hilbe A, Ten-Have T (2009) A comparison of several approaches for choosing between working correlation structures in generalized estimating equation analysis of longitudinal binary data. Stat Med 28(18):2338–2355
SPSS Inc. (2009) SPSS base 17.0 for Windows User's Guide. SPSS Inc., Chicago IL
StataCorp (2011). Stata statistical software: release 12. StataCorp LP: StataCorp, College Station, TX
Strenio J, Weisberg H, Bryk A (1983) Empirical bayes estimation of individual growth-curve parameters and their relationship to covariates. Biometrics, 39:71–86
Tan M, Fang H-B, Tian GL, Wei G (2005) Testing multivariate normality in incomplete data of small sample size. J Multivariate Anal 93(1):164–179
Tang W, He H, Tu X (2012) Applied categorical and count data analysis. Chapman & Hall/CRC, London/Boca Raton
Tsiatis A, Davidian M (2004) Joint modeling of longitudinal and time-to-event data: an overview. Statistica Sinica 14(3):809–834
Tsiatis AA (2006) Semiparametric theory and missing data. Springer series in statistics. Springer, New York
Wan C, Gao L, Li X et al. (2005) Development of the general module for the system of quality of life instruments for patients with chronic disease: items selection and structure of the general module[j]. Chin Ment Health J 11:444–447
Wan C, Tu X, Messing S, Li X, Yang Z, Zhao X, Gao L, Yang Y, Pan J, Zhou Z (2011) Development and validation of the general module of the system of quality of life instruments for chronic diseases and its comparison with sf-36. J Pain Symptom Manag 42:93–104
Wan C, Yang Z, Yang Y (2007) Development of the general module of the system of quality of life instruments for patients with chronic disease: Evaluation of the general module. Chinese J Behavior Medical Sci 16:559–561
Wolfinger R, O'connell M (1993) Generalized linear mixed models a pseudo-likelihood approach. J Stat Comput Simulat 48(3–4):233–243
Zeger S, Liang K, Albert P (1988) Models for longitudinal data: a generalized estimating equation approach. Biometrics, 44:1049–1060
Zhang H, Lu N, C, Feng S, Thurston Y, Xia L, Zhu, and Tu X (2011a). On fitting generalized linear mixed-effects models for binary responses using different statistical packages. Stat Med 30, 2562–2572
Zhang H, Xia Y, R, Chen D, Gunzler W, Tang, and Tu X (2011b). Modeling longitudinal binomial responses: implications from two dueling paradigms. J Appl Stat 38:2373–2390

Chapter 3
Assessment of Diagnostic Tests and Instruments

Hua He, Douglas Gunzler, Yan Ma, and Yinglin Xia

3.1 Introduction

Diagnostic tools and screening instruments are commonly used in clinical and research studies ranging from physical to psychological to social functioning. A diagnostic test not only provides information about a patient's condition, but also influences the health care provider's plan for managing the patient. Early detection of disease often leads to less suffering and quicker recovery, which is especially important for the treatment of cancer patients. Cancer is one of the leading killers in the world, and one of the main reasons of the high mortality rate of cancer patients is that many patients were diagnosed at a very late stage (stage IV). In fact, cancer patients diagnosed at early stage usually have better chance of survival and even complete cure. According to statistics from the National Cancer Institute, the 5-year relative survival rates (comparing to the general population) of pancreatic cancer patients were 22.2 % for localized and 1.8 % for distant patients.

Definitive diagnoses of most malignant cancers must be confirmed by histological examination of the cancerous cells. This requires tissues to be obtained from a biopsy or surgery. Some of the biopsies can only be performed under anesthesia

H. He (✉) • Y. Xia
Department of Biostatistics and Computational Biology, University of Rochester, Rochester, NY 14642, USA
e-mail: hua_he@urmc.rochester.edu; yinglin_xia@urmc.rochester.edu

D. Gunzler
School of Medicine, Case Western Reserve University, Center for Health Care Research & Policy, MetroHealth Medical Center, 2500 MetroHealth Drive, Cleveland, OH 44109, USA
e-mail: dgunzler@metrohealth.org

Y. Ma
Hospital for Special Surgery, Department of Public Health, Weill Medical College of Cornell University, New York, NY 10021, USA
e-mail: yam2007@med.cornell.edu

W. Tang and X. Tu (eds.), *Modern Clinical Trial Analysis*, Applied Bioinformatics and Biostatistics in Cancer Research, DOI 10.1007/978-1-4614-4322-3_3,

and require surgery in an operating room. Thus, it is usually quite expensive and risky for patients, discouraging many to undergo such invasive procedures, especially those at lower risk. Such concerns have generated a great interest to develop less expensive, conveniently administered, and reliable diagnostic tests, using less invasive procedures such as blood tests, X-rays, CT scans, endoscopy, and mammography. As diagnostic tests do not directly detect the pathogens of disease, it is important to assess the accuracy of such tests. For example, a recent study showed that the use of computed tomography (CT) in lung cancer screening may not reduce deaths from the disease, while exposing some individuals to unnecessary invasive treatments (Bach et al. 2007).

The accuracy of diagnostic tests is typically assessed by comparing the test results with the true disease status. For example, patients will be considered as diseased (of the cancer) if such cancerous cells are developed, and as non-diseased otherwise, by histological examinations, a gold standard for detecting cancer. Many diagnostic tests such as mammography for breast cancer have dimensional outcomes and thus may be dichotomized to facilitate clinical use. To provide informative diagnosis, the cut point must be selected carefully so that diagnoses based on such instruments can detect the disease without misidentifying disease-free individuals.

Despite the advance of science and technology in treatment of cancer, there seems no effective treatment for advanced cancer patients in the near future. Many palliative care programs have been developed to help relieve the pain and stress of such patients when they approach the last stage of their lives. For example, depression is common among such patients because of the fear of death and disability associated with such terminal illnesses. Improving the quality of life (QOL) of such terminal cancer patients is the primary goal of palliative care. As QOL encompass a multitude of dimensions pertaining to an individual's emotional, social, and physical wellbeing, including their ability to function in the ordinary tasks of living, no single diagnostic test provides sufficient information about such a multi-faceted construct. For example, the popular medical outcome 36-items short-form health survey (SF-36) contains 36 items (questions) that represent eight domains: Physical Function (PF), Role-Physical (RP), Bodily Pain (BP), General Health (GH), Vitality (VT), Social Function (SF), Role-Emotional (RE), and Mental-Health (MH). These domains may be aggregated into a Mental Component Summary ($\mathrm{MCS} = \mathrm{VT} + \mathrm{SF} + \mathrm{RE} + \mathrm{MH}$) and a Physical Component Summary ($\mathrm{PCS} = \mathrm{PF} + \mathrm{RP} + \mathrm{BP} + \mathrm{GH}$) to help assess the mental or physical health or both conditions.

Unlike diagnostic tests for a single disease condition, the SF-36 is tempting to measure latent constructs such as emotion and social functioning, and as such there typically is no gold standard to validate its external validity. Item-based instruments or questionnaires for measuring conceptual constructs such as the SF-36 are quite popular in research studies and practice, especially in the behavioral and social sciences. Thus, it is of critical importance to understand how such instruments are constructed, their scales are created, and their validities are examined.

Another important consideration for diagnostic instruments is the test–retest reliability. As measurement errors are random, test scores will vary from repeated administrations of the instrument to the same individual. Like external validity, large variations will create problems for replicating research findings and errors for disease diagnoses. Reliability does not imply validity. A reliable instrument measures something consistently, but not necessarily what it is purported to measure. For example, while there are many reliable tests of specific abilities such as different types of cancer and depression, not all of them would be valid for predicting, say, job performance. Reliability indicates precision, while validity speaks to accuracy of diagnosis. For many latent-construct-based instruments such as SF-36 where no gold standard test exists, accuracy can be subject to the judgement call of the expert.

In this chapter, we discuss statistical methods for addressing each of the aforementioned aspects associated with diagnostic tests and instruments.

3.2 Accuracy of Diagnostic Tests

In this section, we discuss assessment of diagnostic tools for binary-outcome tests, mainly the presence and absence of some kind of disease such as breast cancer. We start with native binary tests, i.e., the binary outcome is either the raw output from some diagnostic device such as broken bones or the result based on dichotomizing an underlying latent continuum such as a physician's feeling about the condition of some disease for a patient.

3.2.1 Test Sensitivity and Specificity

For a binary diagnostic test T, the most commonly used indices for its accuracy in diagnosis of a disease of interest are the true positive (negative) rates, i.e., the proportion of positive diagnosis among those diseased (non-diseased) subjects. The true positive rate is also called sensitivity, so by definition sensitivity $= \Pr(T$ positive $\mid$ diseased). The true negative rate is also called specificity, defined by specificity $= \Pr(T$ negative $\mid$ non diseased). In general both sensitivity and specificity need to be high in order for the test to have good performance. Usually, a sensitivity (specificity) of 80 % is considered good. However, one may be weighed more than the other, depending on the purpose of the test. For example, mammographic tests for early detection of breast cancer may focus on high specificity when used for women less than 50 because of the small prevalence rate among such women and associated extremely high false positive rates (see below). Thus, together sensitivity and specificity characterize the accuracy of a diagnostic test.

Youden's index, defined as (sensitivity + specificity)/2, is sometimes used to summarize such accuracy, especially when sensitivity and specificity are viewed as equally important. Other commonly used indices include the positive predictive value (PPV), the probability of a subject with positive test result being actually diseased and the negative predictive value (NPV), the probability of a subject with negative test result being actually non-diseased. A good diagnostic test should be high in both sensitivity and specificity. If the disease prevalence is known, then sensitivity/specificity and PPV/NPV are equivalent mathematically in the sense that one can be computed based on the other. i.e.,

$$\begin{aligned} PPV &= \Pr(\text{diseased T positive}) = \frac{\text{Se} \cdot p}{\text{Se} \cdot p + (1 - \text{Sp})(1 - p)}, \\ NPV &= \Pr(\text{non-diseased T negative}) = \frac{\text{Sp}(1 - p)}{\text{Sp}(1 - p) + (1 - \text{Se})\, p}, \end{aligned} \tag{3.1}$$

where p denotes disease prevalence and Se (Sp) denotes the sensitivity (specificity). Unlike the sensitivity and specificity, which together characterize the quality of the test, PPV and NPV both depend on disease prevalence, making them useful in optimizing the performance of the test for a given situation. For example, when screening for a rare disease, we may have a large percent of false positives if the test specificity is not sufficiently high (Tu et al. 1992). The rate of false positive is

$$\text{Rate of false positive} = 1 - PPV = \frac{(1-Sp)(1-p)}{Se \cdot p + (1-Sp)(1-p)} \approx \frac{1-Sp}{Se \cdot p + 1 - Sp}.$$

For any Sp, the rate above can exceed any value $0 \leq b < 1$ for sufficiently small p. In other words, when screening for a rare disease, a test with sufficiently high specificity (with the exact value depending on p and tolerable rate of false positive) is needed to put the rate of false positive under control. This fact explains why screening is not recommended for a rare disease such as mammographic screening for breast cancer for women less than 50.

Estimation and inference for sensitivity and specificity are straightforward. For example, for a simple random sample from the population, we can estimate the sensitivity and specificity by

$$\begin{aligned} \widehat{Se} &= \frac{\text{number of diseased and positive test subjects}}{\text{number of diseased subjects}}, \\ \widehat{Sp} &= \frac{\text{number of non-diseased and negative test subjects}}{\text{number of non-diseased subjects}}. \end{aligned} \tag{3.2}$$

Likewise, PPV and NPV can be estimated by substituting the parameters in (3.1) by their sample proportions. Note that (3.2) also applies to case-control studies where diseased and non-diseased subjects are sampled separately. In such cases, one may estimate PPV and NPV based on (3.1) if information about the disease prevalence p is available. Since $\widehat{Se}$ and $\widehat{Sp}$ are proportions, we must be mindful about low power for binary outcomes and obtain sufficiently large samples to ensure reliable estimates.

3.2.2 *Receiver Operating Characteristic Curve*

In many applications, binary diagnostic tests are often the result of dichotomizing some continuous or ordinal outcomes such as blood pressure for facilitating clinical research and practice. For example, the Body Mass Index (BMI) is a continuous outcome indicating the amount of body fat based on comparing the weight with the height of an individual. In clinical studies, it is often used as a binary test with easy-to-interpret categories. For adults, an individual with BMI with 25 or above is considered being overweight. Indices such as sensitivity and specificity of binary test can then be applied. However, as BMI is not 100 % correlated with the amount of body fat, it is quite possible that a person with a BMI above 25 is fit and normal. To reduce the chance to erroneously label people as being overweight, one may increase the cut point to make the criterion more stringent, or specific, for overweight. Doing so, however, would increase the error in the other direction by mislabeling overweight subjects as being normal. Before discussing how to select optimal cut points to balance the sensitivity and specificity, let us first look at the relationship between a binary test and its underlying continuous counterpart.

3.2.2.1 Dichotomizing a Continuous Test

Let y denote the continuous test outcome and D the binary indicator with $D = 1$ (0) for the diseased (non-diseased) subject. For ease of discussion and without the loss of generality, we assume that larger values of the continuous test outcome indicate the presence of the disease. Thus, given a cut-off value c, a subject with a test reading $y \geq c$ ($<c$) is more likely to be diagnosed with (without) the disease. Corresponding to c the dichotomized test has the sensitivity and specificity given by

$$Se(c) = \Pr(y \geq c \mid D = 1) = \int_{c}^{\infty} f_1(t)\, dt,$$

$$Sp(c) = \Pr(y < c \mid D = 0) = \int_{-\infty}^{c} f_0(t)\, dt, \qquad (3.3)$$

where $f_k(c) = \Pr(y \leq c \mid D = k)$ denotes the probability density function (PDF) of y, given the subject's disease status $D = k$ $(k = 0, 1)$.

Given a sample of test readings, we can readily estimate the sensitivity and specificity above. Let y_{1i} (y_{0j}) denote the test outcome from the ith (jth) subject from a sample consisting of n_1 diseased (n_0 non-diseased) subjects ($1 \leq i \leq n_1, 1 \leq j \leq n_0$). It follows from (3.3) that the empirical estimates of $Se(c)$ and $Sp(c)$ are given by

$$\widehat{Se}(c) = \frac{1}{n_1}\sum_{i=1}^{n_1} I_{\{y_{1i} \geq c\}}, \quad \widehat{Sp}(c) = \frac{1}{n_0}\sum_{j=1}^{n_0} I_{\{y_{0j} < c\}}, \tag{3.4}$$

where $I_{\{y \geq c\}}$ is a set indicator with the value 1 if $y \geq c$ and 0 otherwise. Given a cut point c, $\widehat{Se}(c)$ ($\widehat{Sp}(c)$) is the sample proportion of the positive (negative) test within the diseased (non-diseased) sample.

With (3.3) and (3.4), we can readily dichotomize a continuous test, and estimate the dichotomized test sensitivity and specificity. For example, by arbitrarily decreasing the threshold value for BMI, we can increase sensitivity and detect any overweight person. However, the improved sensitivity does not translate into more accurate diagnosis, as many normal or even underweight people will be labeled as being overweight. On the other hand, by increasing the cut point, we can reduce and even completely eliminate the error in misclassifying normal and underweight people. But, again doing so will not improve the test either, since more and more overweight individuals will be misdiagnosed as c increases. In other words, the performance of the dichotomized test is determined by the distribution of the continuous outcome and is not manipulatable through changing the cut point. This inherited quality of the dichotomized test is characterized by the Receiver Operating Characteristics or ROC curve.

3.2.2.2 ROC Curve

The Receiver Operating Characteristics (ROC) curve is the most popular approach to characterize the properties, or diagnostic-ability, of a diagnostic test. The ROC curve is the plot of the pair $(1 - Sp(c), Se(c))$ in R^2, the two dimensional Euclidean space, over the theoretical range $R = (-\infty, \infty)$ of the cut point c. If y has no diagnostic value, i.e., $f_1(c) = f_2(c)$, the ROC curve is the straight line $y = x$, called *reference line*. Such a test is the same as blind guesses. If higher test outcomes are associated with increased likelihoods of disease, then the ROC curve will increase from (0,0) to (1,1) and form a concave shape above the reference line, since the slope of the tangent line of the ROC curve $-f_1(c)/f_0(c)$ is a decreasing function (Stewart 2010). The concavity of ROC measures the informativeness of the dichotomized test, the more concave the ROC, the better the performance of the test. Another important property of ROC curve is the invariance under monotone transformation of the test outcome; for any increasing function $h(\cdot)$, the cut point $h(c)$ for $h(T)$ provides the same sensitivity and specificity as that of the cut point of c for T, thus $h(T)$ has the same ROC curve as that of T. This latter feature is quite useful for facilitating modelling and inference about ROC curves, as we discuss below.

ROC curve also provides us a convenient tool to choose cut points to achieve highest diagnostic accuracy. Such optimal cut points often depend on the diagnostic objectives, available resources, and logistic considerations. For example, PPV is quite sensitive to small changes in test specificity with extremely low disease prevalence such as breast cancer for young women. In this case, it is important to calibrate the test in favor of higher specificity. On the other hand, when screening for breast cancer for people over 50, we may want to have higher sensitivity because of the increased risk for women in this age group. In general, optimal cut points are those on the ROC curve that are closest to the point (0,1), and different approaches are developed based on the different closeness criteria. The two mostly used criteria are the average of sensitivity and specificity (Youden index) and the Euclidean distance.

Since ROC curves are determined by the distributions of the test in the diseased and non-diseased populations, a direct approach for modeling of ROC curve is to model these distributions. For example, we may model the two distributions as $f_0(v) \sim N(0,1)$ and $f_1(v) \sim N(\alpha, \sigma^2)$, where $N(\mu, \sigma^2)$ denotes a normal distribution with mean 0 and variance σ^2. Note that we have assumed without the loss of generality the standard normal for $f_0(v)$ because of the invariance of ROC curve under monotone transformation, since any normal distribution can be transformed to the standard normal $N(0,1)$. The sensitivity and specificity under the binormal model can be expressed as: $Se(c) = 1 - \Phi\left(\frac{c-\alpha}{\sigma}\right)$ and $Sp(c) = \Phi(c)$, where Φ denotes the cumulative distribution function of $N(0,1)$. Hence, the normal-based ROC curve is given by

$$x = 1 - \Phi(c), \quad y = 1 - \Phi\left(\frac{c-\alpha}{\sigma}\right), \quad -\infty \leq c \leq \infty,$$

or more succinctly by

$$y = 1 - \Phi\left(\frac{\Phi^{-1}(1-x) - \alpha}{\sigma}\right) = \Phi\left(\frac{\Phi^{-1}(x) + \alpha}{\sigma}\right), \quad 0 \leq x \leq 1.$$

Fig. 3.1 is a plot of binormal ROC curves for several different pairs of (α, σ). Note that the ROC curve for $\sigma \neq 1$ is not concave (part of the curve falls under the diagonal line) due to the artifact of the binormal model. Such ROC curves are called improper in the literature.

3.2.2.3 Area Under ROC Curve

Since the ROC curve is a performance measure for a given test T, we may summarize the information using a single value, the area under the curve (AUC):

$$\mathrm{AUC} = \int_0^1 Se(Sp)\,\mathrm{d}Sp = \Pr(y_1 \geq y_0). \tag{3.5}$$

where y_k above denotes the test outcome for the diseased ($k = 1$) and non-diseased ($k = 0$) subject. In general, higher values of AUC indicate better performance of

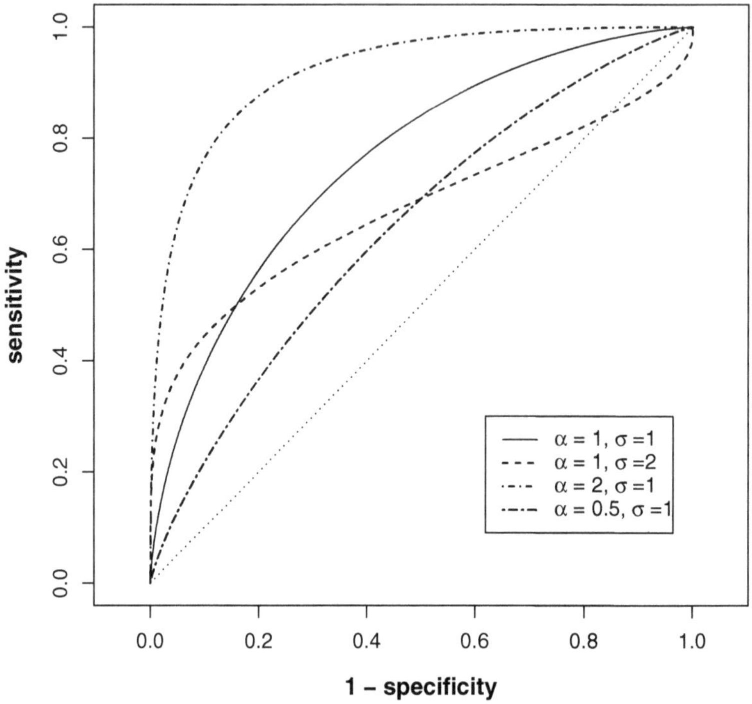

Fig. 3.1 Binormal ROC curves

the test and vice versa. If AUC = 0.5, the ROC curve becomes the reference line and the test is no better than random guesses. On the other hand, if the AUC is close to 1, the test is close to being perfect. Although there is no clear-cut as to how large an AUC is considered good, a rule of thumb in practice is ≥ 0.85 or 0.9.

3.2.2.4 Inference for ROC Curve and AUC

To estimate the ROC curve, we first arrange the pooled $n_1 + n_0$ outcomes y_{1i} and y_{0j} into an ascending order, denoted by $x_1 < \ldots < x_{n_1+n_0}$. The ordered set $\{x_i; 1 \leq i \leq n_1 + n_0\}$ divides the real line R into $n_1 + n_0 + 1$ intervals, namely $(-\infty, x_1], (x_1, x_2], \ldots, (x_{n_1+n_0}, \infty)$. As the points c within each interval yield the same $\widehat{Se}(c)$ and $\widehat{Sp}(c)$, there are at most $n_1 + n_0 + 1$ distinct pairs $(1 - Se(c), Sp(c))$. The empirical ROC curve is the piecewise linear curve connecting the points defined by such pairs.

The same idea applies to tests with ordinal outcomes. However, as the sensitivity and specificity are only defined over a range of discrete values, the empirical ROC is actually a set of points in R^2, rather than a curve. In practice, these points are typically connected with line segments for improved clarity.

By using the empirical estimate for the AUC in (3.5), we obtain an estimate of AUC:

$$\widehat{\text{AUC}} = \frac{1}{n_1 n_0} \sum_{i=1}^{n_1} \sum_{j=1}^{n_0} \left(I_{\{y_{0j} \prec y_{1i}\}} + \frac{1}{2} I_{\{y_{0j} = y_{1i}\}} \right).$$

The above is actually the area under the empirical ROC curve for the continuous test. Readers familiar with the Mann–Whitney two-sample statistic will notice that the above is actually the Mann–Whitney two-sample U-statistic. For ordinal tests or continuous tests with tied test outcomes due to round-off errors, the modified U-statistic is used instead:

$$\widehat{\text{AUC}} = \frac{1}{n_1 n_0} \sum_{i=1}^{n_1} \sum_{j=1}^{n_0} \left[I_{\{y_{0j} < y_{1i}\}} + \frac{1}{2} I_{\{y_{0j} = y_{1i}\}} \right].$$

The empirical estimates of sensitivity, specificity, and ROC curve are distribution-free, as no mathematical model is imposed on the distribution of the test outcome. If $f_k(c)$ can be reasonably modeled using some parametric distribution such as the binormal model discussed earlier, the model-based AUC estimate is more efficient as it takes advantage of the additional information about the structure of $f_k(c)$. Such a parametric approach may not be as restrictive as its first glance, since all these quantities are invariant under monotone transformation, and test outcomes may often be transformed to approximately follow the normal distribution.

For example, when the test outcome is ordinal, the binormal model can still be applied, if the ordinal outcome is viewed as the result of grouping the values of a latent continuous outcome with a set of cut points. For example, consider an ordinal test with m levels, $v_1 < \ldots < v_m$. If we assume the binormal model above for the latent outcomes of the diseased and non-diseased populations, we can find a set of $m - 1$ numbers cut points, denoted by $c_1 < \ldots < c_{m-1}$ such that

$$p_1(v_l) = \Phi(c_j) - \Phi(c_{j-1}), \quad p_0(v_l) = \Phi\left(\frac{c_j - \alpha}{\sigma}\right) - \Phi\left(\frac{c_{j-1} - \alpha}{\sigma}\right), \quad (3.6)$$

$$j = 1, \ldots, m, \quad c_0 = -\infty, \quad c_m = \infty.$$

The binormal model contains two parameters α and σ. For ordinal test outcomes, if there are n_{kj} subjects with disease status k and test outcome j, where $k = 1\ (0)$ denotes the diseased (non-diseased) sample and $j = 1, \ldots, n_k$, then the likelihood under the binormal model is $L = \prod_{k=0}^{1} \prod_{j=1}^{m} p_k^{n_{kj}}(v_j)$ (Dorfman and Alf 1969). Inference about α and σ is readily performed based on the theory of maximum likelihood.

3.2.2.5 Comparison of Multiple AUCs

When several tests are used for diagnosing the same disease, it is often of interest to compare diagnostic ability across the multiple competing tests. In this case, the different tests are applied to the same diseased and non-diseased sample and the outcomes are then used to construct the ROCs and compute the AUCs for the tests under comparison. As the different test results are obtained from the same group of subjects, they are correlated and thus the multiple AUCs are dependent. When empirical AUCs are used, inference about differences across such correlated distribution-free AUC estimates is complex, but can be facilitated by the theory of multivariate U-statistics (DeLong et al. 1988; Kowalski and Tu 2008).

3.2.2.6 ROC Surface for Polytomous Disease Status

Although the classic paradigm of dichotomizing a continuous test is useful for clinical studies and practice, it may not serve well in certain applications. For example, the dichotomized BMI test is easy to administrate, but it is kind of crude to classify subjects into two categories of overweight vs. normal, since not only the normal group is a mixed bag of normal and underweight people, but the subjects within the overweight group may be quite heterogeneous as well. The theory of ROC curve for dichotomized tests can be extended to more than two (ordered) categories such as underweight, normal, and overweight in the BMI test example (Nakas and Yiannoutsos 2004).

To illustrate the basic idea, consider trichotomizing a continuous test into three categories corresponding to three disease status $A < B < C$. Let $c_1 < c_2$ be two cut points dividing test outcomes y into the three categories, $y \leq c_1$ classified as A, $c_1 < y \leq c_2$ as B, and $y > c_2$ as C. As with the standard ROC curve, we may define three quantities below to describe the accuracy of the trichotomized test within the current context:

$$x = \Pr(T < c_1 \mid A), \quad y = \Pr(c_1 < T < c_2 \mid B), \quad z = \Pr(T > c_2 \mid C).$$

When the pair (c_1, c_2) runs over all possible regions, the corresponding points, (x, y, z), form an ROC surface in R^3. Like ROC curves, the volume under the ROC surface has a nice interpretation as the probability of correctly classifying a subject into one of the categories.

3.3 Assessment of Instrument

In the preceding section, we discussed how to assess the accuracy of diagnostic test. As seen from the discussion, such diagnostic tests are often obtained by dichotomizing a continuous or ordinal test outcome. Although such a test outcome is well defined in many studies, it is not always the case. In fact, we are often confronted with the very task to create such an outcome. For example, improving the QOL of patients with advanced incurable cancers and severe chronic illnesses is the primary goal of many treatment programs. As QOL encompass a multitude of dimensions pertaining to an individual's emotional, social, and physical conditions, no single test provides sufficient information about such a multi-faceted construct. Many measures of QOL are constructed by questionnaires or instruments to probe different aspects of the construct. This section focuses on constructing outcomes from such instruments and assessing their validities when used to measure latent constructs such as QOL. We start with a brief introduction of instrumentation by using a simple example to describe how items are created, selected, and grouped to form different facets and domains of the instrument.

3.3.1 Scale Construction

Consider the QOL for lung cancer patients. The process for developing an instrument for this latent construct typically starts with forming a small nominal group of lung cancer patients such as 15 and a small focus group of experts specializing in this topic. First, the focus group discusses and confirms the structure of the instrument, which are usually divided into facets and/or domains such as physical, mental, and social functioning. The nominal group then proposes some possible items under each facet or domain after a careful review of the instrument structure. The items identified are further refined by the focus group through in-depth interviews of the patients in the nominal group. The selected items by the focus group then undergo some pilot testing on a small group of subjects. The data collected are analyzed to confirm and/or revise the within-facet or -domain items for construct validity using statistical modeling tools such as exploratory and confirmatory factor analysis and correlation and regression analysis. We illustrate the process with data from a real study.

The medical outcome study 36-items short-form health survey (SF-36) is a multipurpose, self-report health status instrument. The original English version has been translated and used in more than 40 different countries (Lubetkin et al. 2003; Wan et al. 2007). A study has been recently conducted to evaluate its performance for patients in China with four types of chronic diseases: hypertension, coronary heart diseases, chronic gastritis, and peptic ulcer (Wan et al. 2012). More details about the study are found in Example 1 of Sect. 3.6.

An application of factor analysis to the data yielded eight principal components (eigenvalue >1), accounting for 66 % of the cumulative variance. For example, shown in Table 3.1 are the loadings of the items Q3a-3j on the first factor (PF) and those of items Q4a-4d and Q5a-5c on the second factor (RP), symbolizing the physical functioning and role-physical and -emotional domains, respectively. The loadings from the rest of the items (not shown) also clearly identified their contributions to the remaining domains, mental health, vitality, general health, social

Table 3.1 Factor Analysis of CSF-36

Items	P1	P2	P3	P4	P5	P6	P7	P8
Q3a	0.56	0.30	0.32	0.04	0.07	0.03	0.10	0.04
Q3b	0.79	0.21	0.10	0.08	0.02	0.04	0.08	0.02
Q3c	0.85	0.14	0.11	0.07	0.11	0.07	0.10	0.09
Q3d	0.78	0.18	0.19	0.06	0.09	0.00	0.03	0.02
Q3e	0.75	0.07	0.00	0.22	0.08	0.12	0.18	0.07
Q3f	0.73	0.19	0.10	0.12	0.05	0.01	0.18	0.04
Q3g	0.79	0.18	0.15	0.04	0.15	0.04	0.11	0.08
Q3h	0.86	0.11	0.15	0.05	0.03	0.03	0.00	0.08
Q3i	0.80	0.05	0.00	0.17	0.13	0.16	0.15	0.11
Q3j	0.69	0.03	0.00	0.18	0.14	0.31	0.06	0.05
Q4a	0.18	0.74	0.11	0.01	0.08	0.07	0.10	0.12
Q4b	0.19	0.80	0.09	0.08	0.05	0.08	0.09	0.12
Q4c	0.18	0.82	0.13	0.01	0.01	0.01	0.06	0.07
Q4d	0.22	0.79	0.09	0.08	0.12	0.06	0.00	0.09
Q5a	0.10	0.68	0.12	0.11	0.10	0.26	0.17	0.19
Q5b	0.16	0.75	0.12	0.10	0.10	0.22	0.18	0.12
Q5c	0.07	0.64	0.12	0.22	0.03	0.20	0.27	0.02

functioning, vitality and bodily pain. Thus, the analysis results provide an initial good indication of the construct validity of this CSF-36 when used to assess QOL of the four types of chronic diseases. Results from a formal assessment of the internal validity are presented in Example 1 of Sect. 3.6, along with other performance measures such as external validity and reliability.

3.3.2 *Internal Consistency*

When developing an instrument, it is important to check the coherence of its items, i.e., whether together they measure the latent construct of interest such as depression, social functioning, and personality. This is known as *internal consistency* or *construct validity*. For example, the ten items within the PF of the SF-36 all pertain to physical activities such as walking, running, bending, and kneeling (Ware Jr and Sherbourne 1992). As the total item score of a facet or domain is used as a dimensional scale for the latent construct, it is important that such items are additive within a facet or domain. For example, if larger item scores imply higher on the continuum of the construct, items within a facet must be positively correlated. One way to ensure such internal consistency is to examine all between-item correlations across the within-facet items. This approach, however, is impractical for facets with a large number of items. In addition, it is difficult to assess the strength of item coherence with a large number of pairwise correlations.

3.3.2.1 Spearman–Brown ρ_K

A formal framework for assessing the internal validity is the domain-sampling model. Under this classic model, a measure of the latent construct of interest is the mean item score based on a random sample of K items from a population of items underlying the latent facet (domain). The true value of the latent construct is the limit of the averaged item scores when the number of items approaches infinity, $y_\infty = \lim_{K\to\infty} \frac{1}{K}\sum_{k=1}^{K} y_k$. This is akin to how the population mean of a random variable is defined in statistics. If all items share an equal amount of the common core, i.e., a constant product-moment (PM) correlation $p_1 = Corr(y_k, y_\infty)$ independent of any particular item k, then $p_1 = \sqrt{\overline{\rho}_\infty}$, where $\overline{\rho}_\infty = \lim_{K\to\infty} \binom{K}{2}^{-1} \sum_{(k,l)\in C_2^K} \rho_{kl}$ with $\rho_{kl} = Corr(y_k, y_l)$ and C_2^K denoting the set of $\binom{K}{2}$ combinations of 2 distinct elements (k, l) from the integer set $\{1, \ldots, K\}$ (Tu 2009).

This limit of averaged pairwise correlations $\overline{\rho}_\infty$ is known as the *reliability coefficient* (RC), and p_1 is called the *reliability index* (RI) for a single item. The identity $p_1 = \sqrt{\overline{\rho}_\infty}$ is fundamental to the classic measurement theory, as it expresses the incomputable correlation coefficient involving the latent true score y_∞ as a function of an estimable quantity $\overline{\rho}_\infty$. To minimize measurement errors, the averaged item score $\overline{y}_\cdot = \frac{1}{K}\sum_{k=1}^{K} y_k$, or equivalently the total score, $y_\cdot = \sum_{k=1}^{K} y_k$, is often used in most applications. For convenience, we use the average score in the discussion of internal validity in the rest of the chapter.

The RI p_K for K items is defined as the PM correlation between $\overline{y}_\cdot$ and y_∞. It can be shown that $p_K = \sqrt{\frac{K\overline{\rho}_\infty}{1+(K-1)\overline{\rho}_\infty}} + o(1)$, with the higher order term $o(1) \to 0$ as $K \to \infty$ (Tu 2009). This second important identity, known as the *Spearman–Brown Prophecy* formula, generalizes the first fundamental result for a single item to a measure of *internal consistency* of K items. Accordingly, $\rho_K = \frac{K\overline{\rho}_\infty}{1+(K-1)\overline{\rho}_\infty}$ is called the RC for a facet or domain with k items. Although in theory $-1 \leq \rho_K \leq 1$, $\rho_K \geq 0$ in almost all applications, as items in real studies are generally positively correlated.

3.3.2.2 Cronbach Coefficient Alpha

In addition to the Spearman–Brown ρ_K, another widely used index, indeed more popular, for internal consistency is the *Cronbach coefficient alpha*:

$$\alpha_K = \frac{K}{K-1} \frac{\sum_{(k,l)\in C_2^K} Cov(y_k, y_l)}{\sum_{k=1}^{K}\sum_{l=1}^{K} Cov(y_k, y_l)}, \quad K \geq 2.$$

The Cronbach α_K is generally different from the Spearman–Brown ρ_K, unless the item variance $\sigma^2 = Var(y_k)$ is a constant independent of k. The difference between the two may be small for instruments consisting of items with similar item variances.

Like ρ_K, $0 \leq \alpha_K \leq 1$ in real study applications, with larger values indicating stronger item coherence. In most studies, an acceptable range of α_K is 0.7 or higher. However, we must be mindful about a potential caveat in α_K when using this as a general guideline. Since α_K increases towards 1 as K gets large, it may not be sensible to compare the values of α_K between two scales if numbers of items differ tremendously between the two. Also, we may require a higher α_K for a scale with a large number of items van Zyl et al. (2000).

3.3.2.3 Intraclass Correlation

Another popular validation approach is the intraclass correlation coefficient or ICC. ICC explicitly models the between-instrument variation, hence the name of intraclass correlation, based on the one-way linear mixed-effects model (LMM). Let y_{ik} denote the score of the kth instrument from the ith subject ($1 \leq i \leq n$, $1 \leq k \leq K$). Consider modeling y_{ik} using a single-factor, LMM:

$$y_{ik} = \mu + \lambda_i + \varepsilon_{ik}, \quad \lambda_i \sim N\left(0, \sigma_I^2\right), \quad \varepsilon_{ik} \sim N\left(0, \sigma^2\right), \quad 1 \leq k \leq K. \quad (3.7)$$

Under (3.7), the random effect $y_{i\infty} = \mu + \lambda_i$ is the true latent score, and y_{ik} is the observed version for the ith subject using the kth instrument. Thus, σ_I^2 represents the variation of the latent score in the population, while σ^2 accounts for measurement error.

The ratio or ICC, $\rho_{\text{ICC}} = \frac{\sigma_I^2}{\sigma_I^2+\sigma^2}$, describes the fraction of the variation of the latent construct variable relative to the total variability, with ρ_{ICC} closer to 1 (0) indicating a small (large) measurement error in both instruments. Note that the one-way LMM in (3.7) imposes the assumptions of a common mean and variance for measurement error for both instruments. We can readily revise this model to relax each or both assumptions. For example, the following two-way LMM accommodates potential difference in the mean across the instruments:

$$y_{ik} = \mu_k + \lambda_i + \varepsilon_{ik}, \quad \lambda_i \sim N\left(0, \sigma_I^2\right), \quad \varepsilon_{ik} \sim N\left(0, \sigma^2\right), \quad 1 \leq k \leq K. \quad (3.8)$$

We can test the null, $H_0 : \mu_k - \mu_l = 0$ for all $1 \leq k, l \leq K$ to see if there is any systematic difference between the two instruments.

Under the LMM in (3.7), the PM correlation between any y_{ik} and y_{il} is $\rho_{\text{ICC}} = Corr\left(y_{ik}, y_{il}\right) = \frac{\sigma_I^2}{\sigma_I^2+\sigma^2}$, a constant independent of k and l ($k \neq l$). Thus, the single-item RC is simply $\rho_1 = \rho_{\text{ICC}}$. Further, by correlating y_{ik} and $y_{i\infty}$, we obtain the single-item RI $p_1 = \sqrt{\rho_1}$, the first fundamental result for the relationship between p_1 and ρ_1 derived under the classic domain sampling model. Note that ρ_1 under (3.7) is equal to this ICC, rather than the limit of averaged pairwise PM correlations under the domain sampling model.

3.3.3 *External Validity*

When there is some gold standard, we can assess the external or criterion validity of the instrument by comparing the instrument under investigation directly with the gold standard. We can use a wide range of analytic tools including correlation, regression, and other related statistical modeling techniques.

If both the instrument and gold standard have continuous outcomes, we can assess the closeness between the two instruments using the product—moment (PM) correlation, intraclass correlation coefficient (ICC), the concordance correlation coefficient (CCC), and other robust rank-based correlations such as Spearman's rho. Although the PM correlation is widely used for this purpose, we must be mindful about its limitation in general when used to examine the accuracy of the instrument.

To illustrate, suppose the outcomes from the gold standard and investigative instrument in a hypothetical study with six subjects are $(3,5)$, $(4,6)$, $(5,7)$, $(6,8)$, $(7,9)$, $(8,10)$. Although the outcomes hardly agree, the Pearson estimate of the PM correlation $\widehat{\rho}_{\text{PM}} = 1$. The paradox here is the downward bias in the instrument, sample mean 5.5 for the gold standard vs. 7.5 for the instrument. Thus, although perfectly correlated (precision), the instrument has poor criterion validity (accuracy).

3.3.3.1 Concordance Correlation Coefficient

The CCC addresses the aforementioned flaw associated with the PM correlation. Let y_1 (y_2) denote the score of the gold standard (instrument). The CCC between y_1 and y_2 is defined as

$$\rho_{\text{CCC}} = \frac{2\sigma_{12}}{\sigma_1^2 + \sigma_2^2 + (\mu_1 - \mu_2)^2}. \tag{3.9}$$

where $\mu_k = E(y_k)$, $\sigma_k^2 = Var(y_k)$ and $\sigma_{12} = Cov(y_1, y_2)$. The CCC ρ_{CCC} above ranges between -1 and 1; $\rho_{\text{CCC}} = 1$ (-1) if the two raters completely agree (disagree) and $\rho_{\text{CCC}} = 0$ if y_k are independent. Further, ρ_{CCC} can be expressed as: $\rho_{\text{CCC}} = \rho_{\text{PM}} C_b$, where $C_b = 2\left[\left(\frac{\sigma_1}{\sigma_2}\right) + \left(\frac{\sigma_2}{\sigma_1}\right) + \left(\frac{\mu_1 - \mu_2}{\sqrt{\sigma_1 \sigma_2}}\right)^2\right]^{-1}$ (Lin 1989). As C_b is inversely related to the bias, $(\mu_1 - \mu_2)$, C_b increases as the bias becomes smaller. Thus unlike ρ_{PM} and other popular *association* measures such as Spearman's rho, ρ_{CCC} captures both *accuracy* and *precision*.

We can perform a formal test, $H_0 : \mu_1 - \mu_2 = 0$, for potential systematic bias in the instrument using the paired t-test. As for precision, σ_2^2 is generally larger than σ_1^2 in practice, as it likely contains other sources of variability, in addition to the variation across the subjects. We can also conduct a formal test, $H_0 : \sigma_1^2 = \sigma_2^2$, to see if the instrument is as precise as the gold standard. However, the latter test is technically more challenging, as it involves second-order moments have recently developed a novel approach for facilitating comparisons for such second-order moments.

For many instruments, especially those that measure latent constructs in the behavioral and social sciences, there is generally no gold standard for measuring the latent construct either error-free or with error at a level negligible for all practical purposes. As a result, one often uses established instruments as a reference standard. For example, thanks to its popularity in QOL research, the SF-36 instrument has been translated into many different languages for its more effective use in non-English speaking countries. Although rigorous translation standards such as back-translation are implemented during the translation, it is still important to assess its validity. In many studies, the original English version of SF-36 is generally used as a criterion for external validity. In this case, both the original and translated instruments contain errors in measuring the latent QOL construct. By identifying y_1 and y_2 as the scores from the original and translated instrument, the CCC can again be applied to assess the external validity of the translated instrument. If the original instrument is translated without or with negligible errors, the null $H_0 : \sigma_1^2 = \sigma_2^2$ should hold true when applied to the same study population, though the hypothesis may not be testable because of the language barrier. In general, translation errors may be inevitable and σ_1^2 is likely larger than σ_2^2, with the amount of difference depending on the level of authenticity of the translation.

Note that the use of original instrument as a gold standard for a translated version requires bilingual subjects in the study, which may not be feasible. Other instruments that measure identical or similar latent construct in each of the domains may be used as a reference standard. For example, to help validate the CSF-36 discussed earlier, the investigators used the Quality of Life Instruments for Chronic Diseases-General Module (QLICD-GM), an instrument developed in Chinese for Chinese patients with chronic diseases based on the World Health Organization's definition of Quality of Life and programmed decision procedures (Wan et al. 2005, 2007). In such cases, as the domain mean scores are typically different between the different instruments, the PM correlation ρ_{PM} is more appropriate, since it accounts for such differences. We discuss this issue using a real data example in Sect. 3.6.

3.3.3.2 Kappa Coefficient

For discrete outcomes, either ordinal or categorical, Kappa coefficient κ is the most popular measure of validity when assessing a single instrument against a gold or reference standard. For example, the capacity of friends and family member informants to make judgments about the presence of a mood disorder history for older patients has theoretical, clinical, and public health significance. In a study on depression diagnosis among senior primary care patients (Duberstein et al. 2011), the validity of such informant-reported mood disorder diagnoses was examined using mood disorder diagnoses based on medical chart reviews and data collected in a structured interview with the patient as a gold standard. The instrument under evaluation is the informant's diagnosis about the patient, with a three-level ordinal variable indicating major, minor, and no depression.

Consider two instruments measuring the same construct such as mood disorder, with G ordinal/categorical outcomes: $r_1, r_2, \ldots, r_G$. Note that one of them may be a gold standard, but for our discussion, this does not matter since the validation criteria is the same. Let y_{img} be a binary indicator, with $y_{img} = 1$ if the mth instrument's rating on the ith subject belongs to the gth category and $y_{img} = 0$ otherwise ($m = 1, 2$; $1 \leq i \leq n$). Let

$$\pi_{(g_1, g_2)} = E\left(y_{i1g_1} y_{i2g_2}\right), \quad \phi_{mg_m} = E\left(y_{img_m}\right),$$
$$\pi_{e(g_1, g_2)} = \phi_{1g_1}\phi_{2g_2}, \quad 1 \leq g_1, g_2 \leq G.$$

Cohen's weighted κ is defined as

$$\kappa = \frac{\sum_{g_1=1}^{G} \sum_{g_2=1}^{G} w_{(g_1, g_2)} \left(\pi_{(g_1, g_2)} - \phi_{1g_1}\phi_{2g_2}\right)}{1 - \sum_{g_1=1}^{G} \sum_{g_2=1}^{G} w_{(g_1, g_2)} \phi_{1g_1}\phi_{2g_2}}, \tag{3.10}$$

where the bivariate outcome $\mathbf{g} = (g_1, g_2)^\top$ denotes the range of the two instruments' ratings, and the weight $w_{\mathbf{g}}$ is a known nonnegative function satisfying $0 \leq w_{(g_1, g_2)} \leq 1$ (Cohen 1960; Ma et al. 2008). The index κ ranges between -1 and 1; $\kappa = 1$ (-1) if the two instruments' ratings agree (disagree) perfectly with each other, and $\kappa = 0$ if the two ratings are not correlated.

In general, when the agreement (disagreement) exceeds the chance factor, $\kappa > 0$ $(\kappa < 0)$, with its magnitude indicating the degree of agreement (disagreement). If each of the instruments captures the latent construct to some extent, $\kappa > 0$ as in most validation studies. The unweighted kappa corresponds to the special weight function, $w_{(g_1, g_2)} = 1$ if $g_1 = g_2$, and 0 otherwise, and κ in this case only counts $\pi_{(g, g)}$ along the diagonal:

$$\kappa = \frac{\sum_{g=1}^{G} \left(\pi_{(g, g)} - \phi_{1g}\phi_{2g}\right)}{1 - \sum_{g=1}^{G} \phi_{1g}\phi_{2g}} = \frac{\sum_{g=1}^{G} \pi_{(g, g)} - \sum_{g=1}^{G} \phi_{1g}\phi_{2g}}{1 \quad \sum_{g=1}^{G} \phi_{1g}\phi_{2g}}. \tag{3.11}$$

In comparison, the weighted version tolerates discrepancies between the two instruments as set forth by the weight function ($w_{(g_1, g_2)} < 1$ for $g_1 \neq g_2$ to downweight the contributions from off-diagonal entries). In many applications, especially those with a large number of response categories, discrepancies between observers are inevitable. By judiciously weighting such responses, the weighted version can provide a more useful index of agreement for practical purposes (Ma et al. 2008).

3.3.4 *Test–Retest Reliability*

Upon establishing the validity (internal and external), we next assess the test–retest reliability of the instrument as the final step before it can be used in practice. This reliability aspect of the instrument is concerned with the ability of the instrument to yield identical or similar results when administered to the same subject over different, but closely spaced times. Although there is no rigid rule for the spacing of the times, the general idea is to select a time window to minimize any systematic difference, but long enough to characterize the variability due to repeated administrations of the test.

For example, in the study of CSF-36 discussed in Sect. 3.2.2, each of the 534 study patients was administered the CSF-36 at the time of admission to the hospital. A subsample of 197 patients was randomly selected to take the questionnaire a second time 1–2 days after hospitalization. The time window 1–2 days was adopted in this study to observe test–retest variability without much noticeable bias due to memory recall and changes in QOL in response treatment and care.

All the reliability indices in Sect. 3.3 can be used to assess the test–retest reliability of the instrument. In addition, popular measures of association such as the PM correlation, Spearman's ρ, Kendall's τ_b, and Goodman–Kruskal γ can be applied as well. As discussed earlier, the association measures are insensitive to systematic changes during the repeated administrations of the instrument and as such may be used when systematic differences are expected. For example, a random subsample of 409 patients in the CSF-36 study was administered the questionnaire at discharge (after about 2 weeks of treatment). As the QOL has been improved for the patients thanks to treatments, the mean of each of the domain score at discharge is likely to be different from its counterpart at admission. Association indices may provide more reliable estimate will not detect such treatment effects.

3.3.5 *Inference About Validity Indices*

Inferences for all the indices discussed above have been developed. Most available methods place the multivariate normal on the joint distribution of the items, which are prone to bias when data does not follow the assumed models (Feldt et al. 1987; Donner and Zou 2002; Shrout and Fleiss 1979; McGraw and Wong 1996; Fleiss et al. 2003). More recently, distribution-free alternatives have become available, which provide robust inference regardless of the data distribution (Kowalski and Tu 2008; Ma et al. 2010, 2008, 2010; Lu et al. 2012).

3.4 Real Data Examples

In this section, we illustrate the various topics discussed in the preceding sections with data from some real studies.

Example 1. A recent study carried out at the University of Rochester examined the validity of depression screening tools among postpartum women. A total of

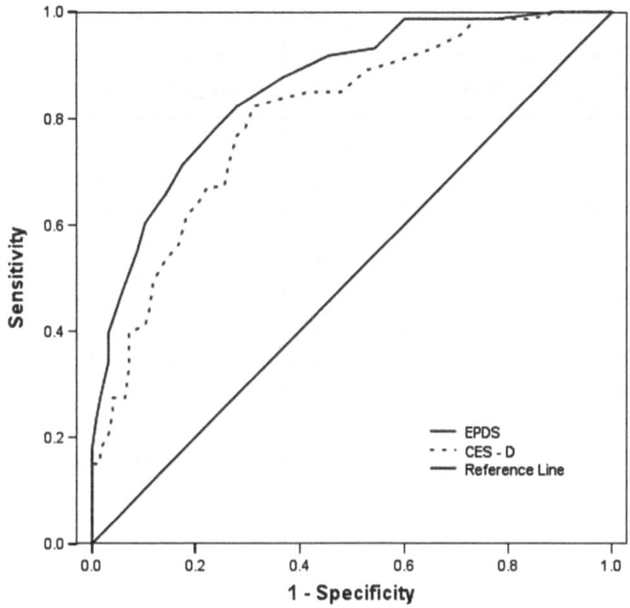

Fig. 3.2 Empirical ROC Curves for EPDS and CES-D

198 women in the study were administered several depression screening tools, including the Center for Epidemiologic Studies Depression Scale (CES-D) and the Edinburgh postnatal depression scale (EPDS), with the latter especially developed for postpartum women. Each item on either instrument is scored using a Likert scale and the total score is used as a continuous outcome for indicating depression. In addition, each subject was interviewed by trained therapists for depression using the Structured Clinical Interview for depression or SCID (Spitzer et al. 1994). As the SCID diagnosis is much more reliable, it generally serves as the gold standard for depression diagnosis. However, as the SCID interview takes several hours, it is expensive and inconvenient to administer. The objective of the study was to investigate the less expensive alternatives such as CES-D and EPDS to see if they would provide reliable depression diagnosis.

Shown in Fig. 3.2 are the empirical ROC curves for the two tests. It is clear that both CES-D and EPDS are informative for detecting depression, as both plots are well above the reference line. In addition, EPDS seems to outperform CES-D slightly, since its ROC curve lays almost entirely above that of CES-D. This may not be unexpected, since CES-D is a screening test for a general population, while EPDS is specifically designed for detecting depression in postpartum women. These findings are quantified in the estimate and inference of AUCs for the two tests.

Show in Table 3.2 are the AUC estimates and associated standard errors and p-values. The p-value in the table corresponds to the null H_0 : AUC $= 0.5$, i.e., the reference line for an uninformative test. These very small p-values indicate that both instruments are quite informative for screening for depression in this study

Table 3.2 Estimates of AUCs, standard errors, and p-values for BDI-II and EPDS for PostPartum Depression Study

Screening instrument	AUC	Std. error	p-value
CES-D/EPDS	0.808/0.867	0.031/0.026	<0.000/0.000

population. The larger AUC estimate for EPDS also confirms the better performance for this test, as suggested by the ROC curves in Fig. 3.2.

Since both ROC curve and AUC estimates show that EPDS is a better test, we formally compared the two empirical AUCs using the SAS PROC LOGISTIC based on the multivariate U-statistic approach discussed in Sect. 3.2.2.5, and obtained a p-value $= 0.0376$. Thus, EPDS does have a significantly better performance in detecting depression in this particular study population.

Example 2. A recent study was conducted to evaluate the performance of a Chinese version of SF-36, or CSF-36, when used to assess health related quality of life (HRQOL) for patients with hypertension, coronary heart diseases, chronic gastritis or peptic ulcer in mainland China. As noted earlier, the SF-36 instrument has eight domains, which can further be aggregated into the Physical and Mental Health subscales.

In this study, there were 534 patients from the aforementioned four disease groups: 157 with hypertension, 133 with coronary heart disease, 124 with chronic gastritis, and 120 with peptic ulcer. The patients ranged in age from 16 to 86, with a mean age of 54.7. After administered the CSF-36 at the time of admission to the hospital, a random subsample of 197 patients were asked to take the questionnaire again 1–2 days after hospitalization, to help evaluate the test–retest reliability of the translated instrument. Another subsample of 409 completed the assessment at discharge to help assess the sensitivity of the instrument to improvement in QOL posttreatment. All study patients were also given the QLICD-GM at admission. This instrument, developed in native Chinese, has 30 items classified into three domains of Physical, Psychological, and Social Functioning, providing criteria for assessing criterion-related validity of the similar domains in CSF-36 (Wan et al. 2005, 2007).

Factor analysis was applied to the data collected from the subjects at the time of admission, and as noted in Sect. 3.3.1, the first eight principal components accounted for 66 % of the cumulative variance. The factor loadings of the items on the eight principle factors, as shown in Table 3.1 for a subset of them, also lend support for the domain structure of the original SF-36. To formally validate internal consistency, Cronbach's α for the eight domains, as well as the two Physical and Mental Health summaries were computed. As shown in Table 3.3, the α values were by and large quite good.

With the data from the subsample of 197, we assessed the test–retest reliability for each domain subscale as well as the Physical and Mental subscales. The difference between the first and second assessments were not statistically significant (paired t-test p-value > 0.05) for all but the PF functioning domain (paired t-test p-value $= 0.039$), indicating negligible effect of treatment and care on this

Table 3.3 Estimates of Cronbach Coefficient Alpha α, and mean, standard deviation (std dev), for SC and DEXA, location (LS) and scale shift (SS), product-moment correlation (PM) and concordance correlation coefficient (CCC) between SC and DEXA for the Penn State Young Women's Health Study

	PF	RP	BP	GH	VT	SF	RE	MH	PCS	MCS
Internal consistency										
α	0.93	0.88	0.54	0.73	0.70	0.62	0.85	0.74	0.88	0.84
Test–retest reliability										
ρ_{PM}	0.76	0.66	0.66	0.80	0.78	0.77	0.68	0.76	0.81	0.82
ρ_{CCC}	0.75	0.67	0.65	0.78	0.80	0.76	0.67	0.74	0.80	0.83
ρ_{ICC}	0.76	0.66	0.67	0.79	0.79	0.77	0.68	0.77	0.79	0.81

Table 3.4 Product–moment correlation (PM) between QLICD-GM and CSF-36 domains

QLICD-GM	CSF-36 domains							
domains	PF	RP	BP	GH	VT	SF	RE	MH
Physical function	0.66	0.34	0.36	0.41	0.47	0.32	0.30	0.28
Psychological function	0.35	0.30	0.28	0.42	0.45	0.43	0.40	0.54
Social function	0.38	0.30	0.26	0.38	0.39	0.33	0.27	0.41

subsample of patients. Shown in Table 3.4 are the different correlation estimates between the two assessments for each domain. As seen, all three indices are quite similar, although by and large the estimates of CCC ρ_{CCC} were smaller than their PM and ICC counterparts. The rather small differences across the three indices reflect the fact that the subjects had similar QOL outcomes for each of the domains over the two assessments, indicating good test–retest reliability for the instrument.

In this study, the QLICD-GM was used for assessing the criterion-related validity of CSF-36. Shown in Table 3.4 are the Pearson estimates of the product–moment correlation (ρ_{PM}) for the similar domains between QLICD-GM and CSF-36, based on the 513 patients who completed both QLICD-GM and CSF-36. It is seen that overall the correlations between the same and similar domains of QLICD-GM and CSF-36 are higher than those between the different and non-similar domains. For example, the Pearson estimate between the Physical Function of QLICD-GM and Physical Functioning of CSF-36 was 0.66, higher than any other estimates involving the Physical Function of QLICD-GM and remaining domains of CSF-36. Additionally, the Psychological Function of QLICD-GM and the Mental Health domain of the CSF-36 had the largest correlation coefficient (0.54). These results confirm the criterion-related validity of CSF-36 to a reasonable degree.

Note that since the mean domain scores between QLICD-GM and CSF-36 are different, the popular validity measures such as CCC and ICC are not applicable, but the PM correlation ρ_{PM} can be used, for it makes allowances for such differences. Thus, although ρ_{PM} is not a validity index in general, it may be used in some cases to provide a quantitative indication about the validity of an instrument under investigation.

Table 3.5 Estimates of mean, standard deviation, location shift (LS), scale shift (SS), product–moment (PM), concordant correlation coefficient (CCC), and test statistics and associated p-values for comparing skinfold calipers and dual energy X-ray absorptiometry for Example 2

Mean (std dev)		LS	SS	PM (std err)	CCC (std err)
$\mu_1\ (\sigma_1)$	$\mu_2\ (\sigma_2)$	$\frac{\lvert\mu_1-\mu_2\rvert}{\sqrt{\sigma_1\sigma_2}}$	σ_1/σ_2	ρ_{PM}	ρ_{CCC}
23.8(0.37)	21.6(0.45)	0.56	0.81	0.80 (0.039)	0.68(0.047)
Hypothesis testing (Test statistic | p-value)					
$\mu_1 = \mu_2$			$\sigma_1^2 = \sigma_2^2$		
64.6 | < 0.01			9.2 | < 0.01		

Example 3. This study examined whether relatively inexpensive skinfold calipers (SC) would provide a reliable alternative for measuring body fat compared to the standard dual energy X-ray absorptiometry (DEXA) (Lloyd 1998). The percentage of body fat was obtained for a group of adolescent girls using both SC and DEXA. The measurements were initially taken at age 12, and then every six months thereafter. For illustration purposes, we focus on the 89 subjects with their first assessment at age 12.5.

Shown in Table 3.5 are estimates of the mean (μ_k) and standard deviation (σ_k), along with the location (LS) and scale shift (SS), for SC ($k = 1$) and DEXA (k=2). The differences in μ_k and σ_k between the two instruments gave some indication of both location and scale shift, which consequently led to quite different Pearson estimates of ρ_{PM} and CCC ρ_{CCC} (0.80 vs. 0.68). A formal test of the null $H_0 : \mu_1 = \mu_2$ for location and $H_0 : \sigma_1^2 = \sigma_2^2$ for scale shift based on the methods of Ma et al. (2010) confirmed the systematic upward bias in mean percent of body fat and underestimation of the variability of body fat across the girls in this age group (see Table 3.5 for the Wald statistic and p-value). Thus, the product–moment (PM) correlation is inappropriate for assessing the validity of SC. Further, the relatively low CCC indicates that SC may not be sufficiently accurate for measuring body fat for girls in this particular age group, substantiating the limitation of skinfold calipers in detecting subcutaneous fat, whereas breast, lower body, and visceral fat are increasing over this age range due to the onset of puberty.

3.5 Software

ROC analysis is now available in most popular statistical software packages such as SAS, SPSS and Stata. For example, one can simply point and click on the interactive session windows in SPSS, and Stata to select disease variables, test outcomes and desirable types of ROC analysis to obtain estimates of ROC and AUC. ROC analysis can also be carried out using SAS PROC LOGISTIC (with SAS 9.2 and newer version). User-developed packages are also available in R.

Most reliability indices have also been implemented in these packages. For example, Cronbach alpha can be computed by SAS PROC CORR. Although some of these are not officially support by SAS yet, user-developed macros are often available for such indices such as CCC and ICC.

3.6 Discussion

In this chapter, we discussed various notions of reliability and statistical models for assessing the different types of reliability. Although the concepts of and statistical methods for the external validity and diagnostic-ability of instruments have been extensively discussed in the literature, most of such discussion has focused on medical devices for detecting often a single dimension, well-defined phenomena such as heart disease and lung cancer. In contrast, discussions of multi-faceted, less clear latent concepts and associated validity issues such as depression and QOL have largely been confined to the literature in the behavioral and social sciences. However, as more effective treatments have been and are being developed to prolong the life of patients suffering cancer and other terminal illnesses, QOL has become an important bench marker for patients care, and more studies have began to include this measure, especially for those focusing on palliative care. In addition, as we move to the next level of health care model by integrating social, behavioral, and lifestyle factors to biological and physical changes and studying the interplay between such biological, behavioral, and societal factors, measures of multidimensional constructs such as physical functioning, mental health, and social support will play an increasingly important role in developing and delivering personalized treatments.

The increasing popularity of including measures of latent health quality constructs such as QOL as study outcomes creates both challenges and opportunities for new statistical models and methods. For example, inferences for Cronbach coefficient α and ICC all involve second-order moments, while most robust methods such as the popular generalized estimating equations typically model the mean, rather than higher-order moments. In recent years, however, numerous effects have been made to address the lack of methods for modeling higher-order moments (Ma et al. 2008, 2010; Lu et al. 2012). Most importantly, most of these approaches are developed by utilizing the Functional Response Models, a new class of models for complex response functions and regression relationships, which in particular provides an effective means to address higher-order moments (Kowalski and Tu 2008).

References

Bach P, Jett J, Pastorino U, Tockman M, Swensen S, Begg C (2007) Computed tomography screening and lung cancer outcomes. J Amer Med Assoc 297(9):953–961

Cohen J (1960) A coefficient of agreement for nominal scales. Educ Psychol Meas 20(1):37–46

DeLong E, DeLong D, Clarke-Pearson D (1988) Comparing the areas under two or more correlated receiver operating characteristic curves: a nonparametric approach. Biometrics 44(3):837–845

Donner A, Zou G (2002) Testing the equality of dependent intraclass correlation coefficients. J Roy Stat Soc D 51(3):367–379

Dorfman D, Alf E (1969) Maximum-likelihood estimation of parameters of signal-detection theory and determination of confidence intervals—Rating-method data. J Math Psychol 6(3):487–496

Duberstein P, Ma Y, Chapman B, Conwell Y, McGriff J, Coyne J, Franus N, Heisel M, Kaukeinen K, Sörensen S, Tu X, Lyness J (2011) Detection of depression in older adults by family and friends: distinguishing mood disorder signals from the noise of personality and everyday life. Int Psychogeriatr 23(04):634–643

Feldt L, Woodruff D, Salih F (1987) Statistical inference for coefficient alpha. Appl Psychol Meas 11(1):93

Fleiss J, Levin B, Paik M (2003) Statistical methods for rates and proportions, 3rd edn. Wiley, New York

Kowalski J, Tu XM (2008) Modern applied U-statistics. Wiley series in probability and statistics. Wiley-Interscience, Hoboken, NJ

Lin L (1989) A concordance correlation coefficient to evaluate reproducibility. Biometrics 45(1):255–268

Lloyd CJ (1998) Using smoothed receiver operating characteristic curves to summarize and compare diagnostic systems. J Amer Stat Assoc 93:1356–1364

Lu N, Gunzler D, Zhang H, Ma Y, He H, Tu X (2012) On robust inference for intraclass correlation coefficients. Technical report, The Department of Biostatistics and Computational Biology, University of Rochester, Rochester, New York

Lubetkin E, Jia H, Gold M (2003) Use of the SF-36 in low-income Chinese American primary care patients. Medical Care 41(4):447–457

Ma Y, Alejandro G, Hui Z, Tu X (2010) A u-statistics-based approach for modeling cronbach coefficient alpha within a longitudinal data setting. Stat Med 29(6):659–670

Ma Y, Tang W, Feng CY, Tu XM (2008) Inference for kappas for longitudinal study data: Applications to sexual health research. Biometrics 64(3):781–789

Ma Y, Tang W, Yu Q, Tu X (2010) Modeling concordance correlation coefficient for longitudinal study data. Psychometrika 75(1):99–119

McGraw K, Wong S (1996) Forming inferences about some intraclass correlation coefficients. Psychol Meth 1(1):30–46

Nakas C, Yiannoutsos C (2004) Ordered multiple-class roc analysis with continuous measurements. Stat Med 23(22):3437–3449

Shrout P, Fleiss J (1979) Intraclass correlations: uses in assessing rater reliability. Psychol Bull 86(2):420–428

Spitzer RL, Gibbon M, Williams JBW (1994) Structured clinical interview for axis I DSM-IV disorders. Biometrics Research Department, New York State Psychiatric Institute.

Stewart J (2010) Calculus: early transcendentals. Brooks/Cole Publishing Company, Pacific Grove, CA

Tu X (2009) The domain-sampling model for measurement errors. Technical report, The Department of Biostatistics and Computational Biology, University of Rochester, Rochester, New York

Tu X, Litvak E, Pagano M (1992) Issues in human immunodeficiency virus (HIV) screening programs. Amer J Epidemiol 136(2):244–255

van Zyl J, Neudecker H, Nel D (2000) On the distribution of the maximum likelihood estimator of cronbach's alpha. Psychometrika 65(3):271–280

Wan C, Gao L, Li X et al. (2005) Development of the general module for the system of quality of life instruments for patients with chronic disease: Items selection and structure of the general module[j]. Chin Ment Health J 11:444–447

Wan C, Jiang R, Tu XM, Tang W, Pan J, Yang R, Li X, Yang Z, Zhang X (2012) The hypertension scale of the system of Quality of Life Instruments for Chronic Diseases, QLICD-HY: a development and validation study. Int J Nurs Stud. 49(4):465–480

Wan C, Yang Z, Yang Y (2007) Development of the general module of the system of quality of life instruments for patients with chronic disease: Evaluation of the general module. Chinese J Behavior Med Sci 16:559–561

Ware Jr J, Sherbourne C (1992) The MOS 36-item short-form health survey (sf-36): I. conceptual framework and item selection. Med Care 30:473–483

Chapter 4
Analysis of Sequential Clinical Trials

W. Jackson Hall

4.1 Introduction

We discuss issues and methods for the analysis of clinical trials that have formal rules for what is commonly referred to as *early stopping*. However, we take a broader view, and include trials having a formal *sequential stopping rule*; the distinction will soon become apparent. Moreover, it is impossible to separate the analysis of such trials from their design, and so we must, of necessity, provide an overview of the design—that is, specification of the stopping rule—of a sequential clinical trial as well.

We focus on some of the most popular sequential design and analysis principles, although sometimes with a nonstandard perspective, and do not attempt to be anything like all-inclusive. This is an active area of statistical research and innovation, and many of the newer methods will need more time to find their appropriate place in the repertoire of clinical trial methodology. Moreover, there are good technical books—we will regularly refer to those by Whitehead (1997) and by Jennison and Turnbull (2000)—and available software to do the basic analyses for sequential trials—we will refer to PEST (2008) and SAS 9.2 (2008), to a recent paper by Whitehead (2011) and to some programs of our own (Hall 2010); others include EAST Version 5 (2007) and S+SeqTrial (2000).

First, a few historical notes: Formal utilization of sequential stopping rules got its early motivation from sampling inspection and quality control activities, first in the *double-sampling plans* of Dodge and Romig (1929) and later in the more elaborate and fully sequential, sampling one-at-a-time, methods of Wald (1947). This work was motivated by quality control needs to reduce the amount of sampling

W.J. Hall (✉)
Department of Biostatistics and Computational Biology, University of Rochester, Rochester, NY 14642, USA
e-mail: Hall@bst.rochester.edu

W. Tang and X. Tu (eds.), *Modern Clinical Trial Analysis*, Applied Bioinformatics and Biostatistics in Cancer Research, DOI 10.1007/978-1-4614-4322-3_4,

(on average) required to meet quality requirements. The earlier double-sampling plans only allowed for a preliminary sample followed by a follow-up sample if results in the preliminary sample were not sufficiently extreme.

Some years later, the same kinds of issues were raised in clinical research, especially clinical trials. But now there was ethical motivation as well as the obvious one of expedience and economy: If the outcome of a clinical trial seems nearly inevitable while it is still in progress, it is important (i) to allow termination to avoid unnecessary experimenting on patients, and (ii) to speed up the opportunity to extend a new successful treatment to control-group patients in the trial, and to the qualifying public at large, or to terminate use of an inferior or even dangerous new treatment. Armitage (1954, 1975) was a major initiator of this work.

All methods have some common ancestry: all grew out of the Neyman–Pearson approach to statistical hypothesis testing, with focus on control on the probabilities of two types of error, rejecting a null hypothesis when it is true and failing to reject when an alternative of particular interest is true. The latter is more often approached as *power*, namely the probability of rejecting, rather than failing to reject, when the alternative is true. Beyond that, there were two "jumping off points" for sequential design and analysis. First, Wald (1947) introduced his *sequential probability ratio tests (SPRTs)* and established their highly efficient properties. However, his designs were open-ended in that there was no upper bound on the amount of sampling that might be required, although truncation of his open ended designs could be done in ways that would not appreciably affect measures of their performance. Various linear-boundary designs (no longer limited to Wald's parallel linear boundaries) were developed, especially triangular stopping boundaries. Wald's designs were optimal in the sense of minimizing the expected amount of sampling required if either null or specific alternative hypothesis was true, compared with any other design with the same (or smaller) error probabilities. Certain triangular designs were found to sacrifice only a small amount of Wald's optimality and yet have a clear upper bound on the amount of sampling required and were efficient (at least approximately) in minimizing the expected amount of sampling at an intermediate hypothesis, where Wald's designs were inefficient (see Anderson (1960), Lorden (1976), Whitehead and Jones (1979) and Hall (1980)). Special cases of such triangular designs, and modifications thereto, as well as other linear boundary designs are the basis for useful clinical trial designs: the aforementioned book by Whitehead, and supporting PEST software, focus on these designs. So this class of designs constitutes modifications of Wald's famous *SPRT*s.

Second, the other large class of designs grew out of modifiying analysis of fixed-sample plans to recognize the fact that the trial might have been stopped early. The context was that of having scheduled one or more interim analyses to evaluate progress of the trial and permitting the possibility of early termination if results were sufficiently extreme. Development started with specification of when these interim analyses would be done, together with rules for termination of the trial at one of these analyses—defined in terms of critical values for test statistics or in terms of nominal significance levels—for potential early stopping at the specified interim analysis times. The maximum amount of observation—a final analysis time if needed—would need to be increased somewhat in order to preserve the significance level and power of the trial. Such interim analyses were deemed important for ethical reasons, as noted above. Early plans of this type were those of Pocock (1977) and O'Brien and Fleming (1979). Later variations allowed a variety of other boundary shapes without quite the same focus on only minor adjustments to the original nonsequential (fixed sample size) design. The aforementioned book by Jennison and Turnbull and recent versions of SAS software (SAS 9.2 2008) provide assistance with the design and analysis of these so-called *group-sequential* trials.

All of the early group-sequential designs were for testing a null hypothesis of "no effect," or "no difference" (between treatments), versus two-sided alternatives—that the new treatment is "superior" or "inferior" to the standard treatment, with equal attention given to each alternative. Such designs are well justified for comparing two treatments "head to head" to find the better of the two, or to conclude insufficient evidence for either conclusion. However, in the context of new versus standard (or placebo) treatment, suitability may be questioned in that ongoing monitoring of the trial gives equal attention to superiority and inferiority, whereas ethical issues may be raised by experimenting on patients in order to distinguish between "inferiority" and "insufficient evidence" conclusions. We recommend asymmetric designs in this latter setting—specifically, using a one-sided design for concluding "superiority" or "insufficient evidence of it" with significance level and power control, with further partitioning of the latter region for conclusions of "insufficient evidence" or "inferiority." Such designs accommodate two-sided p-values for the null hypothesis but may compromise power at particular measures of inferiority. This approach has been put forward by Whitehead (1997, 2011) and his PEST software, and will be further clarified here.

It is often stated that there is a price to be paid for allowing interim analyses, with a sequential stopping rule. This is true only in the sense that there is some risk that the total amount of sampling—data accumulation, more generally—may be increased, relative to a comparable fixed-sample plan. However, the *expected* amount of sampling required—that is, on average over many similar trials for which actual sample sizes would vary since some would stop early—may be greatly reduced; and in any particular trial, the probability of reduced sampling is typically very high. There may be an administrative "price," however, resulting from the uncertain duration of a trial.

In all of the early work, inferences were limited to acceptance or rejection of hypotheses under test. In more recent years, the need for further inference, notably p-values to gauge the strength of the evidence against a null hypothesis, estimation of effect sizes and confidence bounds, and intervals for effect sizes, was recognized and methodology developed. It is this that is our primary focus, but, as stated, we can only present such methodology in the context of the trial design.

Here is an overview: In Sect. 4.2, we give some background and examples of the applied contexts in which sequential trials are used. We give some indication of why sequential sampling affects the methods of inference to be employed. In Sect. 4.3, we describe the most common statistical modeling in such trials and include a summary of sample size calculations appropriate when no sequential rule is to be employed. In Sect. 4.4, we briefly review inference principles derived from p-values in nonsequential settings, since similar principles form the basis for inference in sequential trials. In Sect. 4.5, we describe designs with continuous monitoring—termed *fully sequential designs*—and the associated analysis methods, including adjustments for discrete-time monitoring and p-value-based inferences after reaching a stopping boundary. We start with a simple design in which nothing more than normal probability software is needed for p-value-based inference. We then describe linear-boundary designs, primarily the very efficient *triangular designs*, but these require special software for analysis. In Sect. 4.6, we move on to so-called *group-sequential designs* and their design and analysis; in these, a limited number of interim ongoing analyses are planned or contemplated. We start with 2-stage designs for which the only additional software needed for their analysis is that for bivariate normal probabilities. We then extend to multistage designs and describe methods for design and analysis using SAS software—some simplified methods of Whitehead (2011) as well as those using SAS 9.2. In Sect. 4.7, we briefly describe some approaches to more flexible design construction and analysis, and briefly mention some other relevant topics. In Sect. 4.8, we summarize the general principles and methods of analyzing data from a sequential clinical trial.

4.2 Sequential Clinical Trials and the Need for Special Analysis Methods

We focus on Phase III clinical trials in which a new treatment is to be evaluated relative to a standard (or possibly placebo) treatment. Occasionally, the context may simply be comparison relative to a historical control or to comparison between two new treatments.

Instead of continuing to a fixed time period, or to a fixed number of patient responses, it is often important to allow for earlier termination when trial results warrant it. Typically, patients become available for the trial one-by-one, and consequently trial results become available in a sequential fashion over time. Often, it is simple enough to carry out periodic analyses on available results—possibly by, or for, a *Data Safety Monitoring Committee (DSMB)* and preserving "blindness" for those running the trial—and hence make possible a decision to terminate the trial at such periodic time points.

However, it is dangerous to simply allow stopping "when the results look good" or "when the results appear hopeless" without a clear plan; inferences can be badly biased if stopping rules are not pre-specified and appropriately taken into account when interpreting the data; even interim adjustment to future stopping boundaries should not be based on results of analyses so far.

Readers familiar with *inverse binomial sampling*—sampling Bernoulli trials until a pre-specified number, say s, of successes have been observed—will be familiar with the impact of sequential sampling on standard inference concepts. Here, the distribution of the number of trials required is a negative binomial distribution.

An unbiased estimate of the success probability p is $(s-1)/(n-1)$ rather than the familiar s/n, with s the number of successes in n trials. Note that n, not s, is now random. Still, s/n continues to be the maximum-likelihood estimate (MLE), although biased, and normality and optimality are only large-sample attributes. Hypothesis testing would be based on an underlying *negative binomial* sampling distribution rather than a *binomial* distribution. Even the sample space is affected, being the integers $\{n = s, s+1, s+2, \ldots\}$ rather than $\{s = 0, 1, \ldots, n\}$.

Double-sampling plans are often used in medical research when testing potential agents for activity, and observing a Bernoulli sequence of successes ("active") and failures ("inactive"). A preliminary sample of size n_1, say, is specified. If the number of successes in the first n_1 trials is sufficiently large, say at least s_1, then sampling is continued until a total of n trials have been completed. Inference about the true success ("activity") rate p is carried out only if the trial is not terminated at the first stage, after observing n_1 trials. Methods are devised to adjust inferences about p—whether estimation of it, a confidence interval for it, or a test of a hypothesis about it—that recognize that the count of successes is no longer simply binomial with parameters n and p. In particular, $n = n_1$ when stopping early and, after n observations, $s \geq s_1$. Such sampling schemes allow early stopping only for "futility"—when the available evidence is not very promising. See Simon (1989) for specific treatment of it (and Sect. 4.7). The original plans also allowed stopping after the first sample when evidence was highly favorable Dodge and Romig (1929)

If inference is to be made about a mean, using a sample mean—from a sample of normally distributed observations for definiteness—the same kind of double-sampling scheme raises similar issues. For example if early stopping occurs whenever the sample mean after n_1 observations is far from a hypothesized value—suggesting a strong positive effect or a strong negative effect—then the distribution of the sample mean when sampling does *not* stop early is affected the cumulative sum after all n observations includes the cumulative sum after the first n_1, and since extreme values of the latter would have been removed by early stopping, extreme values after all n observations are not as likely as if early stopping had not been allowed. Thus, the sampling distribution has been affected, having smaller tail areas and consequently less significant p-values, and therefore any frequentist-based inference methods will need to recognize the possibility of early stopping even if it did not occur. This is a special case of a so-called *group-sequential* stopping design; explicit formulas and a numerical illustration for this double-sampling case are given in Sect. 4.6.1.

We allow for more general sampling schemes and associated stopping rules, including those used popularly in clinical trial practice.

4.3 Statistical Models for Sequential Clinical Trial Data

We start with the special case of sampling from two normal (Gaussian) populations. The data come about as follows: Subjects (e.g., patients) become available one-by-one. They are randomized to receive either the new treatment or the control treatment, to receive the new treatment with probability p and the control (standard or placebo) treatment with probability $q = 1 - p$, with p specified as part of the design. Most commonly, $p = q = \frac{1}{2}$ but other choices are possible, such as $p = \frac{3}{5}$ or $p = \frac{2}{3}$, and various blocking and stratification systems may be used to achieve more balance. Choices other than $p = \frac{1}{2}$ may lead to slight losses in statistical efficiency for the primary analysis but may be justifiable by other considerations, such as providing greater experience with the new treatment, or providing a larger sample size in the treatment arm of the study for various arm-specific substudies, or to increase efficiency in the neighborhood of an alternative considered likely *a priori*, or to encourage increased participation if subjects are likely to prefer being in the treatment arm. A response is measured in each subject after a fixed time after recruitment. (Studies involving time to an endpoint, such as death, will be considered later.) Hence, the data consist of a sequence of X's (responses from control patients) and a sequence of Y's (responses from treatment patients). For now, the assumption is that the X's are independent and normally distributed with mean μ and variance σ^2 while the Y's are independent and normally distributed with mean λ and variance τ^2, and independent of the X's.

At the time of any particular data analysis, suppose m X's and n Y's are available. The hypothesis of primary interest is that $\theta \equiv \lambda - \mu$ is 0 versus a specified alternative, either one-sided or two-sided, with power to be controlled at the reference value θ_R (>0), say.

A standard test statistic would be $Z = (\bar{Y} - \bar{X})/SE$ where $\bar{X}$ and $\bar{Y}$ are sample averages and SE is an estimate of the standard error of the numerator, namely $\sqrt{(\hat{\tau}^2/n) + (\hat{\sigma}^2/m)}$ with the ^'s indicating estimates of variances by sample variances. (If τ is assumed to be equal to σ, then a common pooled estimate would be used.) In a fixed-sample-size trial, the null hypothesis would be rejected if the observed value of the test statistic was sufficiently large and a p-value would be the probability in the upper tail of the standard normal distribution beyond the value of the test statistic. For two-sided testing, p would be replaced by twice the smaller of p and $1 - p$. The t-distribution for Z would give similar results unless m and n were small.

The *statistical information* (see below) in the data is quantified by $v \equiv 1/SE^2$, which $= N/[(\tau^2/p) + (\sigma^2/q)] = Npq/(p\sigma^2 + q\tau^2)$ with $N = m + n$, $p \approx n/N$, $q \approx m/N$ and σ and τ estimated. (We use actual proportions m/N, n/N rather than randomization probabilities.) We see that information v is proportional to total sample size N.

For later purposes, we find it convenient to re-express this test statistic in what may be termed the *cumulative-sum form* or *score form*: $S = v \cdot (\bar{Y} - \bar{X})$ $(= Z/SE = Z\sqrt{v})$. Note that S is (approximately) normally distributed with mean θv and variance v. (The statistic S is the *efficient score* for θ; see below.) Also, utilizing the *Central Limit Theorem*, we see that this approximation is valid whatever the true distributions of the data whenever m and n are not too small (assuming finite variances), so any requirement of normality for X and Y is generally unnecessary.

As a simpler example yet, imagine that we are sampling from a single normal population with mean θ and variance σ^2 and we wish to infer about θ. After N observations, the cumulative sum S of the observations, when divided by σ^2, would have variance $v = N/\sigma^2$, the information, and mean $\theta v = N\theta/\sigma^2$. This explains the *cumulative-sum form* terminology.

A statistic with mean proportional to v and variance v behaves like a *Brownian motion* on timescale v and with *drift* the proportionality constant (assuming the covariance, between the statistic at $v = v_1$ and at a later value v_2, is v_1—equivalently, as information increases, the successive values of the statistic have *independent increments*). A sample path varies around a ray from the origin, the drift being the slope. This is the advantage, mathematically, of this cumulative-sum form, which is often found to provide a good approximate model for test statistics as the amount of data, and hence the statistical information, increases. It is clearly true when sampling from a single normal population and readily verified also for the two-normal populations case above. [A Brownian motion $S(v)$ on timescale v and with drift θ has mean θv and variance v; it is normally distributed for each fixed v, and has independent increments.] Upon observing $S(v) = s$ at time v, the MLE of the drift is s/v—whether sampling until a pre-specified v or until a pre-specified level s (when achievable) or until any other (achievable) stopping boundary is reached. As in the inverse binomial sampling case, however, the MLE is no longer necessarily the estimate of choice, as we shall see.

Now let us generalize somewhat more. Consider any statistical model for the two samples of data, with one (numerical) parameter θ of interest and additional nuisance parameters. In the two-normal populations example above, the parameter of interest was the difference between λ and μ, while one of the two means, say μ,—or equivalently $\lambda + \mu$—and the two variance parameters were nuisance parameters, additional parameters required by the model but left unspecified by the null hypothesis of interest and by any alternative at which power is to be controlled. *Scores* for each parameter are defined as the respective derivatives of the logarithm of the joint probability density of the data, that is, the *log likelihood*. *Information* about each parameter is the variance of the corresponding score, and *co-information* the corresponding covariance between scores. (Sample versions of information and co-information are the negatives of derivatives of corresponding scores, and are often used to estimate information.) Evaluating all scores and information terms at the null value of the parameter of interest, say $\theta = 0$, the *effective* or *efficient score* is the score for the parameter of interest, with nuisance parameters replaced by *maximum likelihood estimates* (assuming $\theta = 0$). Estimated effective information (associated with the effective score) needs to be determined from a matrix calculation, namely, the reciprocal of the upper left corner in the inverse of the matrix of sample information and co-information terms. If there is a single nuisance parameter, this leads to effective information equal to information for θ minus the quantity: co-information squared, divided by information for the nuisance parameter. In our two normal populations example, all co-information terms are 0, and hence effective information for $\theta = \lambda - \mu$ is the same as information for θ—there is no loss in information when parameters are *orthogonal* (zero co-information). No worry—software provides all of these quantities.

It turns out, from large-sample considerations, that effective scores behave like a Brownian motion on the effective information timescale. Hereafter, we leave the term "effective" implicit, and simply deal with test statistics in cumulative form that can be justified as being approximately normally distributed, with mean proportional to variance and where variance is (estimated effective) information (and covariance between values at different v's is the variance at the smaller v).

This same model extends to various *semiparametric statistical models* (e.g., Choi et al. 1996); we exemplify this case with trials based on survival analysis below.

Many writers about sequential methods prefer the *standardized test statistic* form Z, the basic version being of the estimated parameter of interest divided by its estimated standard error, as for the difference-between-two-means case above. We prefer, and will mainly use, the *cumulative-sum form*. When observing several points in the (S, v)-plane, associated with interim monitoring, say, the *cumulative-sum* form is especially useful in that such points vary around a ray from the origin with slope θ, and v is the variance of S. However, writing the former as $Z = \hat{\theta}/SE$ and the latter as $S = score = Z/SE$, then $Z = S/\sqrt{v}$ and $SE^2 = 1/v$. With these transformations, we can readily go back and forth between the two representations. The Z form may be thought of as a *Wald test statistic* and the S form as a *score test statistic*. Users need to be careful, when reading literature on sequential clinical trials and when using associated software, about whether the *standardized* or the *score* (*cumulative-sum*) form is being used. Indeed, a third possibility is the *MLE* form, with $\hat{\theta} = S/v = Z \cdot (SE)$.

Stopping rules, along with inference methods upon stopping, will be defined in terms of a sequence of S (or of Z) statistics, based on amounts of information available at successive analyses. We focus on designs which specify when interim analyses are to be performed in terms of pre-specified and increasing amounts of statistical information, generally proportional to total available sample size.

Now we turn to the case of a survival-analysis-based clinical trial. Each treatment has associated with it an (unknown) survival curve $\bar{F}(t)$ with a subscript on $\bar{F}$ indicating treatment or control group (and with $\bar{F} = 1 - F$ with F the distribution function of elapsed time t to endpoint), specifiying the probability of remaining endpoint-free up to a specified time. When the endpoint of interest is death, the term "survival" becomes appropriate, but the term is used more generally. The null hypothesis of interest is that the two survival curves, one for treated subjects and one for control subjects, are identical. Alternatives of interest are most commonly phrased in terms of a *proportional-hazards model*, specifiying that the survival curve—on time scale t of elapsed time from entry in the trial—for the treated group is a power of that for the control group: $\bar{F}_T(t) = \bar{F}_C(t)^{\mathrm{HR}}$. The power, *HR*, termed the *hazard ratio*, is usually written as $HR = \exp(-\theta)$ so that the hypothesis that $HR = 1$—the null hypothesis of identical survival curves—becomes $\theta = -\log(HR) = 0$. The term *hazard* refers to the conditional probability of the endpoint occurring in the next short interval of time, given that it has not yet occurred—sometimes, in the mortality context, referred to simply as the *ongoing risk of death* in those surviving so far. Then *HR* becomes the ratio, treatment:control, of hazards when assuming that the ratio is constant over time. That is, it is a representation of *relative risk*, often found to be adequately constant over the follow-up time periods in a clinical trial. Power computations are usually focused on a particular value for *HR*, such as 0.80 ($\theta_R = -\log(0.80) = 0.2231$), under a proportional-hazards (i.e., constant hazard ratio) assumption—representing a 20 % reduction in ongoing risk when assigned to treatment relative to control.

In a trial, at any particular time point, patients have been followed for varying amounts of follow-up time, from patient-specific recruitment to the current time. Each will either have had an endpoint time (from recruitment) recorded or simply a follow-up time without endpointing recorded, a *censored* time. The *logrank statistic* is the effective score statistic S (e.g., Choi et al. 1996), as described in Whitehead (1997) and in Jennison and Turnbull (2000), and also in SAS manuals. The corresponding information, the variance of the logrank statistic, is now (approximately) proportional to the total number D of endpoints recorded so far, the proportionality constant being pq. Hence, D plays the role of sample size rather than N, the total number of patients recruited so far. The logrank statistic S behaves like a Brownian motion with drift $\theta = -\log(HR)$, and serves as a cumulative-sum form of a test statistic for successive analyses in a sequential trial on the information time scale $v \approx pqD$. Alternatively, S can be converted to *standardized form* Z by dividing by $\sqrt{v}$.

Hence, in a survival-analysis-based trial, interim analyses are scheduled at pre-specified values of accumulated numbers of events, increasing values of D, rather than of N. In practice, there is a need to allow flexibility in the specified values of information at which analyses are to be scheduled (see Sect. 4.7). For now, however, we act as if analyses can be carried out at pre-specified values on the information timescale, or even virtually continuously. To control significance level and power in a survival-analysis trial, a "fixed-sample-size" trial is actually a trial that continues until a pre-specified number D_o of endpoint events occur. Sequential trials are those allowing termination at various values of D. (In practice, we use the estimated variance—provided by software—of the logrank statistic as v, rather than the approximation pqD; just as in other trials, it is *estimated* variance of S that is used.)

How big must a nonsequential trial be? We confine attention to designs with significance level α for testing $\theta \leq 0$ versus $\theta > 0$ and power $1-\beta$ at θ_R (> 0); we say *designs of strength* (α, β) *at* $(0, \theta_R)$. Such designs, when used for testing against two-sided alternatives, have significance level 2α. In order to meet this strength requirement, a standard argument yields

$$v_o = (z_\alpha + z_\beta)^2/\theta_R^2 \tag{4.1}$$

as the needed information v_o, where z_γ is such that the tail area beyond it in a standard normal distribution is γ; that is, $\bar{\Phi}(z_\gamma) = \gamma$ with Φ the standard normal distribution function and $\bar{\Phi} = 1 - \Phi$. The *critical value* for S, or for $|S|$ in standard two-sided testing, is then $z_\alpha\sqrt{v_o}$. For example, in a two-arm clinical trial with randomization probabilities p and q, observation of a quantity with variance σ^2, and θ the difference between means, the needed total sample size is $N_o = v_o\sigma^2/(pq)$ with v_o in (4.1). In a survival-analysis-based trial, the total number of events needed is $D_o = v_o/(pq)$, and sample sizes need to be chosen to make this feasible. This quantity, v_o, plays an important role in designing sequential trials with similar strength requirements.

We must emphasize now that, when drawing any inference from the data when an analysis has been carried out, and a data-based decision to terminate the trial has been made, we must take into account the stopping rule utilized in the trial. As we have seen in Sect. 4.3, it may well affect the sample space and it very definitely affects the probability distribution over the sample space. These play essential roles in the Neyman–Pearson theory of hypothesis testing, the basis for inference procedures discussed herein, and in p-value-based inference following termination of a trial, as well as construction of point and interval estimates.

Before describing some of the popular designs and their associated analyses, for trials with a prescribed sequential stopping rule, we review the inference principles used non-sequentially, as they will be extended to sequential cases in later sections.

4.4 Inference Based on p-Values: Non-sequential Case

We review briefly how all basic elements of inference, following a clinical trial with no provision for early stopping, can be based on an ordering of the sample space and associated p-values. We do this since the same principles will be applied to analysis of data from sequential clinical trials. We assume here that a pre-specified test statistic for the primary analysis is, to an adequate approximation, normally distributed with mean θ, an *effect size*, specified as 0 under a null hypothesis, positive under a one-sided alternative and simply unequal to 0 under a two-sided alternative. Total information is fixed at $v = v_o$.

We first focus on one-sided testing, and afterwards comment on the two-sided case. Associated with the pre-specified test statistic is an ordering of the sample space. Values larger than the observed value of the test statistic are considered *more extreme*—that is, favoring the alternative hypothesis more strongly. A p-value is computed, representing the probability of obtaining a value as or more extreme than that observed under an assumption that the null hypothesis is true. To interpret the result relative to a pre-specified significance level, the null hypothesis is rejected if p is at most α—equivalent to comparing the test statistic with a critical value. Here, both p and α are one-sided. For two-sided testing, the null hypothesis of "no difference" is also rejected if $1 - p \leq \alpha$, now in favor of a negative effect.

For follow-up inferences beyond that based on the null p-value, let us assume, for definiteness, that the test statistic is a score statistic S, assumed (approximately) normal with mean θv_o and variance (information) v_o. Also, define a p-value p_θ for testing a revised null hypothesis that the true effect size is (at most) θ. That is, p_θ is the probability of a result as or more extreme than that observed when the true effect size is θ. In this setting,

$$p_\theta = P_\theta(S \geq s_{\text{obs}}) = \bar{\Phi}\big((s_{\text{obs}} - \theta v_o)/\sqrt{v_o}\big) = \bar{\Phi}(\sqrt{v_o}(\hat{\theta} - \theta)) \tag{4.2}$$

with s_{obs} the observed value of S, v_o the information, $\hat{\theta} = s_{\text{obs}}/v_o$ the MLE, and $\bar{\Phi}(x)$ the upper tail area beyond x in a standard normal distribution. It is essential—for consistency in what follows—that p_θ be increasing in θ, as it clearly is here. One basis for estimating θ, other than relying on maximum likelihood, is to choose as estimate the value $\hat{\theta}_m$ for which $p_{\hat{\theta}_m} = \frac{1}{2}$. Such an estimate is said to be *median-unbiased*: It is just as likely to overestimate θ as to underestimate it. Due to symmetry of the normal distribution, we see from (4.2) that $\hat{\theta}_m = \hat{\theta}$, the MLE. The MLE here is known to be *mean-unbiased* (or simply "unbiased") in that its expectation is precisely the true θ (assuming normality). Hence, the MLE is both mean- and median-unbiased (assuming normality). When extending to sequential designs, the symmetry evident here is lost, and we focus on median-unbiasedness as it is generally more feasible and is p-value based. It also has the advantage of retaining its median-unbiasedness under monotone transformation of parameter and estimate, a property not valid for mean-unbiasedness; hence, in a survival analysis, a median-unbiased estimate $\hat{\theta}_m$ of $\theta = -\log(HR)$ yields $\widehat{HR}_m = \exp(-\hat{\theta}_m)$ as a median-unbiased estimate of the hazard ratio HR. Moreover, median-unbiased estimates are consistent with p-value-based confidence bounds, being 50 % confidence bounds.

To obtain a confidence bound more generally, an upper confidence bound with confidence coefficient $1 - \gamma$ is obtained by equating p_θ to $1 - \gamma$ and solving for $\theta = \hat{\theta}_U$. For a lower confidence bound with confidence coefficient $1 - \gamma$, equate p_θ to γ and solve for $\hat{\theta}_L$. Then $(\hat{\theta}_L, \hat{\theta}_U)$ forms a confidence interval with confidence coefficient $1 - 2\gamma$. In this non-sequential case, these bounds and intervals are easily expressed in terms of the deviate z_γ for which the tail area of the standard normal distribution is γ: $\hat{\theta}_U = \hat{\theta} + z_\gamma/\sqrt{v_o}$ and $\hat{\theta}_L = \hat{\theta} - z_\gamma/\sqrt{v_o}$. But not so in sequential analysis. (Here, z_γ is the familiar 1.96 for a 95 % confidence interval.)

We finally remark that, at the analysis stage, neither β nor the reference value θ_R any longer have relevance; the only relevance of α is that it may serve as a critical value for the null p-value p_0.

We will use each of these methods for inference after a sequential trial, to be described in Sects. 4.5 and 4.6. See also Siegmund (1985) and Whitehead (1997).

4.5 Designs and Analysis with Monitoring in Continuous Time: Fully Sequential Designs

This case, although it has not been especially popular in practice (but see Moss et al. (1996, 2002) and references in Whitehead (1997)), includes a family of designs for which simple explicit formulas are available for various inference quantities, such as significance level and power, expected stopping times, p-values, estimates, and confidence intervals for the primary parameter, and so on. It therefore provides a good vehicle for describing how the principles and methods, described in Sect. 4.4, are extended to inference after a sequential trial. Continuous monitoring provides a good approximation if analyses are carried out very frequently, say after each new observation is recorded, or carried out weekly or monthly in a trial lasting several years. Whether S is based on the difference between sample averages or on a logrank statistic, only S and its estimated variance need to be computed at each analysis, and these only require a reasonably current data file on patient recruitment and withdrawals, treatment assignments, and sample values or endpoint event occurrences. A delay in initiating the first analysis to allow accumulation of sufficient data to obtain a reasonable estimate of information is often advisable and will have little effect on analyses since the very early parts of stopping boundaries have negligible chances of being reached.

We confine attention to *linear-boundary designs* for observation of a Brownian motion process $S(v)$. We assume that analyses will be done very frequently, sufficiently so that we can (for now) act as if information is accumulated in continuous time. There is a pre-specified upper boundary line, $a + c\,v$ $(a > 0)$ (information v); a pre-specified lower boundary, $b + d\,v$ $(b < 0)$, and possibly a vertical boundary at $v = v^o$, required if $c > d$ since the upper and lower boundaries will not otherwise meet. If $c < d$, no vertical boundary is essential; the upper and lower boundaries then form a triangle, and the maximum value for v is the location of the apex of the triangle, namely $v^o = (a - b)/(d - c)$.

The Brownian motion is observed until a boundary is reached. We focus on a one-sided test of $\theta \leq 0$ vs $\theta > 0$. Then, upon reaching the upper boundary, the null hypothesis is rejected in favor of the alternative $(\theta > 0)$, implying superiority of the treatment relative to control. If the lower boundary is reached, the null hypothesis is accepted (not rejected). If there is a vertical boundary and it is first reached, rejection is appropriate whenever the process value exceeds a pre-specified value on the vertical boundary and the null hypothesis accepted otherwise. In two-sided testing, a vertical boundary may be reserved for a "no difference" conclusion and the lower boundary for rejection in favor of $\theta < 0$, or possibly only the early part of the lower boundary so reserved.

What principles guide choice of a particular linear-boundary design? We only summarize briefly since our primary focus is on interpretation of the data once a boundary is reached. Wald (1947) SPRT has parallel-line $(c = d)$ boundaries. He showed, with proper choice of $a,\ b,\ c$, that such boundaries uniquely minimize the *expected* boundary-hitting time, at both $\theta = 0$ and at a target θ_R, among *all* designs—not just those with linear boundaries—that have the same strength (α, β). These designs are open ended, with no upper bound on the possible values of v, but he showed that truncation at a large $v = v^o$ would have no noticeable effects on performance characteristics. Wald's criteria ignored performance characteristics at other parameter (drift) values. The expected stopping time reaches a maximum at some drift value between $\theta = 0$ and θ_R.

In an attempt to reduce the maximum expected stopping time, Anderson (1960) introduced general linear boundaries and derived formulas for the distribution along the boundaries; he showed numerically, in the $\beta = \alpha$ case, that a particular triangular design nearly minimizes the expected stopping time at $\frac{1}{2}\theta_R$—that is, a drift value halfway between null and alternative values—and with only a small increase (relative to an SPRT) in expected stopping time at $\theta = 0$ and θ_R. (Also see Lai (1973), Lorden (1976), Hall (1980), Whitehead (1997).) The probability distribution along linear stopping boundaries was developed in Anderson (1960). More compact formulas appear in Hall (1997). A derivation for a popular special case (with $b = -a$ and a particular choice of the slopes) is in Whitehead (1997).

Before providing details about triangular designs and their analysis, we consider a simpler case, leading to simple formulas for p-values, the basis for analysis and inference. Specifically, consider a linear-boundary design with only upper and vertical linear boundaries (and no lower boundary), an *open bottom restricted design*. (A comparable *open top design* is mentioned in Whitehead (1997), as possibly suitable when early stopping is allowed only for evidence of an unsafe new treatment, but not for an unusually effective new treatment. But these roles could be reversed, as here.) Indeed, simpler still—and without loss of generality by shifting the drift parameter—we first take the upper boundary to be horizontal: $s = a$ for $v \in [0, v^o)$; the vertical boundary is at $v = v^o$ with $s \leq a$. Rejection of the null hypothesis occurs if and only if the horizontal boundary is reached prior to $v = v^o$. There are two design parameters, a and v^o, to be determined by the two strength requirements. It may be shown (see below) that, upon stopping at $(v, s) = (v_{\text{obs}}, s_{\text{obs}})$, $r_{\text{obs}} = s_{\text{obs}}/v_{\text{obs}}$ is the MLE of the drift θ; here $r_{\text{obs}} = a/v_{\text{obs}}$ if $v_{\text{obs}} < v^o$ and $= s_{\text{obs}}/v^o$ if $v_{\text{obs}} = v^o$.

To meet the strength requirements, we need a formula for $P_\theta(V < v^o)$ with V the boundary-hitting (information) time, to be equated, in turn, to α when $\theta = 0$ and to $1 - \beta$ at $\theta = \theta_R$. This probability is well known in the Brownian-motion literature, by a reflection argument, to be

$$P_\theta\big(V < v\big) \;=\; \bar{\Phi}\big((a - \theta v)/\sqrt{v}\big) \;+\; \exp(2a\theta)\,\bar{\Phi}\big((a + \theta v)/\sqrt{v}\big)\,, \qquad (4.3)$$

the probability of reaching level a prior to v when the drift is θ.

Note that, with $\theta = 0$, (4.3) $= 2\bar{\Phi}(a/\sqrt{v})$, so $a/\sqrt{v^o} = z_{\alpha/2}$. Substituting this, θ_R and v^o in (3) yields $\bar{\Phi}(z_{\alpha/2} - \theta_R\sqrt{v^o}) + \exp(2z_{\alpha/2}\theta_R\sqrt{v^o})\,\bar{\Phi}(z_{\alpha/2} + \theta_R\sqrt{v^o}) = 1 - \beta$, to be solved for $\sqrt{v^o}$ (by a search), thereby completing specification of this open bottom restricted design with strength (α, β).

Formula (4.3) also enables computation of the expected stopping time, by (numerically) integrating $P_\theta(V > v)$ from 0 to v^o. This quantity is useful for evaluating the design and comparing it with other design choices.

To define a p-value for this design, the sample space for (v, s) consists of the horizontal segment at level a for $v \leq v^o$ and the vertical segment at $v = v^o$ for $s \leq a$. The obvious ordering is from $-\infty$ on the vertical segment up to $s = a$ and then from $v = v^o$ back to $v = 0$ along the horizontal segment. Equivalently, the sample space is ordered by $r = s/v$, sweeping counterclockwise from $r = -\infty$ at the foot of the vertical boundary up to a/v^o where the boundaries meet, and then onwards to $+\infty$ as v approaches 0. This provides the basis for p-value-based inference. See Fig. 4.1, where a similar open bottom restricted design is shown but with a sloping upper boundary.

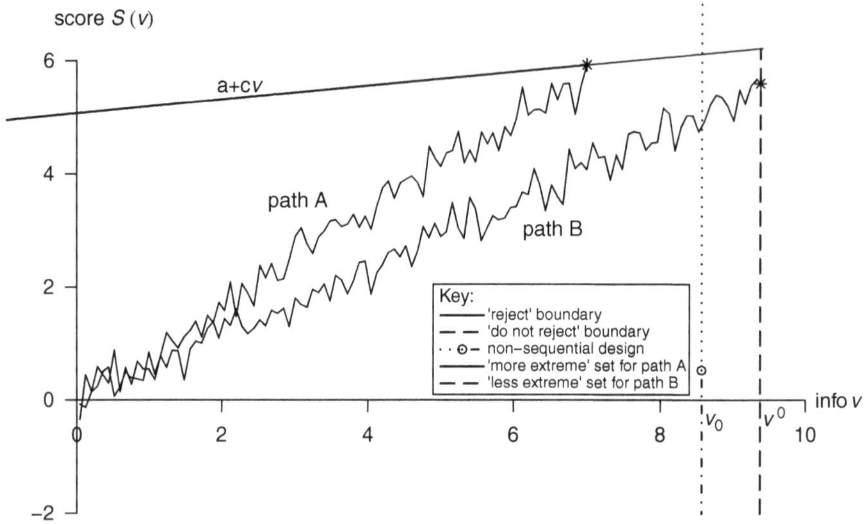

Fig. 4.1 An open bottom restricted design and associated inferences. Paths leading to the upper boundary before the vertical one result in rejection of the null hypothesis in favor of a positive effect; those first reaching the vertical boundary result in failure to reject. Two sample paths are shown: Path A reaches the upper boundary; any path reaching the upper boundary earlier is considered "more extreme," and a p-value represents the null probability of such an event. Path B reaches the vertical boundary; any path reaching the upper boundary, as well as any reaching the vertical boundary at a higher score value, is considered "more extreme". Equivalently, paths reaching the vertical boundary at lower score values are "less extreme." A median-unbiased estimate of the effect size is that parameter (drift) value for which the corresponding p-value is 0.50; a 95 % confidence interval has endpoints which are the parameter values for which the p values are 0.975 and 0.025, respectively. See text

Upon first reaching the upper boundary, at $v = v_{\text{obs}}$, (4.3) gives the p-value p_θ. If instead the vertical boundary is first reached, at $s = s_{\text{obs}}$ ($\leq a$), the probability of less extreme values $s < s_{\text{obs}}$ at v_{obs} is found to be

$$P_\theta\big(V = v^o \text{ and } S \leq s\big) \;=\; \Phi\big((s-\theta v^o)/\sqrt{v^o}\big) - \exp(2a\theta)\,\Phi\big((s-2a-\theta v^o))/\sqrt{v^o}\big)\,. \tag{4.4}$$

In this case, p_θ is found by subtracting (4.4) from unity. These p-value formulas then enable computation of the null hypothesis p-value p_0 and enable solving for $\hat{\theta}_m$, $\hat{\theta}_L$, and $\hat{\theta}_U$ by equating p_θ to $\frac{1}{2}$, $1-\gamma$, and γ, in turn. Solutions can be obtained by simple searches in the neighborhood of $\hat{\theta} = s_{\text{obs}}/v_{\text{obs}}$ and of $\hat{\theta} \mp z_\gamma/\sqrt{v_{\text{obs}}}$.

For a non-sequential design, with information v_o (sample size $\propto v_o$) and MLE $\hat{\theta} = r_{\text{obs}} = s_{\text{obs}}/v_{\text{obs}}$, the p-value p_θ is the first term on the right in (4.3); see (4.2). Hence, the second term in (4.3) is a correction to recognize the possibility of having stopped earlier.

For the case $\theta = 0$, formula (4.3) reduces to $2\,\bar{\Phi}(a/\sqrt{v})$. For comparison, the probability that a normally distributed variable S with mean 0 and variance v

exceeds a is exactly half this. Thus, when stopping at $v < v^o$ with this design, the correct one-sided p-value is twice that computed naively—that is, ignoring the stopping rule and treating the MLE as normally distributed! [Writing (4.3) and (4.4) in terms of the ratio $r = s/v$, they may be differentiated w.r.t. r to derive the likelihood from which r_{obs} may be verified to be the MLE—the second-order derivative w.r.t. r and θ vanishes at $\theta = r$. A simpler and general derivation follows directly from the *Likelihood Ratio Identity* (Siegmund 1985). The "2-fold" fact noted above is a consequence of a standard reflection argument for Brownian motion.]

A more interesting design would have a sloping upper boundary $a + c\,v$, as in Fig. 4.1. Its properties may be derived from those above by replacing θ in (4.3) and (4.4) by $\theta - c$ and replacing s in (4.4) by $s - cv^o$. The design now has three parameters, a, c, and v^o, and hence an additional constraint is needed. One possibility would be to specify v^o/v_o—e.g., as 1.1—with v_o the non-sequential information needed for strength (α, β). Another would be to specify the null probability of rejection at an earlier time, say $\frac{1}{2}v^o$, to be α_1. For the latter, two equations in the two unknowns $a/\sqrt{v^o}$ and $c\sqrt{v^o}$—(4.3) with $\theta = -c$, equated in turn to α_1 and α with $v = \frac{1}{2}v^o$ and v^o, respectively—need solving first and then the revised (4.3) may be used as before to determine v^o. We omit the details, as focus here is on analyzing data from such a trial.

The p-value is again given by the revised (4.3) at v_{obs} if stopping prior to v^o and by subtracting (4.4) (with $\theta \to \theta - c$ and $s \to s - cv^o$) from unity if reaching v^o at $s = s_{\text{obs}}$. Then p_0, $\hat{\theta}_m$, and confidence bounds can be determined as before.

To make the above numerically explicit, suppose a clinical trial is to be carried out, evaluating the difference θ in mean response to treatment relative to control, with null value $\theta = 0$, one-sided significance level 0.05, and power 90 % ($\beta = 0.1$) at $\theta = 1$ unit. Standard deviations of responses are assumed to be 2.0 units, for both treatment and control (but estimated variances are to be used in monitoring and inference). After m and n ($m + n = N$) observations in the respective groups with $m \approx n \approx \frac{1}{2}N$, the MLE would be $\hat{\theta} = \bar{Y}_n - \bar{X}_m$ (in obvious notation), which is approximately normally distributed with mean θ and variance $\approx 16/N$; equivalently, $S = (N/16)\,(\bar{Y}_n - \bar{X}_m)$, which is normal with mean $\theta\,v$ and variance $v = N/16$.

A non-sequential test would reject H_0 if $\hat{\theta} \geq 1.645\sqrt{16/N}$ with N chosen to meet the power requirement; namely, $N_o = 16(1.645 + 1.282)^2 = 137.08$ (or 138) and $v_o = 8.57$. Hence, the critical value for $\hat{\theta}$ would be 0.562.

Now suppose we want to allow early stopping for efficacy, and choose an open bottom design with upper boundary chosen to meet the strength requirements, and, in addition, that $\alpha_1 = 0.01$ at $v = \frac{1}{2}v^o$. It turns out that $a = 5.0716, c = 0.1222$, and $v^o = 9.369$. Hence, $N^o = 16v^o \approx 149.9$, or $N^o = 150$, with $n = 75$ in each arm, an approximately 9 % increase in (maximum) sample size. But the probability of the trial stopping prior to the fixed sample size of 138 is (from the revised (4.3)) 86 % if $\theta = 1$, 37 % if $\theta = \frac{1}{2}$ and 11 % if $\theta = 0$; and the expected value of N is found to be 89 if $\theta = 1$, 129 if $\theta = \frac{1}{2}$, and 148 if $\theta = 0$, being 64 %, 93 %, and 107 %, respectively, of the value $N_o = 138$ required for a non-sequential test. Hence, for the price of a 7 % increase in sample size (on average—and 9 % at most) under null hypothesis conditions, considerable savings are achieved as the effect size increases. Now monitoring would be done after each observation until $s \geq a + c * N/16$ or until $N = 150$.

Suppose now the trial continued to $N = 150$ with $s_{\text{obs}} = 5.6$, resulting in failure to reject the null hypothesis—path A in Fig. 4.1. The true null p-value, calculated as described above, would be 0.054. By contrast, a naive analysis, ignoring the possibility of earlier stopping, would result in $p = 0.034$, and the result incorrectly declared statistically significant.

Suppose instead, at $N = 112$ ($v = 7.0$), $s = 5.93$, just over the stopping boundary—path B in Fig. 4.1. The true and naive null p-values would then be 0.029 (from the revised (4.3)) and 0.013, respectively. For these data, $\hat{\theta}_m$ is found to be 0.750 to be compared with the MLE, or naive estimate, of $s/7.0 = 0.847$.

Although this is not a commonly used design, it has served the dual purpose here of illustrating the effects of a stopping rule on p-values and valid estimates, in contrast to naive analyses that ignore the stopping rule; and the simplicity of the design has led to simple explicit formulas that illustrate how p-value-based inference can be carried out.

For an *open top design*—with stopping early only for safety concerns—formulas (4.3) and (4.4) remain correct if Φ and $\bar{\Phi}$ are interchanged, but (4.4) now gives the probability above s rather than below it.

This methodology can be used for the analysis of any linear-boundary design, using formulas in Hall (1997) and Whitehead (1997). We focus here on the special class of triangular designs featured in Whitehead and in his PEST software, the only ones that have been used in practice; we shall refer to them simply as *triangular designs*:

Observe a Brownian motion $S(v)$ (e.g., score test statistic S) with drift (parameter) θ on information time scale v. Stopping occurs at the first v for which $S(v) =$

$$\frac{a}{r_\beta \theta_R} + \frac{1}{4} r_\beta \theta_R v \quad (upper), \qquad -\frac{a}{r_\beta \theta_R} + \frac{3}{4} r_\beta \theta_R v \quad (lower), \qquad v < v^o = \frac{4a}{(r_\beta \theta_R)^2}, \tag{4.5}$$

with $a = -2\log(2\alpha)$. Rejection of the null hypothesis occurs if the upper boundary is reached at v. Here, α is the one-sided significance level for testing $\theta \leq 0$ versus $\theta > 0$, and the constant r_β is chosen to assure power $1 - \beta$ at θ_R.

Here are a few values: for $\alpha = 0.025$ and $\beta = 0.20,\ 0.10$, or 0.05, respectively, $a = 5.9915$ and $r_\beta = 1.4058,\ 1.2147,\ 1.0904$; for $\alpha = 0.05$, the corresponding values are: $a = 3.2189$ and $r_\beta = 1.3289,\ 1.1277,\ 1.0$.

The slopes and intercepts of the boundaries assure the above significance level and power and, subject to these, to nearly minimizing the expected boundary-hitting time at the θ-value where it is largest, somewhere between 0 and θ_R. The apex of the triangle can be truncated at $0.87v^o$ with negligible effects on properties of the design.

The distribution of the stopping (boundary-hitting) time V along the boundaries of a linear-boundary design is given in Hall (1997), and for this triangular design in Whitehead (1997) (and available in software). These will be needed when analyzing the data upon reaching a boundary; see below.

The expected amount of information required to stop a trial, a so-called "expected stopping time" (measured in information units), when using a triangular design is obtainable from software (PEST or Hall 2010). A triangular design has the possibility of requiring a large value for information (and hence a large sample size)—perhaps as much as 50 % larger than that for a non-sequential trial (for α between 0.025 and 0.05 and β in the range 0.1–0.2). However, the expected information needed to stop is far less than that needed for a non-sequential trial. Specifically, for the $\alpha = 0.025$ case and β at most 0.2, the expected stopping time at the null hypothesis, when compared with v_o for a non-sequential test, is reduced by approximately 50 %, and reduced by at least 25 % at every θ value. Still, these are only expected values, with actual values being quite variable. But from another perspective, the chance of stopping earlier with the triangular design than with a non-sequential design with the same strength can be shown to be 97 % under null conditions and at least 83 % at any θ value (for the $\alpha = 0.025$ case, with comparable savings for other cases). Hence, if the uncertainty of the duration (on an information scale) of a trial can be tolerated, great savings can typically be achieved by using a triangular design, and monitoring the trial very frequently, in near-continuous information time.

These one-sided tests can accommodate two-sided testing, of $\theta = 0$ versus $\neq 0$ with two-sided significance level 2α. Then the null hypothesis is rejected in favor of $\theta < 0$ if the lower boundary is reached prior to v_L, chosen (by software) so that the null probability of reaching the lower boundary prior to v_L is α. For $\alpha = 0.025$ and $\beta = 0.05,\ 0.1,\ 0.2$, respectively, $v_L\theta_R^2 = 2.9112,\ 3.3419,\ 3.9637$; for $\alpha = 0.05$, $v_L\theta_R^2 = 2.0187,\ 2.3549,\ 2.8522$. See Fig. 4.2.

Let us now turn to interpreting data from such a trial. The methods are very much the same as for the *open bottom restricted design* described earlier. The data can again be appropriately summarized by just two numbers, v_{obs} and s_{obs}, the observed information time v upon reaching a boundary and the observed boundary value $s_{\text{obs}} = S(v_{\text{obs}})$.

It turns out, from *sufficient statistics* considerations, that any details of the path leading to the boundary carry no useful information about the drift parameter θ beyond that contained in $(v_{\text{obs}}, s_{\text{obs}})$. Here, these two numbers can be combined, in one-to-one fashion, into a single number, their ratio $r_{\text{obs}} \equiv s_{\text{obs}}/v_{\text{obs}}$, which is the slope of a line segment from $(0, 0)$ to $(v_{\text{obs}}, s_{\text{obs}})$. Hence, r_{obs} is a sufficient statistic. It also has interpretation as the *maximum likelihood estimate (MLE)* of θ. (For notes on verifications, see above.) Note that r_{obs} varies from $-\infty$ to $+\infty$ as the boundary-hitting point moves counterclockwise around the boundary, from $(v, s) = (0, -a/(r_\beta\theta_R))$ along the lower boundary to the apex $(4a/(r_\beta\theta_R),\ 2a/(r_\beta\theta_R))$, and then in reverse direction along the upper boundary to $(0, a/(r_\beta\theta_R))$. Thus, $r = s/v$ for boundary point (v, s) fully describes, and *orders*, the sample space. This counterclockwise ordering may be termed *ML-ordering* ("maximum likelihood"). Similar ordering applies to any linear-boundary design.

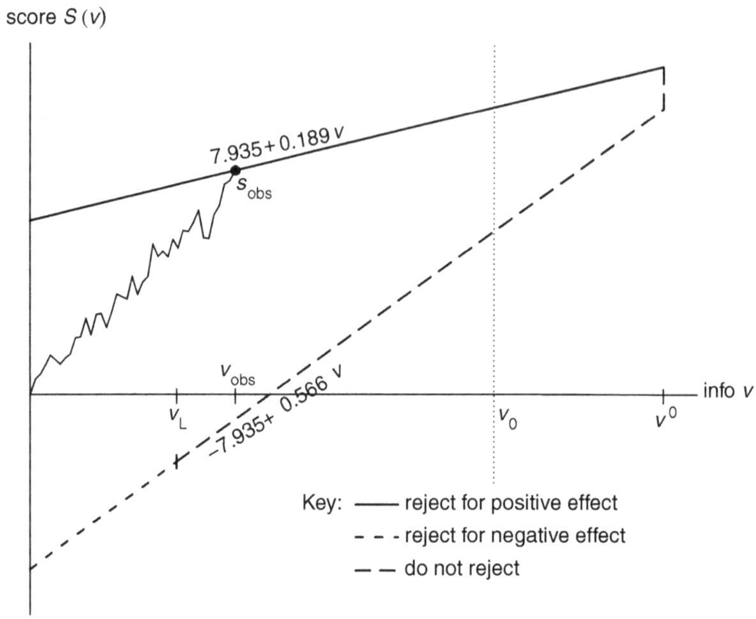

Fig. 4.2 The triangular design of the MADIT trial and associated inferences. This design and sample path, described in the text, was for testing the hypothesis of no difference between treatment and control groups on the risk of survival, with a two-sided significance level of 5 % and power 90 % to detect a hazard ratio of 0.75. The log-rank (score) statistic was plotted against its variance (information) at frequent intervals until a stopping boundary was reached, rejecting in favor of a positive effect if first reaching the upper boundary, in favor of a negative effect if first reaching the early part of the lower boundary, and otherwise leading to an inference of no statistically significant effect. The vertical line indicates the sample space for a non-sequential trial of the same strength. The part of the upper boundary prior to the observed hitting point constitutes the extreme points for defining (one-sided) p-values

Note that a fixed stopping time design (a vertical line at $\pm\infty$, $v = v_o$) is included as a limiting ($a \to \infty$) special case with data $s_{\text{obs}} = S(v_o)$. Then $r_{\text{obs}} = s_{\text{obs}}/v_o$ and ordering by r is equivalent to ordering by s—as would be done upon observing a normally-distributed $S(v_o)$ with mean $\theta\, v_o$ and variance v_o, or when basing inference on an efficient estimate $\hat{\theta}$ with variance $1/v_o$.

Hence, all values of $r\!>\!(<)\ r_{\text{obs}}$ are more (less, respectively) extreme. Equivalently, if stopping on the upper boundary, the earlier part of this boundary is "more extreme," while if stopping on the lower boundary, the earlier part is "less extreme" and hence the later part of the lower boundary and the whole upper boundary is "more extreme."

Thus, if $(v_{\text{obs}}, s_{\text{obs}})$ is on the upper boundary, $p_\theta = P_\theta\{$*upper boundary is reached first and at* $V \le v_{\text{obs}}\}$; if on the lower boundary, $p_\theta = 1 - P_\theta\{$*lower boundary is reached first and at* $V \le v_{\text{obs}}\}$. So p-values are determinable from the distribution of the stopping time V along the boundaries, or equivalently from the distribution of R. Upon hitting the lower boundary with the null p-value at least $1-\alpha$, then v_{obs} must be at most v_L and so, for two-sided testing, $\theta = 0$ would be rejected in favor of $\theta < 0$; the rest of the lower boundary, where $p_0 \in (\alpha, 1-\alpha)$, leads to inference in favor of $\theta = 0$. A two-sided null p-value is defined generally as $2 \cdot [p \wedge (1-p)]$ ($a \wedge b$ stands for the smaller of a and b), and hence consistent with overall rejection when at most 2α.

Consistent with these concepts, a *median-unbiased estimate* $\hat{\theta}_m$ is the solution $\hat{\theta}_m = \theta$ to $p_\theta = \frac{1}{2}$. In particular, if stopping on the upper boundary, $\hat{\theta}_m$ is the effect size for which the probability of stopping on the upper boundary as early or earlier than observed is one-half. This estimate has the interpretation that it is just as likely to underestimate θ as it is to overestimate it. We consider *mean-unbiased estimation* briefly in Sect. 4.7.

To obtain an upper confidence bound $\hat{\theta}_U(r_{\text{obs}})$ for θ, with confidence coefficient $1-\gamma$, we proceed in parallel fashion to the non-sequential case, by solving $p_\theta = 1-\gamma$ for $\theta = \hat{\theta}_U$. Similarly, a lower confidence bound $\hat{\theta}_L$ for θ, with confidence coefficient $1-\gamma$, is found by solving $p_\theta = \gamma$ for $\theta = \hat{\theta}_L$. A confidence interval with confidence coefficient $1-2\gamma$ is $(\hat{\theta}_L, \hat{\theta}_U)$. All of this presumes that p_θ is increasing in θ whatever the observed data—as it may be shown to be for linear-boundary designs; otherwise inconsistencies can occur.

The needed distribution along the boundaries is not so simple as in the open-bottom case, now being an infinite series with each term including normal integrals. Hence, analysis generally requires dependence on software, for the p-value formula and inversion of it—as provided by PEST (2008) or by FORTRAN programs available from Hall's website.

We illustrate with the MADIT trial (Moss et al. 1996) in which an implanted cardioverter defibrillator (ICD) was evaluated relative to conventional medical therapy for reducing mortality associated with ventricular arrhythmias. Patients were recruited and randomized equally to treatment (ICD) and control. The design called for a significance level (one-sided) of 0.025 and 90 % power to detect a hazard ratio (HR) of 0.537—that is $\theta = -\log(HR) = 0$ and $\theta_R = -\log(0.537) - 0.622$. A triangular design was chosen, modified for two-sided ($2\alpha = 0.05$) inference.

For the trial boundaries (4.5), $a = 5.9915$ and $r_\beta = 1.2147$, leading to boundaries $7.935 + 0.189\,v$ and $-7.935 + 0.566\,v$ ($v < 42.1$, but truncated to $v^o = 37.0$). For two-sided inference, the early part, $v \leq v_L = 3.3419/\theta_R^2 = 8.64$, of the lower boundary served as a rejection region for a negative effect. Hence, the trial would require at most $4v^o = 148$ events (deaths) but, with high probability, would end after a much smaller number. (Recall that $v \approx pqD$ with $p = q = \frac{1}{2}$ here.) A non-sequential design with the same strength (0.025, 0.10) would require information (4.1) $v_o = 27.17$, or 109 deaths (see Fig. 4.2).

Recruitment began in December, 1990, and continued throughout the trial. Monitoring was done monthly (once ten deaths had been reported) by a statistical assistant who reported results periodically to the DSMB statistician. A (stratified) logrank statistic was plotted against its (estimated) variance v, using SAS survival-analysis software.

Analysis of the MADIT trial: P-values, estimates, and 95 % confidence intervals for θ and for $HR = -\log(\theta)$

	1-sided p	$\hat{\theta}_m$	$\hat{\theta}_L$	$\hat{\theta}_U$	$\widehat{HR}_m$	$\widehat{HR}_L$	$\widehat{HR}_U$
Correct inferences	0.0042	0.786	0.204	1.361	0.456	0.256	0.815
Naive inferences	0.0017	0.842	0.280	1.405	0.431	0.245	0.756

On March 18, 1996, after 51 deaths had been reported in the 196 enrolled patients, the upper stopping boundary was reached (see Fig. 4.2 for a facsimile of the path, and the cited publication for the actual path), at $(v_{\text{obs}}, s_{\text{obs}}) = (12.037, 10.210)$. An emergency conference was called by the DSMB, leading to a recommendation for terminating the trial, which was done immediately. The trial therefore required less than half the number of deaths needed for a non-sequential design, but there was some small risk of requiring a longer trial, with up to a 36 % increase in the number of deaths needed. The big savings achieved was due to finding a hazard ratio estimate smaller than hypothesized; see below.

Software (Hall 2010, and confirmed by PEST) yielded:

As is typical in sequential trials that reject a null hypothesis, the correct sequential analysis is (usually slightly) more conservative than the naive one that ignores the stopping rule, as is seen in the table above. More details appear in Moss et al. (1996) and in Hall and Liu (2002).

In practice, observation cannot be made in truly continuous time. Siegmund (1979) created an adjustment for the simple linear boundary (*open bottom*) case above, which Whitehead and Stratton (1983) adapted to general linear boundaries; see Whitehead (1997) for a full description. In short, if an analysis is done at time v_k and the last previous analysis (if any) was done at time $v_k - \Delta$, then the upper boundary (for S) at v_k should be reduced by $0.583\sqrt{\Delta}$ and the lower boundary increased by the same amount. Graphs in continuous time for a triangular design give what Whitehead calls a "Christmas-tree" effect to triangular boundaries; see examples in his book. If analyses are reasonably close together (in information time v), this generally works well; otherwise, the resulting design is like a group-sequential one and should be analyzed accordingly (see Sect. 4.6).

For general linear boundaries, similar methodology applies; see Whitehead (1997) and Hall (1997). This includes the *restricted designs* of Armitage (1957), with a non-decreasing linear upper boundary and a symmetrically defined non-increasing lower boundary, and truncated; see Whitehead (1997).

4.6 Designs and Analysis with Monitoring in Discrete Time: Group-Sequential Designs

We now deal with the case of analyses being carried out at a limited number of specific times, with time measured on the information scale. In contrast to development of continuous-time monitoring designs, these were originally developed as adaptations of fixed-sample-size (non-sequential) designs to adjust inference methods to recognize that early stopping was a possibility. As a consequence of this possibility, the (approximate) normality of test statistics will have been disturbed, just as was found in the simple linear-boundary case treated in Sect. 4.5.

Although data may be viewed as if collected in "batches" or "groups," from analysis to analysis, actual analyses are carried out at each specified time based on the data accumulated to that time; the term "group-sequential" should not be taken too literally.

4.6.1 Two-Stage Designs

We start with 2-stage designs. As usual, we consider a trial based on a score statistic S with variance = information = v, and S normally distributed with mean θv (before imposition of any stopping rule). We assume, as earlier, that the design is to be constrained to have one-sided significance level α for testing $\theta \leq 0$ versus $\theta > 0$ with power $1 - \beta$ at θ_R (> 0).

After an initial observation period (or sample), with information pre-specified as v_1, observation is terminated and the null hypothesis is rejected in favor of $\theta > 0$ if $s_{\text{obs}} = S(v_1) \geq U_1$ or terminated and "not rejected" if $\leq L_1$ ($< U_1$). Otherwise, observation continues until the accumulated information reaches a pre-specified v_2. Then, the null hypothesis is rejected if $s_{\text{obs}} = S(v_2) \geq U_2$ and otherwise not rejected. For two-sided alternatives, the "not rejected" region may be partitioned into "reject in favor of $\theta < 0$" and "not significantly different from 0" subregions; in symmetric designs ($L_1 = -U_1$), the "no difference" conclusion would be limited to $L_2 < S(v_2) < U_2$, and power $1-\beta$ would also be guaranteed at $-\theta_R$. See Fig. 4.3.

The design specifies v_1, U_1, L_1, v_2, U_2 in a way to satisfy the *strength* requirements. There remain three (two for symmetric designs) free parameters, thus allowing a wide variety of design choices. For example, we may specify the ratio v_1 to v_2 and perhaps the null probabilities of stopping at stage 1. (The latter is in the spirit of *α-spending*; see Sect. 4.7.) Or v_2 might be chosen to be a given percentage larger than the value v_o required for a one-stage (non-sequential) design with the same strength; see (4.1). Or one might vary one design parameter to obtain a reduced expected stopping time under the null or alternative hypothesis. Or the "shape" of the boundaries might be specified; see Sect. 4.6.2.

The data in such a trial may be summarized by the sufficient statistic $(v_{\text{obs}}, s_{\text{obs}})$ upon stopping, with $v_{\text{obs}} = v_1$ or v_2 and $s_{\text{obs}} = S(v_{\text{obs}})$. (But the ratio s/v is no longer 1–1 in (v, s) and hence not sufficient.) Moreover, with $S_i = S(v_i)$, S_1 and $S_2 - S_1$ are statistically independent, so that (S_1, S_2) is bivariate normal with mean vector $(\theta v_1, \theta v_2)$, variances v_1 and v_2 and covariance v_1.

Use of the s and v scales is best for plotting and for description, as plotted data points should vary around a ray from the origin. But for computations, it is often more convenient to transform to the $z = s/\sqrt{v}$ scale.

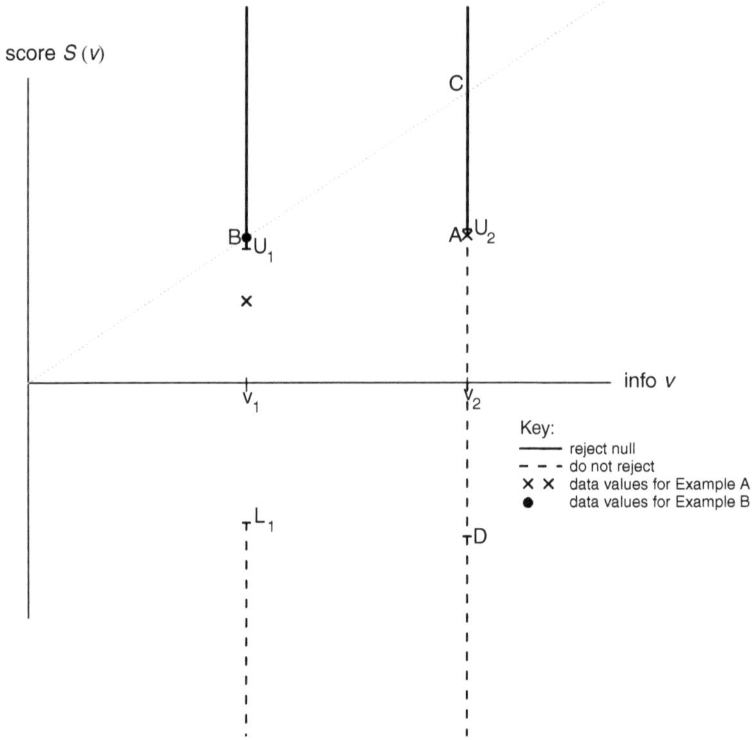

Fig. 4.3 A 2-stage design and two associated examples. A 2-stage design of strength (0.025, 0.10) at (0, 1) (see text) is shown, stopping at either v_1 or v_2 with stopping and decision boundaries marked by L_1, U_1, and U_2. Rejection occurs above U_1 and above U_2, while failure to reject occurs below L_1 and below U_2; this latter boundary may be partitioned at D for two-sided inference, rejecting for a negative effect if below L_1 at v_1 or below D at v_2. Two examples, A and B, of data are indicated. For Example A, $S(v_1)$ was between the boundaries (at **X**) with the value at v_2 indicated by **A** in the "do not reject" region. Extreme values for these data are those above (v_1, U_1) together with those above (v_2, A). For Example B, extreme values are those above (v_1, B) and, for ML-ordering, also those above $(v_2, C = B\, v_2/v_1)$. Corresponding inferences are given in the text

Then $(Z_1, Z_2) = (S_1/\sqrt{v_1}, S_2/\sqrt{v_2})$ (if unrestricted) has a bivariate normal distribution with means $\theta\sqrt{v_i}$ $(i = 1, 2)$, unit variances, and correlation coefficient $\rho = \sqrt{v_1/v_2}$. The means may alternatively be expressed as $\rho\tau$ and τ, respectively, with $\tau = \theta\sqrt{v_2}$ and $\rho\tau = \theta\sqrt{v_1}$. We will use the notation $\Phi_\rho(x, y)$ for the bivariate normal distribution function with 0 means, unit variances, and correlation coefficient ρ.

To design a trial to meet strength requirements, we need the power function $pow(\theta)$, the probability of rejection, being the sum of the probabilities of rejection at v_1 and at v_2. Expressing the boundaries in terms of the Z-statistics, they become $\ell_1 = L_1/\sqrt{v_1}$, $u_1 = U_1/\sqrt{v_1}$, and $u_2 = U_2/\sqrt{v_2}$. Then $pow(\theta) = p_1(u_1 - \rho\tau) +$

$p_2(\ell_1-\rho\tau, u_1-\rho\tau, u_2-\tau)$ with $p_1(u_1) = P_0(S_1 \geq U_1) = P_0(Z_1 \geq u_1) = 1-\Phi(u_1)$ and $p_2(\ell_1, u_1, u_2) = P_0(S_1 \in (L_1, U_1), S_2 \geq U_2) = P_0(Z_1 \in (\ell_1, u_1), Z_2 \geq u_2) = \Phi(u_1) - \Phi(\ell_1) + \Phi_\rho(\ell_1, u_2) - \Phi_\rho(u_1, u_2)$. The last step follows from $P_0(Z \geq z_1, Z_2 \geq z_2) = 1 - P_0(Z_1 < z_1) - P_0(Z_2 < z_2) + P_0(Z_1 < z_1, Z_2 < z_2)$. (A Venn diagram will help.) Upon adding p_1 and p_2, the $\Phi(u_1)$ terms cancel, yielding $pow(0) = 1 - \Phi(\ell_1) + \Phi_\rho(\ell_1, u_2) - \Phi_\rho(u_1, u_2)$. Similarly, the power at θ is

$$pow(\theta) \;=\; 1 - \Phi(\ell_1 - \rho\tau) + \Phi_\rho(\ell_1 - \rho\tau, u_2 - \tau) - \Phi_\rho(u_1 - \rho\tau, u_2 - \tau) \quad (4.6)$$

with τ and ρ defined above. The strength requirements are then: $pow(0) = \alpha$ and $pow(\theta_R) = 1 - \beta$.

There are five design parameters and just two constraints. It is convenient to impose the additional constraint that the v values be in a fixed ratio, say $v_1/v_2 = \rho^2$ (using notation already introduced), and two additional null hypothesis constraints: e.g., specify the null probabilities of reaching each of the stage-1 boundaries. Then the three parameters ℓ_1, u_1, and u_2 can be determined from the three null constraints, and the remaining parameter, say v_2, can be determined by the power constraint (4.6) at $\theta = \theta_R$ ($\tau = \tau_R = \theta_R\sqrt{v_2}$).

Specifically, setting $P_0(S_1 \geq U_1) = P_0(S_1 \leq L_1) = \alpha_1$, we find $1 - \Phi(u_1) = \Phi(\ell_1) = \alpha_1$, and hence $u_1 = z_{\alpha_1} = -\ell_1$. The α constraint (from (4.6)) then becomes $1 - \alpha_1 + \Phi_\rho(-z_{\alpha_1}, u_2) - \Phi_\rho(z_{\alpha_1}, u_2) = \alpha$ or $\Phi_\rho(z_{\alpha_1}, u_2) - \Phi_\rho(-z_{\alpha_1}, u_2) = 1 - \alpha_1 - \alpha$. This equation determines u_2, requiring a search over u_2-values near z_α. The power (4.6) at τ_R, when equated to $1 - \beta$, then determines τ_R and thence v_2; then $v_1 = \rho^2 v_2$. Finally, $U_i = u_i\sqrt{v_i}$ and $L_1 = \ell_1\sqrt{v_1}$. This can all be readily accomplished with software providing univariate and bivariate normal distribution function capability—e.g., SAS functions *probnorm*$(z) = \Phi(z)$ and *probbnrm*$(z_1, z_2, \rho) = \Phi_\rho(z_1, z_2)$—and carrying out two searches to solve the two strength equations, in turn for u_2 and τ_R, as described above. We illustrate this below.

The expected amount of information required to reach a termination boundary is simply $E_\theta(V) = v_1\, P_\theta(V = v_1) + v_2\, [1 - P_\theta(V = v_1)]$ with $P_\theta(V = v_1) = P_\theta(S_1 \leq L_1) + P_\theta(S_1 \geq U_1)$, resulting in

$$E_\theta(V) \;=\; v_2 \;-\; (v_2 - v_1)\big[\Phi(u_1 - \rho\tau) - \Phi(\ell_1 - \rho\tau)\big]\Big|_{\rho\iota - \theta\sqrt{v_1}}. \quad (4.7)$$

This is useful in evaluating and comparing various choices of the design parameters.

We now turn to analyzing data from such a 2-stage trial however v_1 and v_2 and the stopping boundaries L_1, U_1, and U_2 were chosen. The sample space is the union of two disjoint sets in the (v, s) plane, namely $\Omega_1 \equiv \{v = v_1, s \notin (L_1, U_1)\}$ and $\Omega_2 \equiv \{v = v_2, s \in (-\infty, \infty)\}$. We must choose an ordering (in the positive direction) of the sample space. In contrast to the linear-boundary fully-sequential case, there is no obvious single ordering. We consider two alternative possibilities, termed *stagewise*, or simply *SW*, and *maximum-likelihood (ML)* orderings. For SW-ordering, points (v, s) in $\Omega_1 \cup \Omega_2$ are ordered vertically along the line segment at v_1 from $s = -\infty$ to L_1, then vertically along $v = v_2$, from $-$ to $+\infty$, and finally vertically along the line segment at v_1 from $s = U_1$ to $s = \infty$. Hence, (v, s) is more extreme than (v, s')

if $s > s'$, and every point (v_1, s) with $s \geq U_1$ ($\leq L_1$, respectively) is more (less, respectively) extreme than any point (v_2, s). It is as if ordering is defined vertically at v_1, but with the whole vertical line at v_2 substituted for the continuation interval (L_1, U_1) at v_1.

For ML-ordering, points in the sample space are increasingly ordered by the ratio $\hat{\theta} = s/v$, whether $v = v_1$ or v_2. For example, a very large s-value at v_2 may be considered more extreme than an s-value just over the U_1 boundary at v_1.

The p-value function p_θ, once $(v_{\text{obs}}, s_{\text{obs}})$ has been observed, is the probability, when the parameter is θ, of obtaining results as or more extreme (in the positive direction) than $(v_{\text{obs}}, s_{\text{obs}})$. For the *SW-ordering*, we readily find (whether reaching an upper or a lower boundary)

$$p_\theta = \begin{cases} P_\theta\{S_1 \geq s_{\text{obs}}\} & \text{if } v_{\text{obs}} = v_1 \\ P_0\{S_1 \geq U_1\} + P_\theta\{L_1 < S_1 < U_1,\ S_2 \geq s_{\text{obs}}\} & \text{if } v_{\text{obs}} = v_2. \end{cases} \tag{4.8}$$

In the notation introduced above and, in addition, $\hat{\tau} = \hat{\theta}\sqrt{v_2}$, the null p-value is therefore given by

$$p_0 = \begin{cases} 1 - \Phi(\rho\hat{\tau}) & \text{if } v_{\text{obs}} = v_1 \\ 1 - \Phi(\ell_1) + \Phi_\rho(\ell_1, \hat{\tau}) - \Phi_\rho(u_1, \hat{\tau}) & \text{if } v_{\text{obs}} = v_2; \end{cases} \tag{4.9}$$

p_θ is similar but with $\rho\tau$ subtracted from arguments of Φ and from the first argument of Φ_ρ and τ subtracted from the second argument of Φ_ρ.

Note that when stopping at v_1, the p-value formula (4.8) is the same as if the possibility of continuing to v_2 was ignored. However, when stopping at v_2, this p-value is quite different—sometimes smaller, sometimes larger—from the "naive" one, namely $P_\theta(S_2 \geq s_{\text{obs}}) = 1 - \Phi(\hat{\tau} - \tau)$, that ignores the possibility of having stopped at v_1.

Now let us consider *ML-ordering*. As stated, all values (v, s) in $\Omega_1 \cup \Omega_2$ are increasingly ordered according to the ratio s/v. Specifically, if stopping at (v_1, s_1), (v_1, s)-values with $s > s_1$ are again more extreme, but also (v_2, s) values with $s/v_2 > s_1/v_1 = \hat{\theta}$ are more extreme; and if stopping at (v_2, s_2), s-values at v_1 (that are in the sample space, i.e., $\notin (L_1, U_1)$) are more extreme if $s/v_1 > s_2/v_2 = \hat{\theta}$.

A formula for the null ML p-value may be derived very much as the second formula in (4.9) unless $\hat{\theta}v_1 < L_1$ ($\rho\hat{\tau} < \ell_1$), in which case it is easier to derive a formula for the probability of being less extreme and then subtracting from unity. These arguments lead to the following formula for the null ML p-value, covering all cases:

$$p_0 = 1 - \Phi(\ell_1 \wedge \rho\hat{\tau}) + \Phi(u_1) - \Phi(u_1 \vee \rho\hat{\tau}) + \Phi_\rho(\ell_1, \hat{\tau}) - \Phi_\rho(u_1, \hat{\tau}), \tag{4.10}$$

where the notation $a \wedge b$ stands for the smaller of a and b and $a \vee b$ stands for the larger of the two. Again, p_θ is similar but with $\rho\tau$ or τ appropriately subtracted as described after (4.9) above. These orderings are contrasted in Fig. 4.4, exemplifying two sets of data with identical ML-ordered p-values but differing SW-ordered p's.

It turns out that, in principle, the ML p-value $>$ or $<$ the SW p-value when stopping at v_1 according as the upper or lower boundary was reached; and if stopping

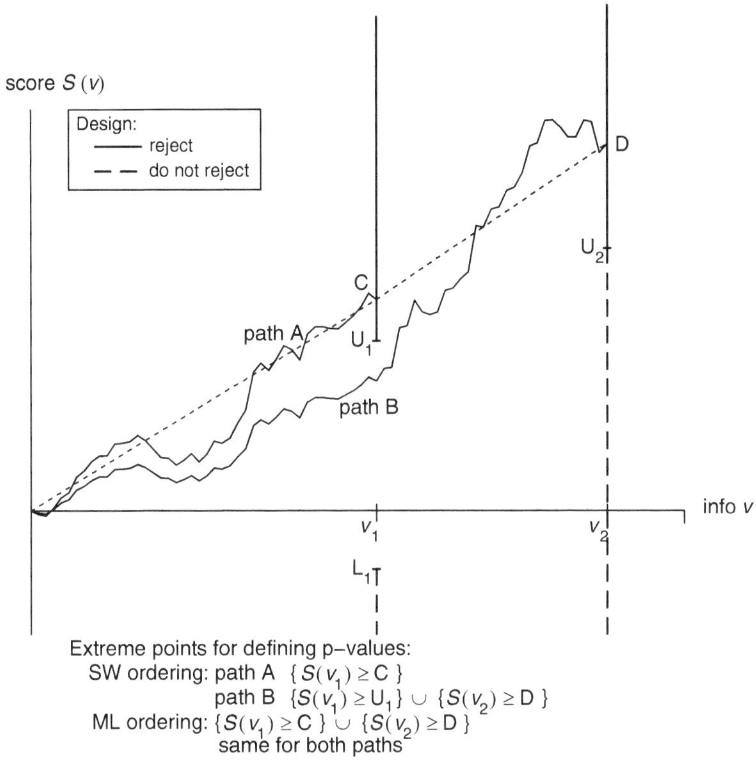

Fig. 4.4 A 2-stage design with comparison of SW- and ML-orderings for inference. A 2-stage design is illustrated along with two examples of sample paths, plotted in near-continuous time although in practice only values at $v = v_1$ and possibly also at v_2 are needed. Path A reaches a stopping boundary (for rejection) at the first stage, so the trial would be terminated at v_1. Path B passes between the boundaries at the first stage, requiring a second stage of observation. The two paths lead to the same MLE $= S(v)/v$ (indicated by the dotted line connecting the origin, **C** and **D**). For ML-ordering of the sample space, these two paths are equally extreme, with extreme points being those above **C** at v_1 and those above **D** at v_2. For SW-ordering, the extreme points for path A are only those above **C** at v_1 while those for path B include all points above U_1 at v_1 along with those above **D** at v_2

at v_2, the ML p-value $<$, $=$ or $>$ the SW p-value according as $\hat{\theta} v_1 > U_1$, between the boundaries, or $< L_1$; however, the differences are often negligible. However, the ML p always differs from the naive one (possibly negligibly). Each of these p-value definitions is consistent with the significance level formula in that each null p is $<$ or $> \alpha$ according as whether or not rejection occurs.

It may be shown that each of these orderings leads to p_θ's increasing in θ, needed for consistent p_θ-based estimates and confidence bounds.

These p-value formulas, whichever ordering is pre-specified, may then be used to compute, in turn, the p-value for the null hypothesis p_0 (which can be converted to a two-sided p in the usual way), a median-unbiased estimate $\hat{\theta}_m$ by solving $p_\theta = \frac{1}{2}$

Two examples of inference in a two-stage design (see Fig. 4.3)

Method	(v, s)	Two-sided p	$\hat{\theta}_m$	$\hat{\theta}_L$	$\hat{\theta}_U$
2-stage[a]	(10.70, 6.3)	0.058	0.584	−0.021	1.185
Naive[b]	(10.70, 6.3)	0.058	0.589	−0.010	1.188
2-stage SW	(5.35, 6.2)	0.0073	1.159	0.312	2.006
2-stage ML	(5.35, 6.2)	0.0073	1.063	0.305	1.788

[a]Results for ML and SW orderings agree in this example, to the decimal places given
[b]Naive analyses ignore the 2-stage stopping rule. The estimate $\hat{\theta}_m$ is then the MLE
In the second example, SW and naive agree since stopping was at the first analysis

for $\theta = \hat{\theta}_m$, and confidence bounds, each with confidence $1 - \gamma$, by equating p_θ, in turn, to $1 - \gamma$ and to γ, and confidence interval with confidence coefficient $1 - 2\gamma$.

We now consider an example, with $\alpha = 0.025$, $\beta = 0.1$ at $\theta = 1$, $v_1 = \frac{1}{2}v_2$ ($\rho = 0.7071$) and $\alpha_1 = 0.005$. To design the trial, first note that $\ell_1 = -u_1 = -z_{0.005} = -2.5754$. Next, we use SAS code to search over u_2 values (in the vicinity of $z_{0.025}$) for which $\Phi_\rho(2.5754, u_2) - \Phi_\rho(-2.5754, u_2) = 1 - 0.005 - 0.025 = 0.970$, finding $u_2 = 2.0028$. We then search over τ-values ($> \sqrt{v_o}$) to solve $1 - \Phi(-2.5754 - 0.7071\,\tau) + \Phi_\rho(-2.5754 - 0.7071\,\tau, 2.0028 - \tau) - \Phi_\rho(2.5754 - 0.7071\,\tau, 2.0028 - \tau) = 0.9$, finding $\tau = 3.2709$ and thence $v_2 = (3.2709/1.0)^2 = 10.70$; finally, $v_1 = \frac{1}{2}v_2 = 5.35$, completing the design specifications. Now $U_1 = u_1\sqrt{v_1} = 5.96 = -L_1$ and $U_2 = u_2\sqrt{v_2} = 6.55$. This design appears in Fig. 4.3.

This design may be compared with a 1-stage design with strength (0.025,0.1) for which $v_o = (1.960 + 1.282)^2 = 10.51$. Hence, by increasing the 1-stage design sample size to a maximum sample size 1.8 % larger, the possibility of reducing the needed sample size by 49 %, whenever stopping at the first of two stages, has been achieved, without any change in the strength! The chance of stopping at $v_1 = 5.35$ is $\Phi(\theta\sqrt{v_1} - u_1) = \Phi(2.313\,\theta - 2.5754)$, which = 0.010 at $\theta = 0$, 0.078 at $\theta = 0.5$, and 0.397 at $\theta = 1$. The expected value of V is, from (4.7), 10.65, 10.28, and 8.58 for the three θ-values just considered, representing a 1.3 % increase, a 2.2 % decrease, and a 18 % decrease, respectively, from 10.51.

To analyze data from the trial, suppose first $(v_{\text{obs}}, s_{\text{obs}}) = (v_2, 6.30)$. Then, for SW-ordering, the null p-value is found from (4.9) to be 0.029. For ML-ordering, we use (10) and find the same value (to three decimal places). To find $\hat{\theta}_m$, we return to (4.9) and (4.10), respectively, with $\rho\tau$ and τ appropriately inserted, and use SAS code to search over τ-values (near $\hat{\theta}\sqrt{v_2}$) to find $\hat{\tau}_m$ for which $p_\theta = \frac{1}{2}$; then $\hat{\theta}_m = \hat{\tau}_m/\sqrt{v_2}$. We proceed similarly to find confidence bounds. Results are in the table below, along with comparable naive values, obtained by ignoring the 2-stage nature of the design. Another example, now with $(v_{\text{obs}}, s_{\text{obs}}) = (v_1, 6.2)$, is also summarized there.

Both examples are illustrated graphically in Fig. 4.3. In each of these examples—reflecting the fact that the stopping rule illustrated here is quite conservative—the p-values did not differ from the naive ones, but small differences are not uncommon;

the estimates and confidence bounds show slight differences from the naive ones. Indeed, whenever stopping occurs at the first analysis, SW and naive analyses are identical, but not so for ML-ordering analyses as they recognize the possibility of not stopping at the first analysis.

Further consideration of 2-stage designs, and with detailed SAS programs, appears in Whitehead (2011), but limited to SW-ordering. Of course, tables and software for multistage designs (see below) also apply.

4.6.2 *MultiStage Designs*

We now suppose the total number of analyses planned is $K \geq 2$, still requiring a one-sided significance level of α and power $1 - \beta$ at a specified θ_R. A design is specified by up to $3K$ parameters, allowing for as many as K analyses, with each of the first K-1 analyses allowing the possibility of termination of the trial or continuation, but with termination certain at the Kth analysis. Analyses are scheduled at information times, that is, when accumulated information has reached specific values $v_1, \ldots, v_K$. Associated with each analysis "time" are two critical values, L_i and $U_i > L_i, i < K$, and a single value U_K at time $v_K = v^o = v\text{-}max$ (or set $L_K = U_K^-$). The trial ends at any analysis time $v_i < v_K$ if the score (cumulative-sum) statistic S is $\geq U_i$ or $\leq L_i$, with rejection of the null hypothesis in favor of $\theta > 0$ if $\geq U_i$. Reaching the lower boundary L_i results in "failure to reject."

For a symmetric two-sided design, $L_i = -U_i, i = 1 \ldots, K$, and rejection occurs the first time v_i that either L_i or U_i is reached—in favor of $\theta < 0$ or > 0, respectively—while a conclusion to not reject the null hypothesis can be reached only at v_K, and then whenever $|S(v_K)| < U_K$. (Some symmetric designs permitting earlier stopping in favor of the null hypothesis are mentioned briefly in Sect. 4.7.) However, any one-sided design may be modified for two-sided inference, but without the same power control at negative values.

To convert from cumulative sum (score) monitoring to Z-statistic monitoring, the critical values U_i and L_i are replaced by $u_i = U_i/\sqrt{v_i}$ and $l_i = L_i/\sqrt{v_i}$. We continue to focus on cumulative-sum scaling when describing designs since paths of the statistic vary around a straight line (unless the assumed model is invalid).

For one-sided designs, there are $3K$-1 design parameters to specify and only two constraints on them to meet strength requirements. For symmetric two-sided designs, there are $2K$ design parameters and still just two constraints. The standard solution to this abundance of parameters is (i) to specify the analysis times up to a single free parameter (see below) and (ii) to specify "shapes" for the array of needed upper and lower critical values, again up to a single free parameter. Then the two free parameters are chosen to meet strength requirements.

For (i), a standard way has been to equally space the information times, writing $v_i = (i/K)\,v^o$ with $v^o = v\text{-}max$, yet to be determined. But other pre-determined spacings are allowable, and may well be deemed more suitable; indeed, if $K \geq 5$, say, it may not be considered appropriate to allow stopping with so little information as would be available at $v_1 = v^o/K$. So equal spacing may be replaced by specifying all ratios v_i/v^o with v^o yet to be determined. An equivalent possibility is to specify the ratios $r_2, \ldots, r_K$ with $r_i \equiv v_i/v_1$ with v_1 yet to be determined (Whitehead 2011). For *(ii)*, a variety of boundary "shapes" for the array of critical values have been suggested and studied. (An alternative method, to be mentioned briefly later, is to specify the rates at which Type I, and possibly also Type II, errors are accumulated as v increases.) Then, by specifying boundary shapes up to a single free parameter, it often turns out that this single boundary-shape-parameter can be solved for to meet a significance level requirement; then the single analysis-time parameter (v^o or v_1, with all ratios to it already specified) can be determined to meet the power requirement—very much as in the $K = 2$ case considered earlier. Hence, two one-parameter "searches" are required, but software or tables are available for common choices.

Four families ("shapes") of group-sequential designs considered in the literature are listed below. The first two families are primarily one-sided designs; they can be adapted for two-sided testing as noted earlier. The other two are symmetric two-sided designs, with equal control on power at positive and negative parameter values, and allowing a conclusion of "no significant difference" only at the maximum information time. Even for two-sided designs, it is easier to describe p-values and associated estimates and confidence intervals if we focus on one-sided formulations.

We describe the designs by expressing boundary shape as a function of $t = v/v^o$ with $v^o = v_K$ (or v-*max*), so that t varies from 0 to 1. Each family has a proportionality parameter C, as well as v^o; C can be determined solely by the α requirement and then v^o determined by the β requirement. Any other parameters (e.g., Δ, b) will need user choice, possibly trying different values and checking corresponding properties—e.g., expected information to reach termination of the trial. To convert shapes $upper(t)$ and $lower(t)$ to stopping boundaries for $S(v)$ at $v = v_i$, or for $Z(v)$ at v_i,

$$U_i = \sqrt{v^o}\, upper(t), \quad u_i = upper(t)/\sqrt{t}, \quad t = v_i/v^o\,,$$

and similarly for L_i, ℓ_i and $lower(t)$. For each family, $upper(t)/t$ is decreasing in t and $lower(t)/t$ is increasing in t, required to assure consistency of p-value-based inferences.

4.6.2.1 One-Sided Families of Boundary Shapes

triangular family: $upper(t) = C\,(1+t), \quad lower(t) = -C\,(1-3\,t)\,.$

power family ($\Delta < 1,\, b > 0$): $upper(t) = C\,t^{\Delta}, \quad lower(t) = C[(1+b)\,t - b\,t^{\Delta}]$.

Tables of $C_1 = C$ and $C_2 = b\,C$ are given in Jennison and Turnbull's book for $\Delta = -0.5,\ -0.25,\ 0.0,\ 0.25$, equally-spaced v_i's, $\alpha = 0.05$, $\beta = 0.2,\ 0.1$; they restrict attention to b-values that result in $v^o = [(1+b)\,C]^2$.

4.6.2.2 Symmetric Two-Sided Families of Boundary Shapes

$$lower(t) = -upper(t) \quad \text{and a } \textit{vertical} \text{ boundary at } t = 1.$$

restricted $(b \geq 0)$: $upper(t) = C\,(1 + b\,t)$.

Choice of b resulting in $v^o = (4b\,C)^2$ or $(2b\,C)^2$ have been considered for fully sequential designs; see the end of Sect. 4.5. The case $b = 0$ is included below.

power family $(\Delta \in [0, \frac{1}{2}])$: $upper\ (t) = C\,t^{\Delta}$.

The special cases $\Delta = \frac{1}{2}$ and $\Delta = 0$ are the designs of Pocock (1977) and of O'Brien and Fleming (1979), respectively—the first group-sequential designs to be introduced and investigated. Designs with intermediate Δ values were introduced by Wang and Tsiatis (1987). Tables of C and of $R = v^o/v_o$ appear in Jennison and Turnbull's book for equally spaced v_i's, $2\alpha = 0.01,\ 0.05,\ 0.10$, and $\beta = 0.2,\ 0.1$; here, v_o is the information required for strength (α, β) in a non-sequential $(K = 1)$ design.

The *triangular* and *restricted* designs are discrete-time analogs of linear-boundary designs of Sect. 4.5 (set $t = v/v\text{-}max$ to clarify); the other two families of designs have no fully sequential analogs (as formulas for Brownian motion with these curved stopping boundaries are not available; moreover, linear boundaries are highly efficient there). Examples of three of these shapes—triangular, Pocock and O'Brien-Fleming—for $K = 4$ appear in Fig. 4.5.

A *Pocock design* is quite aggressive, in that boundaries—$\pm C\sqrt{v_i}$ at v_i on the cum-sum scale—for early stopping are not as wide apart as in the other symmetric designs, making early stopping more likely. As a consequence, the increase in the maximum sample size (or information level) needed to accommodate the possibility of early stopping is somewhat greater—e.g., an 18 % increase if $2\alpha = 0.05$, power is set at 90 % and $K = 4$; for this case, the critical values are proportional to $C = 2.361$ instead of the familiar 1.960, appropriate when $K = 1$. By contrast, an *O'Brien-Fleming design* has constant critical values—$\pm C\sqrt{v^o}$—and is quite conservative, with only small chances of stopping at early analyses. For the example above, only a 2.2 % increase in maximum sample size is required, and the constant critical value has $C = 2.024$ instead of 1.960. The other symmetric power designs are intermediate to these two. For these designs, since the maximum information required increases with K, it is often stated that this increase over the $K = 1$ case is a penalty to be paid for allowing early stopping, sometimes referred to as "multiple looks" at the ongoing data. However, v^o is only the *maximum* possible information time (sample size) and the *expected* information required to reach a boundary, or the *expected* sample size, may be substantially reduced compared with the non-sequential $K = 1$ case—as was seen in the 2-stage design case.

When using a triangular (or a one-sided power-family) design, v-max may appear large compared with v_o for a non-sequential design. Yet, expected values of V upon stopping tend to be much reduced—very much as in the fully-sequential case of Sect. 4.5—with the later interim analyses rarely being required.

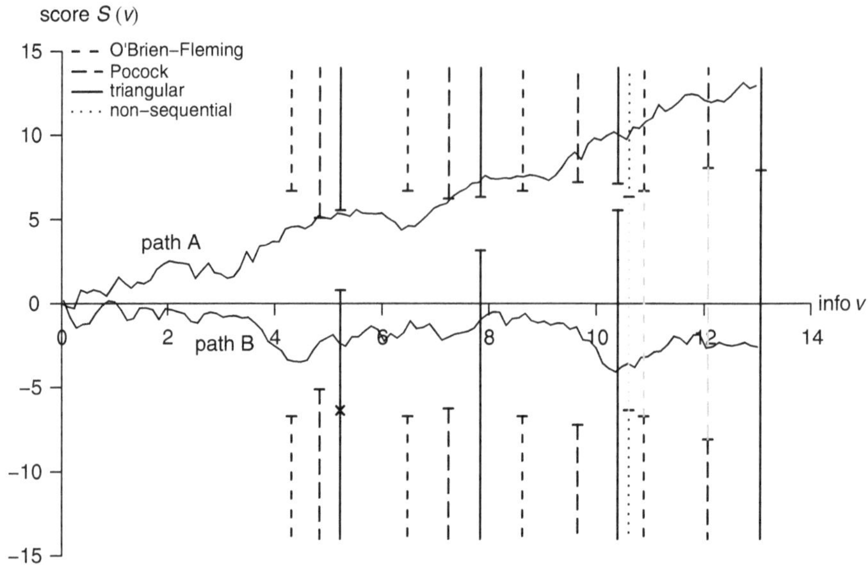

Fig. 4.5 Three 4-stage designs and two associated examples. Three 4-stage designs are illustrated, along with the corresponding 1-stage (non-sequential) design, each of strength (0.025, 0.1) at (0, 1). Two examples of sample paths are shown, in near-continuous time although at most four analyses with their observation points are required; each design requires a different schedule of analysis times. Path A would stop at stage 1 for the Pocock design, at stage 2 for the triangular design and at stage 3 for the O'Brien–Fleming design. Path B would stop at stage 1 for the triangular design—which makes little attempt to distinguish between evidence of a negative effect and of no effect—and at stage 4 for each of the other designs. The **X** on the lower triangular boundary at stage 1 partitions it to convert to a two-sided design (with reduced power for negative effects relative to the others). For details and corresponding inferences, see tables in the text

There are some other variations in the literature, but analysis principles and methods—which we now describe—are all very much the same.

The probability of reaching a particular boundary at stage k is given by a multivariate normal probability of a certain k-dimensional rectangle ($\ell_i < Z_i < u_i,\ i < k;\ Z_k > u_k;$ *all* Z_i*'s correlated*). Specialized software is needed. SAS 9.2 software enables construction of designs for any of the shapes above (and some others); see the Appendix for an example of its use. Tables for each of the power-family shapes are in Jennison and Turnbull's book, as noted above.

Whitehead (2011) gives a simple computing algorithm for designing trials of specified strength and shape and with all ratios $r_i = v_i/v_1$ specified. His method requires only simple SAS codes, specifically the SAS code *seq* which is part of the SAS Interactive Matrix Language *PROC IML* and included in versions of SAS prior to 9.2. He shows how to search first for the proportionality parameter C to achieve significance level α and then to search on v_1 to achieve power $1 - \beta$ at θ_R. Here is his method:

To express shape in terms of the ratio r instead of t, set $t_i = v_i/v^o = r_i/r_K$. Let *domain* represent the $2 \times K$ matrix array with $\sqrt{r_K} \cdot lower(t)/r_i$ at $t = r_i/r_K$ making up the first row and similarly for the second row with *lower* replaced by *upper*. (Recall that $\ell_i = L_i/\sqrt{v_i}$, $u_i = U_i/\sqrt{v_i}$ and set $\ell_K = u_K - 10^{-8}$; Whitehead writes $a = C\sqrt{r_K}$.) A (K-1)-vector *tspace* has elements $(r_2 - r_1, \ldots, r_K - r_{K-1})$. Then the SAS command

call seq(prob, domain) tspace = tspace;

yields a $3 \times K$ matrix, *prob*, with row vectors f, g, and h. The ith element of f is the null probability of stopping at v_i with $S(v_i) \le L_i$; the ith element of g is the null probability of this same event or the event of continuing beyond v_i (and hence $S(v_i) < U_i$), and the ith element of h is the null probability of continuing beyond v_{i-1} (and $h_1 = 1$). It follows that $h_{i+1} = g_i - f_i$, and $h_i - g_i = p_i$, say, the null probability of rejecting at stage i. Therefore, the overall null probability of rejecting is

$$p_0 = \sum_{i=1}^{K}(h_i - g_i) = \sum_{i=1}^{K} p_i \,. \tag{4.11}$$

To have significance level α (one-sided), this last formula must $= \alpha$, and so C (or a) can be searched over until the value α obtains.

To determine the remaining parameter v_1, modify the *domain* matrix by subtracting $\theta_R\sqrt{v_1}\,r_i$ from the ith element of each row and repeat the computation of (4.11), searching over v_1 until the result is $1 - \beta$. This completes the design specification.

The resulting expected amount of information at termination, when the true parameter is θ, is given by inserting $\theta\sqrt{v_1}\,r_i$ in the revised *domain*, and computing $\sum_{i=1}^{K}(f_i + h_i - g_i)\,v_i$. This may be useful, not only for assessing properties of the design, but possibly for adjusting other parameters in the chosen shape.

We now turn to p-value-based inference for a multistage design however it was constructed. The methods described for the $K = 2$ case again apply. The sample space now consists of score values in two semi-infinite (vertical) line segments at each v_i for $i < K$—values leading to stopping at v_i—and an unrestricted (vertical) line at $v = v_K = v^o$. This space is that of the pair of statistics $(v_{\text{obs}}, s_{\text{obs}}) = (v_i, s_i)$ with $s_i = S(v_i) \le L_i$ or $\ge U_i$ for $i < K$ and s_K unrestricted. Several orderings have been proposed and examined in the literature, but we again restrict attention to just two—SW and ML orderings—the only two with considerable versatility in meeting our analysis goals.

For SW-ordering, points in the sample space at $v = v_i$ are ordered by the s coordinate. And all points (v_i, s) with $s \ge U_i$ ($\le L_i$, respectively) are more (less, respectively) extreme than points with $v > v_i$. An advantage of this ordering is that it does not depend at all on the schedule of future analyses nor on critical (boundary) values at such analyses. This provides some flexibility when designs cannot be adhered to precisely; more about this later.

Let $p_{i\theta}$ represent the probability of rejecting at stage i when the parameter is θ—that is, the probability of S_j being between L_j and U_j for each $j < i$ and $\ge U_i$

at stage i. Let $\hat{p}_{i\theta}$ be defined similarly but with U_i replaced by $s_{\text{obs}} = \hat{\theta} v_i$, recalling that $\hat{\theta} = s_{\text{obs}}/v_{\text{obs}}$—that is, the probability of not stopping prior to v_i and $S_i \geq \hat{\theta} v_i$.

For SW-ordering when $v_{\text{obs}} = v_j$, we therefore find

$$p_\theta = \sum_{i<j} p_{i\theta} + \hat{p}_{j\theta} . \tag{4.12}$$

For ML-ordering, first suppose $\hat{\theta} v_i > L_i$ for every $i < K$. (This implies more than that a lower boundary was never reached.) Let j be the smallest i for which $\hat{\theta} v_i \geq U_i$, if ever, and otherwise let $j = K$. We then find

$$p_\theta = \sum_{i<j} p_{i\theta} + \sum_{i=j}^{K} \hat{p}_{i\theta} . \tag{4.13}$$

Now suppose $\hat{\theta} v_i \leq L_i$ for some $i < K$. For this case, it is simplest to compute the probability of "less extreme than observed" and afterwards subtract it from unity to obtain the p-value. Due to the symmetry (about a vector of zeroes) of the joint distribution of the S_i's after subtracting their expectations, we obtain

$$1 - p_\theta = \sum_{i<j} p_{i\theta}^{-} + \sum_{i=j}^{K} \hat{p}_{i\theta}^{-}$$

where $p_{i\theta}^{-}$ and $\hat{p}_{i\theta}^{-}$ are defined as before (4.12) but with $(L_i, U_i, \theta, \hat{\theta})$ replaced by $(-U_i, -L_i, -\theta, -\hat{\theta})$.

As in the $K = 2$ case, these two kinds of p-values rarely agree precisely—only when stopping at v_K with $\hat{\theta} v_j \in (L_j, U_j)$ for every $j < K$. However, differences are usually small. Both differ from the naive p-value, except when stopping at the first analysis and using SW ordering.

Whitehead's method (described earlier) can be adapted to compute these p-values. For SW-ordering, with $(v_{\text{obs}}, s_{\text{obs}}) = (v_j, s_j)$, use the revised domain with θ appropriately subtracted as above, with K reduced to j, u_j replaced by $\hat{u}_j = \hat{\theta}\sqrt{v_j}$ and $\ell_j = \hat{u}_j - 10^{-8}$. For ML-ordering, the computing is not so straightforward. Steps like those above must be carried out for each $i \geq j$, using i terms in each row of *domain* with the first $i - 1$ being boundary values as before and with the ith term $\hat{u}_i = \hat{\theta}\sqrt{v_i}$ and $\hat{\ell}_i = \hat{u}_i - 10^{-8}$. The p-value is then the sum of $g_i - h_i$ from the first ($i = j$) step, and then $g_i - h_i$ from each of the additional steps ($i > j$) added on one-by-one. Estimation and confidence bounds each require a search over θ of the corresponding p_θ-values.

Alternatively and more simply, SAS 9.2 software will provide ML-ordering p-values and inferences—estimates and confidence intervals—derived from them, as well as those derived from SW-ordering. It is only necessary to input the actual design parameters, the v_i's and corresponding boundary values; see the Appendix for an example. (FORTRAN programs are available in Hall 2010).

Characteristics of the 4-stage designs in Fig. 4.5

	Stage	v	L	U
O'Brien–Fleming	1	4.314	−6.700	6.700
	2	6.472	−6.700	6.700
	3	8.629	−6.700	6.700
	4	10.786	−6.700	6.700
Pocock	1	4.828	−5.096	5.096
	2	7.241	−6.241	6.241
	3	9.655	−7.206	7.206
	4	12.069	−8.057	8.057
Triangular	1	5.224	0.793	5.55
	2	7.836	3.171	6.342
	3	10.447	5.549	7.135
	4	13.059	7.928	7.928
Non-sequential		10.507	−6.353	6.353

Tests of strength (0.025, 0.10) at (0, 1) with v-values proportional to 2, 3, 4, 5

Examples of three group-sequential designs appear in Fig. 4.5 and in the table below, a triangular design and two symmetric power-family designs ($\Delta = 0$ and $\frac{1}{2}$). Each allows for up to three interim analyses (K=4), with 90 % power to detect θ=1 at a one-sided significance level of $\alpha = 0.025$. In each, $v_i/v^o = (i+1)/5$, $i = 1, 2, 3, 4$—as if there were to be five equally spaced information times for analyses but with the first one omitted; hence, $r_i = (i+1)/2$. The corresponding non-sequential design is included for comparison.

From the table and the figure, the potential economies of the triangular design are apparent in comparison with two-sided designs, but at sacrifice of some power at negative θ-values. For it, the probability of stopping earlier than the non-sequential test is at least 86 %, and the expected information required for termination is everywhere at least 20 % less than that needed for a non-sequential design.

Fig. 4.5 also includes two examples of sample paths, constructed by simulation under assumptions $\theta = 0.9$ and $\theta = -0.2$ for illustration. They are shown in near-continuous time, but in practice data only need to be analyzed at the information values specified by the specific design being used. We emphasize that schedules are in *information* time, not calendar time. Notice that path A just reaches the first Pocock boundary; it reaches the second triangular boundary and third O'Brien–Fleming boundary. Path B, generated under near-null conditions, reaches the first triangular boundary but only boundary 4 for the other designs.

The table below gives specifics for each design and the corresponding inferences that would be made for each path in Fig. 4.5:

Each design leads to somewhat different inferences, but they are based on differing amounts of information. The latter is reflected in the magnitude of p-values and in the width of confidence intervals. For path A, only small differences are seen

Details for the designs and analyses for the two examples in Fig. 4.5. Simulated data (see Fig. 4.5)

True θ	Design shape	Stopped at stage	s_{obs}	Ordering*	p_0	$\hat{\theta}_m$	$\hat{\theta}_L$	$\hat{\theta}_U$
0.9	O'Brien–Fleming	3	7.537	SW	0.008	0.851	0.167	1.525
				ML	0.008	0.827	0.160	1.481
				Naive	0.005	0.873	0.206	1.541
	Pocock	1	5.212	SW	0.009^-	1.080	0.188	1.972
				ML	0.009^+	0.979	0.169	1.770
				Naive	0.009^-	1.080	0.188	1.972
	Triangular	2	7.400	SW	0.010	0.895	0.150	1.614
				ML	0.010	0.873	0.146	1.580
				Naive	0.004	0.944	0.244	1.645
	Non-sequential		9.902		0.001	0.942	0.338	1.547
−0.2	O'Brien–Fleming	4	−3.243	***	0.838	−0.301	−0.897	0.296
	Pocock	4	−2.626	***	0.775	−0.218	−0.782	0.347
	Triangular	1	−2.494	***	0.862	−0.477	−1.335	0.380
	Non-sequential		−3.827		0.881	−0.364	−0.969	0.240

*Naive inference ignores the stopping rule. The naive estimate is the MLE $= s_{obs}/v_{obs}$, with v-values in previous table. For the second example, SW, ML, and naive values agree

between SW and ML orderings, but results for each differ somewhat from the naive analyses; for path B, both methods agree with naive analyses.

Any one-sided group-sequential design can be adapted for two-sided testing, just as were the triangular designs in Sect. 4.5. Again, simply define a two-sided p-value for the null hypothesis as $2 \cdot [p_0 \wedge (1 - p_0)]$. This implicitly implies rejection of the null hypothesis if the one-sided p-value is "large" (in the chosen ordering) as well as when "small"—that is, when a "less extreme" part of the lower boundary region is first reached, thereby permitting distinction between "insufficient evidence" and "inferior" conclusions. Interpreting the triangular design analysis for the second example in this way, we see that the p-value against negative alternatives would be 0.14, suggesting little evidence in favor of a negative effect.

As a third example, reflect path A in Fig. 4.5 around the axis, representing a sample path for $\theta = -0.9$. It would again reach the first Pocock boundary but would also reach the first triangular boundary, with $s_{obs} = -5.427$ and $p = 0.991$, leading to strong evidence ($p = 0.009$) in favor of a negative effect, and with $\hat{\theta}_m = -1.039$ (naive, SW and ML all agree). For the two-sided designs, $(p_0, \hat{\theta})$ would equal $(1 - p_0, -\hat{\theta})$ found for path A.

However, in standard one-sided designs, power against negative θ-values is limited. For the triangular design above, rejecting for $\theta < 0$ if $S(v_1) \leq -z_\alpha v_1/\sqrt{v_1} = -4.533$ converts the design for two-sided alternatives. Power is then 63 % at $\theta = -1.0$ and 90 % at $\theta = -1.43$, whereas the two-sided designs have 90 % power at $\theta = -1.0$. See Fig. 4.5.

It is often necessary to allow a freer schedule of analyses. This, as well as designs allowing early stopping for futility, will be briefly considered in Sect. 4.7.

4.7 Some Variations: Flexibility, Unbiased Estimation, Inner-Wedge Designs, Overrunning, Secondary Analyses, Binary Data

We briefly mention six additional topics.

4.7.1 Flexibility of Design and Analysis

For a group-sequential design, fixing analysis times at pre-specified values of information is often difficult or inconvenient, especially for survival-analysis-based monitoring where information is proportional to the total event count. More often, DSMB committees like to schedule reviews based on calendar time. To this end, various methods have been proposed that allow some flexibility in the design and associated analysis; see Lan and Zucker (1993); Lan and DeMets (1989), DeMets and Lan (1994), Proschan (1999) and Jennison and Turnbull (2000). Some compromise in precise preservation of significance level and power is incurred. And to avoid bias, any scheduling of analysis times needs to be done without access to results of earlier interim analyses.

When deviations from scheduled information times are not large, the boundary (critical) values can be adjusted using the shape formulas given in Section 6B, and the properties of the design will be approximately preserved. If the trial continues to a Kth analysis, it would be prudent to adjust the scheduled v_K and U_K to restore the desired strength exactly.

An alternative approach is based on shapes associated with chosen *α-spending functions*, which specify the growth rates of the accumulated probability of rejecting the null hypothesis as interim analyses progress, relative to the accumulated information, assuming the null hypothesis to be true. More extreme versions leave analysis times, and even the pre-specified number (K) of them, quite flexible. For one-sided designs, *β-spending functions* may also be needed. Shapes determined by such spending functions differ from the shapes defined in Section 6B, but rough correspondences have been determined; see references.

Still, when analyzing data at an interim analysis—and using SW ordering—all that matters is having available a record of information times at which earlier analyses were carried out and critical (boundary) values imposed at those times, however they were chosen (so long as not dependent on arm-specific data). Then SW p-values can be constructed for any parameter value θ, enabling terminal inference based on p for the null hypothesis, and median-unbiased estimation of θ and associated confidence intervals can be constructed. ML-ordering can only be used in an approximate way, by relying on estimates of future analysis information times and their associated critical values (except at v_K). See Chap. 7 in Jennison and Turnbull (2000).

For continuous monitoring, the "Christmas-tree" adjustment to linear boundaries (end of Sect. 4.5) allows for discrete-time monitoring at times not pre-specified. Analysis can be carried out by Whitehead's approximate method (PEST software) or by applying group-sequential software (SAS 9.2), entering the actual analysis (information) times and the adjusted boundaries.

There is current interest in development of *adaptive designs*, designs with stopping boundaries adjusted in response to results of earlier analyses. The purpose is to modify the power requirements to a new choice of specific alternative, chosen on the basis of interim trial results. These typically entail some sacrifice of the original significance level and power requirements. See Jennison and Turnbull (2006) and references therein.

4.7.2 Unbiased Estimation

We have described construction of median-unbiased estimates, and noted that maximum-likelihood estimates are biased, both median-biased and mean-biased. If an upper stopping boundary is first reached, the MLE tends to overestimate the parameter (the Brownian-motion drift) and to under-estimate it if a lower boundary is first reached.

Mean-unbiased estimates have been developed, both for fully-sequential designs and for group-sequential designs, but formulas are difficult and require some programming for computation. Liu and Hall (1998, 1999) provide discussion of this problem and give needed references.

Median-unbiased estimates have several advantages: they are computed in the same way that confidence intervals are computed, and hence are compatible with them, and the median-unbiasedness is preserved under monotone transformation of the parameter—for example, exponentiating a median-unbiased estimate of a parameter yields a median-unbiased estimate of the exponentiated parameter. This latter property is especially valuable in survival analysis studies, where the hazard ratio is the exponentiated regression coefficient for treatment effect. It is common to accompany an unbiased estimate with its estimated variance, to provide guidance on its reliability. However, interpretation is usually relative to a normal distribution; but estimates in sequential analysis are not normally distributed. Moreover, a variance may be misinterpreted in that it gauges variability over trials that may have stopped much earlier or much later, and not just trials with the same sample size (or same amount of information) as the actual one.

For these reasons, mean-unbiased estimates are rarely quoted in sequential clinical trials. A median-unbiased estimate is deemed most appropriate, although an MLE may also be quoted as part of a naive analysis (that ignores the stopping rule). The MLE is the same, whether the stopping rule is taken into account or ignored. However, there is no known theoretical support for an MLE in a sequential trial; it has no known optimality, is biased, and typically without even approximate normality, and the naive SE may be quite misleading.

Unbiased estimates when observing a single Bernoulli sequence with a sequential stoppping rule were developed and described in Girschick et al. (1946).

4.7.3 Inner-Wedge Designs

For two-sided testing, a disadvantage of the standard (*symmetric power* and *restricted* families) group-sequential designs is that they do not allow early stopping for the null hypothesis, that is, for "insufficient evidence that $\theta \neq 0$." Various *inner-wedge* designs have been introduced to make this feasible, dating back to a procedure of Sobel and Wald (1949), another due to Armitage (1957) and later the *double-triangular designs* of Whitehead (1997). Viewing boundaries in continuous time, these designs replace the terminal vertical boundary by a triangle or "wedge" with vertex at an earlier information time; described another way, the boundary is the overlay of two triangular-like test boundaries, one pointing "northeast" and the other pointing "southeast," the latter one being the mirror image about the abscissa of the former one. See Whitehead and Todd (2004) for some published applications, as well as fuller description and support.

Such designs are primarily appropriate, however, only if there is equal interest in distinguishing between the hypotheses $\theta > 0$ versus $\theta = 0$ and between $\theta < 0$ versus $\theta = 0$. Such applications are infrequent. In an inner-wedge design, once the inner boundary begins, one of the three hypotheses, either $\theta < 0$ or $\theta > 0$, has been ruled out; suppose the latter. Then, thereafter, the trial continues only in order to choose between $\theta = 0$ and $\theta < 0$. In the context of a clinical trial of a new treatment against a standard one, this would seem indefensible, from both ethical and economic perspectives, in that the trial continues in order to distinguish between a "useless" treatment and an "inferior" one.

For these designs, the stopping region at later analysis times is the union of three intervals. This can cause difficulties with p-value-based inference; see Jennison and Turnbull (2000), Sect. 8.5. For alternative, but asymmetric, two-sided designs that allow early termination for the null hypothesis, see the remarks in Sects. 4.5 and 4.6 about adapting one-sided designs for two-sided testing, as done in Whitehead (1997) for triangular designs.

4.7.4 Underrunning and Overrunning

Underrunning refers to settings in which a trial is stopped without having reached a decision boundary, necessitatng interpretation of the available data although still in a "continuation region" (but the term "underrunning" is used differently in the Jennison and Turnbull book). A similar need arises for interim inferences while a trial continues—e.g., by a DSMB. At such times, p-values, estimates, and confidence intervals can be constructed just as if a boundary had been reached. For SW ordering, the current information value is treated as defining an unrestricted vertical boundary. ML-ordering requires no adjustment unless the current analysis is not at a prescribed analysis time, in which case the original design needs to be augmented by this additional analysis time for this calculation. DSMB's usually manage with naive p-values unless a decision regarding termination is to be made.

Overrunning occurs when some additional data become available after the trial has been stopped, due to inertia in the data-collection system. In group-sequential monitoring, the usual analysis method is what Whitehead now terms the *deletion method*: If a trial stops at analysis $k < K$, simply replace the analysis that led to stopping by an analysis that incorporates additional overrunning data, as if it were the scheduled kth analysis. Now the path may or may not reach a stopping boundary but should be interpreted just as in an underrunning analysis. If the trial continues beyond the $(K - 1)$th analysis, then a final analysis can await complete final data. This deletion method is, however, an *ad hoc* one, without formal justification, and resulting p-values and inferences based on them no longer have precise interpretation. Two other methods have been proposed; see Hall and Liu (2002) and Hall and Ding (2008). Relative merits of all three are discussed in the latter reference and in other references there.

4.7.5 Secondary Inferences

After completion of a clinical trial, and computation of a p-value for the primary hypothesis and construction of estimates and confidence intervals for the primary parameter, there are often secondary inference questions. Perhaps the primary analysis was a regression analysis allowing for a main effect for sex along with a treatment effect, and now we want to estimate the sex effect, or test whether it is zero. Or, whether or not a main effect for sex was allowed for in monitoring, we now want to add an interaction term in the model and make inferences about it, that is, about a differential treatment effect across sexes. This latter situation is somewhat in the spirit of goodness-of-fit testing, of whether the model used for monitoring was inadequate in a particular way.

If an appropriate statistic for the secondary inference is correlated with that for the primary inference, with correlation likely to be stable over the course of the trial, then some fairly simple adjustments can be made to lessen or remove bias in secondary inferences. See Whitehead (1997), Liu et al. (2000), Whitehead et al. (2000), and Liu and Hall (2001). However, it can often be argued that such dependence may vary over the course of a trial, especially in regression parameter settings in which any trends in the type of subjects recruited over time may induce such variation.

Lai et al. (2009) developed secondary inference methods for parameters allowed for in the primary monitoring, using *hybrid re-sampling methods*. Hall and Yakir

(2003) deal with the case of inference about parameters that expand the primary model. However, each of these methodologies are quite complex and not available in standard software packages. As a consequence, standard clinical trial practice is to ignore effects of monitoring when carrying out secondary inference, that is, carry it out "naively." When primary inference differs little from naive inference that ignores the stopping rule, it provides some support for ignoring the stopping rule for secondary inference; otherwise, secondary inferences need to be interpreted with some caution.

4.7.6 *Binary Response Data*

For a clinical trial of treatment versus control with binary responses, or indeed for a single Bernoulli data sequence, large-sample approximations justify score-based Brownian-motion approximations. A difficulty arises in that variability depends on the same parameter(s) as do the primary hypotheses. For further discussion, see Jennison and Turnbull (2000). Alternatively, for a trial based on a single Bernoulli sequence, possibly using historical controls, exact sequential test procedures are feasible. Here, testing is whether the success probability p is $\leq$ or $>$ p_0.

For this, a simple choice is a *curtailed sampling* stopping rule as a sequential modification of a fixed-sample-size (n_o) test—that is, stop sampling whenever the number of successes s reaches the critical value s_o for s in the non-sequential design ("reject if $s \geq s_o$") and whenever the number of failures f reaches $n - s_o + 1$ ("do not reject"). This design has the identical strength of the non-sequential one, and rarely requires the full sample size n_o. Modifications can be made to increase the chances of early stopping and still preserve the strength, as follows. The curtailed plan on a (f, s)-lattice has a horizontal stopping-for-rejection boundary at level s_o and a vertical stopping-for-acceptance boundary at $f_o = n_o - s_o + 1$. Replace the horizontal boundary by an increasing (or non-decreasing) curve with a smaller intercept, and replace the vertical boundary likewise by a curve with less slope and a smaller intercept. For any specific boundary, it is elementary to compute the probability of reaching any boundary point, by counting the possible paths from (0,0) to that point and multiplying by $p^s(1-p)^f$, with p the success probability. In this way, the boundary can be modified, perhaps a small amount at a time and by trial and error, in ways that preserve the strength of the non-sequential design, but also have increased chances for early stopping and have smaller expected sample size required for stopping. (We omit details.)

This approach was used in Feldman et al. (2004) for testing $p \leq 0.25$ with strength (0.10,0.10) at (0.25,0.60). A curtailed version of a binomial design with this strength (boundaries at $s_o = 6$ and at $f_o = 8$) requires between 6 and 13 observations. This trial used a triangular design (boundaries $f = -12 + 3s$ and $f = 7 + (2/7)s$), requiring between 4 and 15 observations, with a slight reduction in the expected number of observations.

Sampling in batches ("groups") might also be considered. For some 2-stage Bernoulli-trial designs, popular in Phase-II trials, see Simon (1989). For unbiased estimation in Bernoulli trial designs, see Girshick et al. (1946).

4.8 Summary of the Essential Principles of Analysis: Final Comments

We summarize the essential rules when analyzing a fully-sequential or a group-sequential trial. One essential requirement is that, at any time that an analysis is carried out for consideration of ending the trial—an analysis that makes use of treatment-arm assignments—a pre-specified rule for stopping must be adhered to. Most popular designs require pre-specified stopping boundaries based on a monitoring (test) statistic with properties enabling modeling by Brownian motion approximation; see Sect. 4.3. There are exceptions for other specific modeling, such as with binary data.

Each interim analysis only requires computation of the monitoring statistic based on the accumulated data—here assumed to be a *score* or *cumulative-sum* statistic—its (estimated) variance—*information*—and corresponding boundary values for the statistic determining whether stopping or continuation is in order. If ML-ordering is to be used, the schedule (information times) of future analyses and the corresponding critical values are also needed.

Next, a sample-space ordering needs to have been pre-specified. If fully-sequential, or nearly so, there is only one natural ordering. If group-sequential, stagewise (SW) ordering is the most versatile. If the schedule of analysis times and associated critical values is well fixed, with only minor variations expected in the future, maximum-likelihood (ML) ordering may be chosen instead. In particular, ML-ordering, in contrast to SW-ordering, entertains the possibility that potential future data may represent "more extreme" results than do current data that led to stopping, and also that current data may be "more (or less) extreme" than some earlier possibilities that would have been considered sufficiently extreme to stop the trial. Orderings other than SW and ML may well lead to inconsistencies in p-value-based inferences. Two-sided designs with an inner wedge can also lead to inconsistencies, but adapting one-sided designs for two-sided testing avoids these difficulties and allows smaller sample sizes, but with some loss in power to distinguish between null and "negative effect" conclusions.

The chosen ordering enables definition of associated one-sided p-values for any hypothesized value of the parameter. The p-value for the null hypothesis may be doubled, or, if $> \frac{1}{2}$, subtracted from unity and then doubled, to enable consistent two-sided testing. Comparing with a pre-specified significance level enables acceptance or rejection of the null hypothesis in a way consistent with the earlier interim analyses and, if ML-ordering, with the schedule and boundaries for future analyses.

Equating the one-sided p-value function—p_θ for hypothesis θ—to $\frac{1}{2}$ and solving for $\theta = \hat{\theta}_m$ yields a median-unbiased estimate $\hat{\theta}_m$ of θ. Such an estimate is equally likely to underestimate or to overestimate the true θ. Similarly, equating p_θ to γ and, in turn, to $1 - \gamma$, yields endpoints for a confidence interval for θ having confidence coefficient $1 - 2\gamma$. (These p-value-based methods limit designs to those for which p_θ is monotone in θ, including all designs considered herein.) Mean-unbiased estimation requires special computations and is not so commonly carried out.

It is well to also carry out naive inference—as if the stopping time (information) had been pre-specified, without any opportunity for earlier or later stopping of the trial—for comparison. This naive analysis (incorrectly, in general) treats the observed score statistic as if normally-distributed with mean θv and variance v = information—or treats $\hat{\theta}_{\mathrm{MLE}}$ as normally-distributed with mean θ and variance $1/v$—in spite of, and ignoring, limitations imposed by the stopping rule. Results of analyses that recognize the stopping rule tend to be moderated relative to such a naive analysis, with p-values closer to $\frac{1}{2}$, estimates closer to null values, and confidence intervals wider. But differences are often slight. The degree of agreement or disagreement between correct inferences and naive ones provides a basis for judging potential effects of the stopping rule on secondary analyses, for which methods formally recognizing the stopping rule are limited.

The p-function typically involves multidimensional integrals and/or infinite series. A fully-sequential design with only an upper or only a lower linear boundary, together with a vertical boundary, requires no special software, but other linear-boundary designs do. A 2-stage group-sequential design requires bivariate normal integral software. Multistage group-sequential designs can be analyzed using only the SAS *seq* command and a few simple searches, but fuller SAS software (SAS 9.2) carries out such analyses completely.

Some flexibility in the stopping rule can be accommodated in interim analyses for possible termination of the trial. The significance level can often be preserved, by evaluation of a correct p-value for the null hypothesis, but power may be compromised.

Late data can be accommodated in one of three ways, the choice of which needs to be pre-specified. Secondary analyses, for the most part, can only be carried out naively—that is, ignoring the stopping rule—but induced biases are typically small.

Fully-sequential designs, although rarely used, are the most efficient in terms of controlling the expected duration of a trial. Their usage only requires that data files be kept current throughout the trial and that analyses can be carried out by someone who can keep results concealed from those persons running the trial—as is required anyhow whenever scheduled analyses are done for a DSMB. A disadvantage is that maximum possible duration is longer than in other designs, leading to complications in planning and financing, but trials will tend to end earlier, many with significant savings on average.

For examples with details of designs, see the books by Whitehead (1997) and Jennison and Turnbull (2000) and SAS 9.2 manuals. Rarely have details about stopping rules and monitoring been included in medical publications of trials, but the three MADIT trials of and associates (Moss et al. 1996, 2002, 2009) are exceptions—the first two with fully-sequential linear-boundary designs and the last with a symmetric power family design.

Acknowledgments The services of Xiao Zhang with SAS programming, the Appendix, and the figures are gratefully acknowledged.

Appendix: Sample SAS Programs for Group-Sequential Design and Analysis

```
ods graphics on;

proc seqdesign altref=1

bscale=score

errspend

pss

plots=ASN

stopprob(CREF=0);

design

method(upperalpha)
   =triangular(tau= 1)
method(upperbeta)
   =triangular(tau=0.3)

nstages=4

info=cum(2 3  4 5)

alt=upper

stop=both

alpha=0.025

boundarykey=both ;

run;

ods graphics off;
```

Requests graphics for sequential design analysis.

Specifies the alternative reference values θ_1.

Specifies the statistic scale to be displayed in the boundary. Options are MLE, SCORE, STDZ and PVALUE.

Displays a plot of the cumulative error spending for the boundaries.

Displays power and expected sample size under various hypothetical reference values.

Displays a plot of the average sample numbers

Displays cumulative stopping probabilities under various hypothetical references. The power and expected sample sizes under hypotheses $\theta = c_i \theta_1$ are displayed, where θ_1 is the alternative reference and c_i are the values specified in the CREF=option.

Specifies methods for the upper α and upper β boundary values. Options are WHITEHEAD, ERRFUNCGAMMA, OBF, POC, POW, TRI....etc,

Specifies the number of stages for the design.

Specifies relative information level for all stages in the design. The INFO = EQUAL option specifies equally spaced information levels, and the INFO = CUM specifies cumulative relative information levels.

Specifies the type of alternative hypothesis in the design.

Specifies the condition of early stopping for the design.

Specifies the Type I error probability α.

Specifies types of errors to be maintained in the resulting boundary.

```
ods graphics on;
data boundary;
input _Stop_$1-4 _Alt_$5-10
_Scale_ $11-16 _InfoProp_
_Info_ NullRef Bound_UB
Bound_UA _Stage_ AltRef_L
AltRef_U ;
datalines;
both upper score 0.4 5.224 0
-5.550 -0.793 1 -5.224 5.224
both upper score 0.6 7.836 0
-6.342 -3.171 2 -7.836 7.836
both upper score 0.8 10.448 0
-7.135 -5.549 3 -10.448 10.448
both upper score 1.0 13.06 0
-7.928 -7.928 4 -13.06 13.06
;
run;
proc print data=boundary;
run;

data info3;
input variable$1-4 _scale_
$5-10 estimate _Stage_ ;
datalines;
xiao score 5.3945 1
;

run;
proc print data=info3;
run;

procseqtest
boundary=boundary

parms( testvar=xiao)= info3

nstages=4

order=stagewise

boundaryscale=score
;

run;
```

Input boundary information.

Input data.

Specifies boundary information

Names the SAS data set containing the parameter estimates and their associated standard errors

Specifies number of the stages

Specifies the ordering of the sample space. Options are LR, MLE, STAGEWISE
Options are MLE, SCORE, STDZ, PVALUE.

References

Anderson TW (1960) A modification of the sequential probability ratio test to reduce the sample size. Ann Math Statist 31:165–197

Armitage P (1954) Sequential tests in prophylactic and therapeutic trials. Q J Med 23:255–274

Armitage P (1957) Restricted sequential procedures. Biometrika 44:9–56

Armitage P (1975) Sequential medical trials, 2nd edn. Blackwell, Oxford, UK

Choi S, Hall WJ, Schick A (1996) Asymptotically uniformly most powerful tests in semiparametric and parametric models. Ann Statist 24:841–861

DeMets DL, Lan KKG (1994) Interim analysis: The alpha spending function approach. Statist Med 13:1341–1352

Dodge HF, Romig HG (1929) A modified method of sampling inspection. Bell Syst. Tech. J. 8:613–631

EAST Version 5 (2007) Cytel Statistical Software and Services, Cytel Inc., Cambridge, MA
Feldman AM, Klein H, Tchou P, Murali S, Hall WJ, Mancini D, Boehmer J, Harvey M, Heilman S, Szymkiewicz S, Moss AJ, for the WEARIT and BIROAD Investigators and Coordinators (2004) Use of a wearable defibrillator in terminating tachyarrhythmias in patients at high risk for sudden death: results of WEARIT/BIROAD. PACE 27:4–9
Girshick MA, Mosteller F, Savage LJ (1946) Unbiased estimates for certain binomial sampling problems with applications. Ann Math Statist 17:13–23
Hall WJ (1980) Sequential minimum probability ratio tests. In: Chakravarti IM (ed) Asymptotic theory of statistical tests and estimation, pp. 325–350. Academic Press, New York
Hall WJ (1997) The distribution of Brownian motion on linear stopping boundaries. Sequential Anal 16:345–352, Addendum in: 17:123–124
Hall WJ (2010) Website: http://www.urmc.rochester.edu/biostat/people/faculty/hall.cfm
Hall WJ, Ding K (2008) Sequential tests and estimates after overrunning based on p-value combination. In: Clarke B, Ghosal S (eds) Pushing the limits of contemporary statistics: contributions in honor of Jayanta K. Ghosh, Institute of Mathematical Statistics Collections, vol 3. pp. 33–45
Hall WJ, Liu A (2002) Sequential tests and estimators after overrunning based on maximum-likelihood ordering. Biometrika 89:699–707
Hall WJ, Yakir B (2003) Inference about a secondary process after a sequential trial. Biometrika 90:597–611
Jennison C, Turnbull BW (2000) Group sequential methods with applications to clinical trials. Chapman-Hall/CRC, Boca Raton.
Jennison C, Turnbull BW (2006) Adaptive and nonadaptive group sequential tests. Biometrika 93:1–21
Lai TL (1973) Optimal stopping and sequential tests which minimize the maximum sample size. Ann Statist 1:659–673
Lai TL, Shih M-C, Su Z (2009) Tests and confidence intervals for secondary endpoints in sequential clinical trials. Biometrika 96:903–915
Lan KKG, DeMets DL (1989) Group sequential procedures: Calendar versus information time. Statist Med 8:1191–1198
Lan KKG, Zucker DM (1993) Sequential monitoring of clinical trials: the role of information and Brownian motion. Statist in Med 12:753–765
Liu A, Hall WJ (1998) Minimum variance unbiased estimation of the drift of Brownian motion with linear stopping boundaries. Seq Anal 17:91–107
Liu A, Hall WJ (1999) Unbiased estimation following a group-sequential test. Biometrika 86:71–78
Liu A, Hall WJ (2001) Unbiased estimation of secondary parameters following a sequential test. Biometrika 88:895–900
Liu A, Tan M, Boyett JM, Xiong X (2000) Testing secondary hypotheses following sequential clinical trials. Biometrics 56:640–644
Lorden G (1976) 2-SPRT's and the modified Kiefer-Weiss problem of minimizing an expected sample size. Ann Statist 4:281–291
Moss AJ, Hall WJ, Cannom DS, Daubert JP, Higgins SL, Klein H, Levine JH, Saksena S, Waldo AL, Wilber D, Brown MW, Heo M; for the Multicenter Automatic Defibrillator Implantation Trial Investigators (1996). Improved survival with an implanted defibrillator in patients with coronary disease at high risk for ventricular arrhythmia. N Engl J Med 335:1933–1940
Moss AJ, Zareba W, Hall WJ, Klein H, Wilber DJ, Cannom DS, Daubert JP, Higgins SL, Brown MW, Andrews ML, for the Multicenter Automatic Defibrillator Implantation Trial-II Investigators (2002) Prophylactic implantation of a defibrillator in patients with myocardial infarction and reduced ejection fraction. N Engl J Med 346:877–883
Moss AJ, Hall WJ, Cannom DS, Klein H, Brown MW, Daubert JP, Estes M, Foster E, Greenberg H, Higgins SL, Pfeffer MA, Solomon SD, Wilber D, Zareba W, for the MADIT-CRT Trial Investigators (2009) Cardiac-resynchronization therapy for the prevention of heart-failure events. N Engl J Med 361:1329–1338

O'Brien PC, Fleming TR (1979) A multiple testing procedure for clinical trials. Biometrics 35:549–556
PEST Version 4 (2008) Software. http://www.maths.lancs.ac.uk/department/research/statistics/mps/pest
Pocock SJ (1977) Group sequential methods in the design and analysis of clinical trials. Biometrika 64:191–199
Proschan MA (1999) Properties of spending function boundaries. Biometrika 86:466–473
SAS Version 9.2 (2008) Software. SAS Institute Inc., Cary, NC
S+SeqTrial (2000) Software. MathSoft, Data Analysis Products Division, Seattle
Siegmund D (1979) Corrected diffusion approximations in certain random walk problems. Adv Appl Probab 11:701–719
Siegmund D (1985) Sequential analysis: tests and confidence intervals. Springer, New York
Simon S (1989) Optimal two-stage designs for phase II clinical trials. Controlled Clinical Trials 10:1–10
Sobel M, Wald A (1949) A sequential decision procedure for choosing one of three hypotheses concerning the unknown mean of a normal distribution. Ann Math Statist 16:243–258
Wald A (1947) Sequential analysis. Wiley, New York
Wang SK, Tsiatis AA (1987) Approximately optimal one-parameter boundaries for group sequential trials. Biometrics 43:193–200
Whitehead J (1997) The design and analysis of sequential clinical trials, 2nd edn. (revised), Wiley, New York
Whitehead J (2011) Group sequential trials revisited: simple implementation using SAS. Stat Meth Med Res 20:635–656
Whitehead J, Jones DR (1979) The analysis of sequential clinical trials. Biometrika 66:443–452
Whitehead J, Todd S (2004) The double triangular test in practice. Pharm Statist 3:39–49
Whitehead J, Todd S, Hall WJ (2000) Confidence intervals for secondary parameters following a sequential test. J Roy Stat Soc B 62:731–745
Whitehead J, Stratton I (1983) Group sequential clinical trials with triangular continuation regions. Biometrics 39:227–236

Chapter 5
Dynamic Treatment Regimes

Min Qian, Inbal Nahum-Shani, and Susan A. Murphy

5.1 Introduction

Recent research (see Lavori and Dawson 2000, 2004) stresses the need to take into account patients' heterogeneity in need for treatment when developing intervention programs. In order to improve patient care the type of treatment and the dosage should vary by patients. Additionally, in many cases, the need for treatment may change over time, yielding repeated opportunities to adapt the intervention. For example, patients with mental illnesses (e.g., depression, drug-abuse, alcoholism, etc) often respond differently to treatment and also tend to experience repeated cycles of cessation and relapse (see, e.g., McLellan 2002; Fava et al. 2003). Therefore, the clinical management of mental illnesses requires that clinicians make a sequence of treatment decisions, where the first step is aimed at stabilizing the patient and the following steps are directed to preventing relapse in the long term. Dynamic treatment regimes operationalize this sequential decision making. A dynamic treatment regime is a sequence of decision rules, one per treatment decision, that provide the mechanism by which patient's values on key characteristics, called tailoring variables, are translated into dosage amount and type. Instead of

M. Qian (✉)
Department of Biostatistics, Columbia University
722 W. 168 St. 6th Floor, New York, NY, 10032, USA
e-mail: mq2158@columbia.edu

I. Nahum-Shani
Institute for Social Research, University of Michigan
426 Thompson Street, Room 3336, Ann Arbor, MI 48106, USA
e-mail: inbal@umich.edu

S.A. Murphy
Department of Statistics, University of Michigan
439 West Hall, 1085 S. University Ave., Ann Arbor, MI 48109, USA
e-mail: samurphy@umich.edu

W. Tang and X. Tu (eds.), *Modern Clinical Trial Analysis*, Applied Bioinformatics and Biostatistics in Cancer Research, DOI 10.1007/978-1-4614-4322-3_5,

delivering the same type and dosage of treatment to every patient, a dynamic treatment regime assigns different treatment types/dosages across patients and within each patient across time according to the patient's values on tailoring variables. The term "dynamic treatment regimes" is also known as "adaptive treatment strategies" (Lavori and Dawson 2000; Murphy 2005a), "multistage treatment strategies" (Thall et al. 2002; Thall and Wathen 2005), "treatment policies" (Lunceford et al. 2002; Wahed and Tsiatis 2004, 2006) or "individualized treatment rules" (Petersen et al. 2007; van der Laan and Petersen 2007). All are aimed at constructing a sequence of decision rules that when implemented will produce the best long-term outcome.

Better understanding of dynamic treatment regimes can be gained by considering the following example. This example demonstrates a sequential decision-making problem in the area of clinical science. It will be used throughout the chapter.

Addiction Management Example

Suppose in planning the treatment for alcohol-dependent patients we are particularly interested in making two critical decisions. First we must decide what is the best initial treatment for an alcohol-dependent patient. For example, we may consider two possible treatment options: opiate-antagonist Naltrexone (NTX) and Combined Behavioral Intervention (CBI; Miller 2004). Second, we must decide what is the best subsequent treatment for non-improving patients (i.e., nonresponders) and improving patients (i.e., responders). For example, if a patient is a nonresponder to NTX, we need to decide whether to augment NTX with CBI (NTX + CBI) or switch the treatment to CBI. If a patient is a nonresponder to CBI, we need to decide whether to augment CBI with NTX (CBI + NTX) or switch to NTX. If a patient is a responder, we will refer the patient to a 12-step program (Alcoholics Anonymous 2001), but we need to decide if it is worthwhile to augment the 12-step program with Telephone Disease Management (TDM; Oslin et al. 2003) for an additional period of six months.

The above example is inspired by the ExTENd trial conducted by David Oslin from the University of Pennsylvania (personal communication) and the COMBINE trial conducted by COMBINE Study Research Group (2003). In the context of this example, two simple dynamic treatment regimes may be:

- Regime A: Treat patients with NTX first; then provide CBI for nonresponders and refer to the 12-step program for responders.
- Regime B: Treat patients with CBI first; then provide NTX+CBI for nonresponders and refer to the 12-step program for responders.

The above two dynamic treatment regimes tailor the subsequent treatment to each patient using the patient's response status to the initial treatment. More complex dynamic treatment regimes may use a patient's pretreatment information (e.g., medical history, severity of addiction, and co-occurring disorders) to choose the initial treatment and/or use intermediate outcomes (e.g., the patient's response status, side effects, and adherence to the initial treatment) to choose the subsequent treatment.

In order to further clarify the concept of dynamic treatment regimes, in the next section we use the potential outcome framework to define dynamic treatment regimes and the optimal dynamic treatment regime. In Sect. 5.3, we introduce the SMART design proposed by Lavori and Dawson (2000, 2004) and Murphy (2005a) and discuss the motivation behind it. In Sect. 5.3.3, we compare the SMART design with adaptive experimental designs. In Sect. 5.4, we discuss commonly used methods for developing the optimal dynamic treatment regime using data collected from a SMART trial. More specifically, we focus on Q-learning which is a well-known regression-based method (Murphy 2005b). In Sect. 5.5, we discuss other analyses related to the Q-learning algorithm. Finally, we summarize the chapter and discuss challenges and open questions in Sect. 5.6.

5.2 Potential Outcomes Framework

Potential outcomes were introduced in Neyman (1923) to analyze causal effects of time-independent treatments in randomized studies. Extensions of Neyman's work to the analysis of causal effects of time-independent treatments in observational studies can be found in Rubin (1978). A formal theory of causal inference was proposed in Robins (1986, 1987) to assess the direct and indirect effects of time-varying treatments from experimental and observational data. In this section, we use potential outcome models to provide a framework for developing the optimal dynamic treatment regime. Later we explain how inference concerning potential outcomes can be made, using data from experimental trials.

For simplicity, assume there are only two decision points. The extension to multiple decision points is straightforward. Denote the treatment decision at the initial and secondary stage by a_1 and a_2, respectively. Let S_1 denote the pretreatment information. For each fixed value of the treatment sequence (a_1, a_2), we conceptualize potential outcomes denoted by $S_2(a_1)$ and $Y(a_1, a_2)$, where $S_2(a_1)$ is the intermediate outcome (ongoing information) that would have been observed prior to the secondary decision point if the initial treatment assignment were a_1, and $Y(a_1, a_2)$ is the primary outcome (large values are desirable) that an individual would have if he/she were assigned the treatments (a_1, a_2). In this context, a *dynamic treatment regime* is a sequence of decision rules, (d_1, d_2), where d_1 takes S_1 as input and outputs a treatment a_1 and d_2 takes $(S_1, a_1, S_2(a_1))$ as input and outputs a treatment a_2.

Let $\mathscr{A}$ be the collection of all possible treatment sequences. Then the set of all potential outcomes is $\{(S_1, S_2(a_1), Y(a_1, a_2)) : (a_1, a_2) \in \mathscr{A}\}$ (S_1 is included for completeness). The mean primary outcome for regime (d_1, d_2), denoted by $\mu_{(d_1, d_2)}$, is defined as the average primary outcome that would be observed if the entire study population were assigned (d_1, d_2). Mathematically, this is

$$\mu_{(d_1,d_2)} = E[Y(a_1, a_2)_{a_1=d_1(S_1),a_2=d_2(S_1,a_1,S_2(a_1))}], \tag{5.1}$$

where the expectation is taken with respect to the multivariate distribution of $(S_1, S_2(a_1), Y(a_1, a_2))$ for the treatment sequence determined by (d_1, d_2). The goal is to develop a dynamic treatment regime that leads to the maximal $\mu_{(d_1,d_2)}$ (as compared to other possible dynamic treatment regimes). This dynamic treatment regime is called the *optimal dynamic treatment regime*. Note that (5.1) can be written as a repeated expectation

$$\mu_{(d_1,d_2)} = E\left[E\left[E[Y(a_1,a_2)|S_1, S_2(a_1)]_{a_2=d_2(S_1,a_1,S_2(a_1))}|S_1\right]_{a_1=d_1(S_1)}\right].$$

If we knew the distribution of the potential outcomes $(S_1, S_2(a_1), Y(a_1, a_2))$ for each treatment pattern $(a_1, a_2) \in \mathscr{A}$, classical algorithms based on backwards induction (dynamic programming) (Bellman 1957) could be used to find the optimal sequence of decision rules. The optimal secondary decision rule d_2^* maximizes the mean primary outcome at the second decision point, i.e.,

$$d_2^*(s_1, a_1, s_2) \in \arg\max_{a_2} E[Y(a_1, a_2)|S_1 = s_1, S_2(a_1) = s_2].$$

Note that we use "$\in$" in the above formula since there may be multiple treatments that maximize the mean primary outcome given $S_1 = s_1$ and $S_2(a_1) = s_2$. The optimal initial decision rule d_1^* then chooses the treatment that maximizes the mean primary outcome given that d_2^* is followed at the second decision point, i.e.,

$$d_1^*(s_1) \in \arg\max_{a_1} E\left[\max_{a_2} E[Y(a_1,a_2)|S_1, S_2(a_1)]\middle|S_1 = s_1\right].$$

The sequence of decision rules (d_1^*, d_2^*) is the optimal dynamic treatment regime. And $\mu_{(d_1^*,d_2^*)}$ is the optimal mean primary outcome. Note that the treatment options at each decision point may depend on a patient's ongoing information and/or previous treatment. In the addiction management example, the treatment options at the second decision point depend on a patient's initial treatment and/or whether the patient had improved or not. In the above formulae the maximization at each decision point is taken over all possible treatment options at that point.

In general the multivariate distribution of the potential outcomes for each of the possible treatment patterns is unknown; thus we cannot directly use the arguments given above to construct the optimal dynamic treatment regime. Accordingly, Murphy (2005a) proposed experimental trials by which data can be obtained and used for the formulation of decision rules. In the next section, we introduce this Sequential Multiple Assignment Randomized Trials (SMART) and discuss the motivation for developing this approach for the formulation of dynamic treatment regimes.

5.3 SMART Design

In a SMART trial, each subject may be randomly assigned to treatments several times. More specifically, this is a multistage randomized trial, in which each subject progresses through stages of treatments and is randomly assigned to treatments at each stage. This type of design was first introduced by Lavori and Dawson (2000) and was named "biased coin adaptive within-subject" (BCAWS) design. TenHave et al. (2003) compared the BCAWS design with other designs. Lavori and Dawson (2004) discussed practical considerations for the design. Murphy (2005a) proposed the general framework of the SMART design.

Trials in which each subject is randomized multiple times have been widely used, especially in cancer research (see, e.g., CALGB study 8923 for treating elderly patients with primary acute myelogenous leukemia (Stone et al. 1995)). Precursors of SMART trials include the CATIE trial for antipsychotic medications in patients with Alzheimer's (Schneider et al. 2001) and STAR*D for treatment of depression (Lavori et al. 2001; Fava et al. 2003). In recent years, a number of SMART trials have been conducted. These include phase II trials at MD Anderson for treating cancer (Thall et al. 2000), the ExTENd trial concerning alcohol dependence by David Oslin from the University of Pennsylvania (personal communication), and the ADHD trial conducted by William Pelham from the University at Buffalo, SUNY (personal communication).

To make the discussion more concrete we consider a SMART trial based on our addiction management example (see Fig. 5.1). In this trial, each subject is randomly assigned to one of two possible initial treatments (CBI or NTX). Then in the next two months clinicians record heavy drinking, adherence, side effects, and other intermediate outcomes. If at any time during this two-month period the subject experiences a third heavy drinking day, he/she is classified as a nonresponder to the initial treatment. As soon as the subject is classified as a nonresponder he/she is re-randomized to one of the two subsequent treatments, depending on his/her initial treatment assignment: NTX+CBI or CBI alone for NTX nonresponders; NTX+CBI or NTX alone for CBI nonresponders. However, if the subject is able to avoid more than two heavy drinking days during the two-month period, he/she is considered as a responder to the initial treatment. In this case, the subject is re-randomized to one of the following two possible treatments for an additional period of six months: 12-step program or 12-step program+TDM. The goal of the study is to maximize the number of non-heavy drinking days over a 12-month study period.

This experimental approach is motivated by several disadvantages of traditional "single-stage" experimental designs discussed in the following section.

5.3.1 Motivation for the SMART Design

Most randomized trials are used to compare single-stage treatments. To ascertain the best treatment sequence, an alternative approach to SMART is to conduct

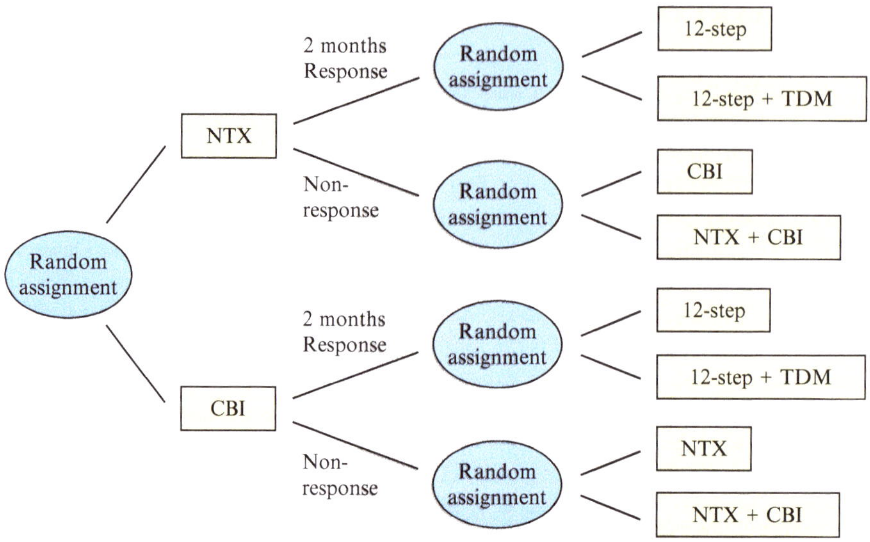

Fig. 5.1 The SMART design for the addiction management study

multiple randomized trials; each trial compares available treatment options at each stage based on results from previous trials and/or based on historical trials and the available literature. For example, instead of the SMART trial for our addiction management study, the researcher may conduct two single-stage randomized trials. The first trial compares the initial treatments (CBI versus NTX). Based on the results of the first trial, the researcher chooses the best treatment and moved on to the second trial where all subjects are initially offered with the chosen treatment and then responders are randomized to one of the two possible conditions: 12-step program or 12-step program+TDM; nonresponders are randomized to one of the two possible conditions: augment current treatment with another type of treatment or switch to another type of treatment. However, this approach has at least three disadvantages as compared to a SMART trial when used to optimize dynamic treatment regimes.

First, this approach may fail to detect possible delayed effects that are cases where a treatment has an effect that is less likely to occur unless it is followed by a particular subsequent treatment. For example in the addiction management study, the first trial may indicate that NTX performs the same as CBI on average. And, the second trial, in which NTX was chosen to be the initial treatment based on the results from the previous trial (since CBI is more expensive than NTX), may indicate that CBI along performs as well as NTX+CBI for nonresponders (and avoids further side effects due to NTX for nonresponders who experience side effects) and the 12-step program is as effective as 12-step+TDM for responders. In that case, the researcher may conclude that regime A described in Sect. 5.1 is the best treatment sequence. However, it is possible that NTX was found to be as effective as CBI during the first trial because for many subjects CBI works best if used over a longer period of time. If the researcher had chosen CBI instead of NTX as the initial treatment for the second trial, then NTX+CBI might be much more effective than NTX alone for nonresponders and the 12-step program may be sufficient for responders. That is, regime B described in Sect. 5.1 might have been more effective than regime A though NTX may initially appear to be as effective as CBI. Hence using the single-stage experimental approach may reduce the ability of the researcher to detect the delayed effects and eventually led to the wrong conclusion as to the most effective regime.

Second, although the results of the first trial may indicate that one treatment is initially less effective than the other, the former treatment may elicit diagnostic information that permits the researcher to better match the subsequent treatment to each subject, and thus improve the primary outcome.

Third, subjects who enroll and remain in the single-stage trials may be different than those who enroll and remain in a SMART trial. This is a type of cohort effects (see Murphy et al. 2007). Consider a one-stage randomized trial in which CBI is compared with NTX. For subjects who are not improving, there are no other options besides dropping out of the study. However, in a SMART trial nonresponding subjects know that their treatments will be altered. Consequently nonresponding subjects may be less likely to drop out from a SMART trial relative to a one-stage randomized trial. Thus the choice of initial treatment based on the single-stage approach may be based on a sample that is less representative of the study population relative to the SMART trial.

From the above discussion, we see that conducting separate trials for different stages and examining treatment options at each stage separately from other stages may fail to detect delayed effects and diagnostic effects and may result in deleterious cohort effects. As a result, the developed sequence of treatment decisions may not be the best choice. This has led researchers to consider designs in which each individual is randomized multiple times, one per critical decision, so as to be able to link different stages in the decision process, and improve the ability to develop dynamic treatment regimes in which sequential treatments are synergetic.

5.3.2 Design Aspects

Denote the observable data for a subject in a SMART trial by (S_1, A_1, S_2, A_2, Y), where S_1 and S_2 are the pretreatment information and intermediate outcomes, A_1 and A_2 are the randomly assigned initial and secondary treatments, and Y is the primary outcome of the subject, respectively. For example, in the addiction management study (see Fig. 5.1), S_1 may include addiction severity and comorbid conditions, S_2 may include the subject's response status, side effects and adherence to the initial treatment, and Y may be the number of non-heavy drinking days over the 12-month study period. Under Robins' consistency assumption (i.e., a subject's treatment assignment does not affect other subjects' outcomes; see Robins and Wasserman (1997)), the potential outcomes are connected to the subject's data by $S_2 = S_2(A_1)$ and $Y = Y(A_1, A_2)$.

The treatment randomization probabilities in a SMART trial are allowed to depend on past history (i.e., past information and treatment). That is the randomization probabilities for A_1 and A_2 may depend on S_1 and (S_1, A_1, S_2), respectively. Thus data from a SMART trial satisfy the "sequential ignorability" assumption (i.e., at each decision point the current treatments are assigned independently of potential future responses to treatment, conditional on the history of treatments and response to date (Robins 2004)). Under the "sequential ignorability" assumption, the conditional distributions of the potential outcomes are the same as the corresponding conditional distributions of the observable data. That is,

$$P(S_2(a_1) \leq s_2 | S_1 = s_1) = P(S_2 \leq s_2 | S_1 = s_1, A_1 = a_1)$$

and

$$P(Y(a_1, a_2) \leq y | S_1 = s_1, S_2(a_1) = s_2) = P(Y \leq y | S_1 = s_1, A_1 = a_1, S_2 = s_2, A_2 = a_2).$$

Thus the mean primary outcome of a dynamic treatment regime, (d_1, d_2), can be written as a function of the multivariate distribution of the observable data:

$$\mu_{(d_1,d_2)} = E\left[E\left[E[Y | S_1, A_1, S_2, A_2 = d_2(S_1, A_1, S_2)] | S_1, A_1 = d_1(S_1)\right]\right].$$

Hence we can evaluate the effect of a dynamic treatment regime or estimate the optimal dynamic treatment regime using data from a SMART trial. In particular, define the Q-function at the secondary decision point by

$$Q_2^*(s_1, a_1, s_2, a_2) = E[Y | S_1 = s_1, A_1 = a_1, S_2 = s_2, A_2 = a_2], \tag{5.2}$$

and the Q-function at the initial decision point by

$$Q_1^*(s_1, a_1) = E\Big[\max_{a_2} Q_2^*(S_1, A_1, S_2, a_2) | S_1 = s_1, A_1 = a_1\Big]. \tag{5.3}$$

Each Q-function measures the quality of the current treatment for patients with the specified past information and treatment assignments. Then the optimal decision rule at each decision point chooses treatment that maximizes the corresponding Q-function.

To power a SMART trial, we need to specify a primary research question. This research question may concern some components of dynamic treatment regimes (e.g., testing the main effect of the second-stage treatment) or the whole regimes (e.g., comparing effects of two dynamic treatment regimes). A good primary research question should be both scientifically important and helpful in developing a dynamic treatment regime. For example in the addiction management study an interesting primary research question would be "on average what is the best subsequent treatment for responders to initial treatment." That is, we want to compare the mean primary outcomes of two groups of responders (12-step versus 12-step+TDM). Standard test statistics (Hoel 1984) and sample size formula (Jennison and Turnbull 2000) for a large sample comparison of two means can be used in this case. Define the standardized effect size δ as the standardized difference in mean primary outcomes between two groups (Cohen 1988), i.e.,

$$\delta = \frac{E(Y|\text{Response}, A_2{=}\text{12-step}) - E(Y|\text{Response}, A_2{=}\text{12-step+TDM})}{\sqrt{[\text{Var}(Y|\text{Response}, A_2{=}\text{12-step}) + Var(Y|\text{Response}, A_2{=}\text{12-step+TDM})]/2}}.$$

Let γ denote the overall initial response rate. Suppose the randomization probability is $1/2$ for each treatment option at the secondary decision point. Standard calculation yields a sample size formula for the two-sided test with power $1-\beta$ and size α:

$$n = 4(z_{\alpha/2} + z_\beta)^2 \delta^{-2} \gamma^{-1},$$

where $z_{\alpha/2}$ and z_β are the standard normal $(1-\alpha/2)$ percentile and $(1-\beta)$ percentile, respectively. To use the formula, one needs to postulate the overall initial response rate γ.

Alternatively researchers may be more interested in primary research questions that are related to dynamic treatment regimes. In this case, Murphy (2005a) advocated that the primary research questions should involve the comparison of two dynamic treatment regimes beginning with different initial stage treatment options. This would allow researchers to decide which of the possible initial stage treatment options are worthy of further investigation. In the addiction management study, we may want to compare regime A with regime B defined in Sect. 5.1. Test statistics and sample size formulae for this type of research question have been provided in Murphy (2005a) and Oetting et al. (2007). In the following, we review the formulae in the context of our example.

Let $p_1(a_1|S_1)$ and $p_2(a_2|S_1, A_1, S_2)$ be the randomization probability at the initial and secondary decision point, respectively. For any dynamic treatment regime of interest (d_1, d_2), assume

$$P(p_1(d_1|S_1)p_2(d_2|S_1, A_1, S_2) > 0) = 1. \tag{5.4}$$

Assumption (5.4) implies that treatments specified by regimes A and B at any decision point for any given past history (i.e., past information and treatments) have positive probabilities of being assigned. Murphy et al. (2001) showed that an unbiased estimator of $\mu_{(d_1,d_2)}$ (the mean primary outcome for regime (d_1, d_2)) is

$$\hat{\mu}_{(d_1,d_2)} = \frac{\mathbb{P}_n\left[\frac{1_{A_1=d_1(S_1)} 1_{A_2=d_2(S_1,A_1,S_2)}}{p_1(d_1|S_1)p_2(d_2|S_1,A_1,S_2)} Y\right]}{\mathbb{P}_n\left[\frac{1_{A_1=d_1(S_1)} 1_{A_2=d_2(S_1,A_1,S_2)}}{p_1(d_1|S_1)p_2(d_2|S_1,A_1,S_2)}\right]}, \tag{5.5}$$

where n is the sample size, $\mathbb{P}_n f = \sum_{i=1}^n f(X_i)/n$ in which X_i is a vector of observations for the ith subject and f is a given function, and 1_Ω is an indicator function which equals 1 if event Ω occurs and 0 otherwise. A consistent estimator of the variance of $\sqrt{n}\hat{\mu}_{(d_1,d_2)}$ is

$$\hat{\tau}^2_{(d_1,d_2)} = \mathbb{P}_n\left(\left[\frac{1_{A_1=d_1(S_1)}1_{A_2=d_2(S_1,A_1,S_2)}}{p_1(d_1|S_1)p_2(d_2|S_1,A_1,S_2)}(Y - \hat{\mu}_{(d_1,d_2)})\right]^2\right).$$

The comparison of regime A with regime B can be obtained by comparing the subgroup of subjects in the trial whose treatment assignments are consistent with regime A with the subgroup of subjects in the trial whose treatment assignments are consistent with regime B. Note that there is no overlap between these two subgroups since a subject's initial treatment assignment can be consistent with only one of the regimes (A or B). The test statistic

$$Z = \frac{\sqrt{n}(\hat{\mu}_{\mathrm{A}} - \hat{\mu}_{\mathrm{B}})}{\sqrt{\hat{\tau}^2_{\mathrm{A}} + \hat{\tau}^2_{\mathrm{B}}}} \tag{5.6}$$

has an asymptotic standard normal distribution under the null hypothesis $\mu_{\mathrm{A}} = \mu_{\mathrm{B}}$ (Murphy 2005a). The standardized effect size for addressing this question is defined as $\delta = (\mu_{\mathrm{A}} - \mu_{\mathrm{B}})/\sqrt{(\sigma^2_{\mathrm{A}} + \sigma^2_{\mathrm{B}})/2}$, where σ^2_{A} and σ^2_{B} are the variances of the primary outcomes for regimes A and B, respectively. Suppose the randomization probability for each treatment option is 1/2 at each decision point. Equation (10) in Murphy (2005a) implies that $Var(\sqrt{n}\hat{\mu}_A) = 4\sigma^2_{\mathrm{A}}$ and $Var(\sqrt{n}\hat{\mu}_B) = 4\sigma^2_{\mathrm{B}}$ in large samples. Using a large sample approximation, the required sample size for the two-sided test $\left(H_0 : \mu_{\mathrm{A}} = \mu_{\mathrm{B}} \quad \text{v.s.} \quad H_1 : \mu_{\mathrm{A}} - \mu_{\mathrm{B}} = \delta\sqrt{(\sigma^2_{\mathrm{A}} + \sigma^2_{\mathrm{B}})/2}\right)$ with power $1 - \beta$ and size α is

$$n = 8(z_{\alpha/2} + z_\beta)^2\delta^{-2}.$$

A detailed derivation of the sample size formula in a similar context can be found in Oetting et al. (2007). Oetting et al. (2007) also discussed additional research questions and the corresponding test statistics and sample size formulae under different working assumptions. A web application that calculates the required sample size for sizing a study designed to discover the best dynamic treatment regime using a SMART design for continuous outcomes can be found at http://methodologymedia.psu.edu/smart/samplesize.

Formulae for the randomization probabilities that create equal sample sizes across all dynamic treatment regimes have been provided in Murphy (2005a). This was motivated by the classical large sample comparison of means for which, given equal variances, the power of a test is maximized by equal sample sizes. Let $k_1(S_1)$ be the number of treatment options at the initial decision point with pretreatment information S_1 and $k_2(S_1, A_1, S_2)$ be the number of treatment options at the secondary decision point with past history (S_1, A_1, S_2), respectively. Murphy's formulae yield

$$p_2(a_2|S_1, A_1, S_2) = k_2(S_1, A_1, S_2)^{-1}$$

$$p_1(a_1|S_1) = \frac{E[k_2(S_1, A_1, S_2)^{-1}|S_1, A_1 = a_1]^{-1}}{\sum_{b=1}^{k_1(S_1)} E[k_2(S_1, A_1, S_2)^{-1}|S_1, A_1 = b]^{-1}}. \quad (5.7)$$

If k_2 does not depend on S_2, the above formulae can be directly used. In our example, there are two initial treatment options and two secondary treatment options for each subject, i.e., $k_1(S_1) = 2$ and $k_2(S_1, A_1, S_2) = 2$ for all possible combinations of (S_1, A_1, S_2). Thus (5.7) yields a randomization probability of $1/2$ for each treatment option at each decision point. In general working assumptions concerning the distribution of S_2 given (S_1, A_1) are needed in order to use the formulae. See Murphy (2005a) for more details.

Some principles and practical considerations are as follows (for more details see Lavori and Dawson (2004), Murphy (2005a) and Murphy et al. (2007)). First, Murphy (2005a) proposed that the primary research question should consider at most simple dynamic treatment regimes so as to simplify the sample size formulae. In our addiction management study, we consider regimes where the initial decision rule is a constant (i.e., does not depend on an individual's pretreatment information) and the secondary decision rule depends only on the individual's initial treatment and his/her response status. Second, both Lavori and Dawson (2004) and Murphy (2005a) pointed out that, when designing the trial, the class of treatment options at each decision point should be restricted (only) by ethical, scientific, or feasibility considerations. Lavori and Dawson (2004) demonstrated how to constrain treatment options and thus decision rules using the STAR*D example (Lavori et al. 2001; Fava et al. 2003). Yet, Murphy (2005a) warns against undue restriction of the class of the decision rules. Our addiction management example reflects this notion. Although we might have reason to believe that non-adherent nonresponders to NTX should receive different treatment from adherent nonresponders to NTX, we do not provide these two groups different treatment options if we are uncertain that such restriction is necessary. Finally, the SMART trial should be viewed as one trial among a series of randomized trials intended to develop and/or refine a dynamic treatment regime. It should eventually be followed by a randomized control trial that compares the developed regime and an appropriate control (Murphy 2005a; Murphy et al. 2007). Note that like traditional randomized trials, SMART trials may involve standard problems such as dropout, incomplete assessments, etc.

5.3.3 SMART Design Versus Adaptive Experimental Designs

The SMART design introduced in the previous section involves stages of treatment. Some adaptive experimental designs also utilize stages of experimentation (Berry

2002, 2004). However, the SMART design is quite different from adaptive experimental designs.

An adaptive experimental design is "a multistage study design that uses accumulating data to decide how to modify aspects of the study without undermining the validity and integrity of the trial" (Dragalin 2006). Chow and Chang (2008) summarized various types of adaptive experimental designs. For example, a response adaptive design modifies the randomization schedules based on prior subjects' observed data at interim in order to increase the probability of success for future subjects (see, e.g., Berry et al. 2001). A group sequential design allows premature stopping of a trial due to safety, futility, and/or efficacy with options of additional adaptations based on results of interim analysis (see, e.g., Pampallona and Tsiatis 1994). A sample size reestimation design involves the recalculation of sample size based on study parameters (e.g., revised effect size, conditional power, nuisance parameters) obtained from interim data (see, e.g., Banerjee and Tsiatis 2006). In general, the aim of adaptive experimental designs is to improve the quality, speed, and efficiency of clinical development by modifying one or more aspects of a trial.

With the above definition, the difference between standard SMART design and adaptive experimental designs is rather straightforward. In a SMART design, each subject moves through multiple stages of treatment. On the other hand, in most adaptive experimental designs each stage involves different subjects. That is, each subject only participates in one stage of treatment. In both cases randomization occurs at each stage. The goal of a SMART trial is to develop a dynamic treatment regime that could benefit future patients. Many adaptive experimental designs (e.g., response adaptive randomization) try to provide most efficacious treatment to each subject in the trial based on the knowledge at the time that subject is randomized. In a SMART trial the design elements such as the final sample size, randomization probabilities, and treatment options are specified prior to conducting the trial. On the other hand, in an adaptive experimental design, the final sample size, randomization probabilities, and treatment options may be altered during the conduct of the trial.

There are some studies in which the adaptive experimental design was combined with the SMART design. For example, Thall et al. (2002) provided a statistical framework for an "outcome adaptive design" in a multicourse treatment setting in which two SMART trials are involved. Each trial used one half of the subjects. If the data from the first trial show a particular treatment sequence to be inferior to the others within a subgroup of subjects, then that treatment sequence option is dropped within that subgroup in the second trial. At the end the best treatment sequence for each subgroup is selected. Thall and Wathen (2005) considered a similar but more flexible design where the randomization criteria for each subject at each stage depend on the data from all subjects previously enrolled. Thall and his colleagues were able to apply such a strategy because subject outcomes in each SMART trial are observed quickly. In many other settings, obtaining subject's outcomes may take a long time (e.g., 12 months in our addiction management study). Thus, adaptation based on interim data is less feasible. How to optimally combine adaptive experimental design with the SMART design is worthy of further investigation.

5.4 Optimal Dynamic Treatment Regimes

In the previous section we discussed issues concerning experimental data and some primary analyses for developing dynamic treatment regimes. In the current section we discuss useful methods for estimating the optimal dynamic treatment regime using data from SMART trials.

5.4.1 *Simple Dynamic Treatment Regimes*

In some cases researchers would like to consider relatively simple dynamic treatment regimes. For example in the addiction management study, the primary research question may be the comparison of two simple dynamic treatment regimes (regimes A and B). Each regime specifies one initial treatment for all patients and assigns one treatment for initial treatment responders and another for nonresponders. There are totally eight such simple dynamic treatment regimes (see Fig. 5.1).

When there are only a few dynamic treatment regimes, we can estimate the mean primary outcome for each regime using the estimator in (5.5) and select the best one. Consider regime A in the addiction management example. When the randomization probability for each treatment option at each decision point is $1/2$, condition (5.4) is satisfied. $\hat{\mu}_{\mathrm{A}}$ is simply the average of Y over subjects whose treatment assignments are consistent with regime A. The estimation of μ_A can be improved by considering doubly robust estimator (Robins 2000), which may result in smaller variance (see Murphy et al. 2001).

5.4.2 *Dynamic Treatment Regimes Involving Covariates*

In general, there may be a large number of possible dynamic treatment regimes. The initial treatment decision may depend on pretreatment information S_1. The secondary treatment may vary according to an individual's pretreatment information S_1, initial treatment A_1, and ongoing information S_2. In fact, there could be infinite number of dynamic treatment regimes. For example, suppose there are two treatment options, 1 and -1, at each decision point and the data are collected from a SMART trial in which the randomization probability is $1/2$ for each treatment option at each decision point. Often one uses summaries of past history (i.e., past information and treatment assignments) to form decision rules. Let H_{11} be a vector summary of S_1 and H_{21} be a vector summary of (S_1, A_1, S_2), respectively. We may be interested in finding the best dynamic treatment regime of the form $\{d_1(H_{11}) = sign(\psi_{10} + H_{11}^T\psi_{11})$ and $d_2(H_{21}) = sign(\psi_{20} + H_{21}^T\psi_{21})\}$, where ψs are the parameters, $sign(x) = 1$ if $x > 0$ and -1 otherwise. ψ_{10} is the main effect of the initial treatment. Each component in ψ_{11} is the interaction effect of the initial treatment and the corresponding component in H_{11}. In our addiction management example, H_{11} could be a two-dimensional vector including addiction severity and an indicator of presence/absence of a co-occurring disorder, and H_{21} could be a four-dimensional vector including the initial treatment, response status, adherence to the initial treatment, and a measure of side effects. In this case, (5.5) is equivalent to

$$\hat{\mu}_{(d_1,d_2)} = \frac{\mathbb{P}_n\left[1_{A_1(\psi_{10}+H_{11}^T\psi_{11})>0} 1_{A_2(\psi_{20}+H_{21}^T\psi_{21})>0} Y\right]}{\mathbb{P}_n\left[1_{A_1(\psi_{10}+H_{11}^T\psi_{11})>0} 1_{A_2(\psi_{20}+H_{21}^T\psi_{21})>0}\right]}. \tag{5.8}$$

Selecting the dynamic treatment regime (i.e., the ψs) that maximizes (5.8) is computationally intractable since the objective function (5.8) is nonconcave in the ψs. An additional problem is that, if H_{11} and H_{21} are of high dimension, the regime that maximizes (5.8) is subject to overfitting the data and may yield a poor mean primary outcome among all dynamic treatment regimes under consideration. This is because we try to maximize $\hat{\mu}_{(d_1,d_2)}$ instead of $\mu_{(d_1,d_2)}$ over (d_1, d_2). Denote the maximizer of $\hat{\mu}_{(d_1,d_2)}$ by $(\tilde{d}_1, \tilde{d}_2)$. When the sample size is relatively small as compared to the complexity of the dynamic treatment regimes, the resulting regime $(\tilde{d}_1, \tilde{d}_2)$ may fit the data well but the regime may not be close to the regime that maximizes $\mu_{(d_1,d_2)}$.

An alternative approach to estimate the optimal dynamic treatment regime is Q-learning, which can be viewed as a generalization of regression to multistage decision making. This method estimates the optimal decision rules by learning the Q-functions defined in Sect. 5.3.2. There are many variants of Q-learning (Watkins 1989; Sutton and Barto 1998; Ormoneit and Sen 2002; Lagoudakis and Parr 2003; Ernst et al. 2005). Below we review Q-learning with function approximation as described in Murphy (2005b).

Let $\mathcal{Q}_1$ be the approximation space for the initial stage Q-function Q_1^* defined in (5.3) and $\mathcal{Q}_2$ be the approximation space for the second-stage Q-function Q_2^* defined in (5.2), respectively. For example, assume there are two treatment options at each decision point; $a_1 \in \{-1, 1\}$ and $a_2 \in \{-1, 1\}$. We may consider linear approximation spaces for the Q-functions,

$$\mathcal{Q}_2 = \Big\{ Q_2(h_{20}, h_{21}, a_2; \boldsymbol{\theta}_2) = \phi_{20} + h_{20}^T\phi_{21} + (\psi_{20} + h_{21}^T\psi_{21})a_2 : \\ \boldsymbol{\theta}_2 = (\phi_{20}, \phi_{21}^T, \psi_{20}, \psi_{21}^T) \in \Theta_2 \Big\}$$

$$\text{and} \quad \mathcal{Q}_1 = \Big\{ Q_1(h_{10}, h_{11}, a_1; \boldsymbol{\theta}_1) = \phi_{10} + h_{10}^T\phi_{11} + (\psi_{10} + h_{11}^T\psi_{11})a_1, \\ \boldsymbol{\theta}_1 = (\phi_{10}, \phi_{11}^T, \psi_{10}, \psi_{11}^T) \in \Theta_1 \Big\} \tag{5.9}$$

where h_{20} and h_{21} are vector summaries of (s_1, a_1, s_2), h_{10} and h_{11} are vector summaries of s_1, and Θ_1 and Θ_2 are the parameter spaces. Note that we use upper case letters to denote random variables or data for subjects in the SMART trial and lower case letters to denote the values of the variables.

Since Q_2^* is the conditional mean of Y given past information and treatment assignments (S_1, A_1, S_2, A_2), we can estimate the second-stage parameter $\boldsymbol{\theta}_2$ using least squares,

$$\hat{\boldsymbol{\theta}}_2 = \arg\min_{\boldsymbol{\theta}_2 \in \Theta_2} \mathbb{P}_n[Y - Q_2(H_{20}, H_{21}, A_2; \boldsymbol{\theta}_2)]^2.$$

Similarly, since Q_1^* is the conditional mean of $\max_{a_2} Q_2^*$ given (S_1, A_1), a least squares estimator of the initial stage parameter $\boldsymbol{\theta}_1$ (with Q_2^* estimated by $Q_2(H_{20}, H_{21}; \hat{\boldsymbol{\theta}}_2)$) is

$$\hat{\boldsymbol{\theta}}_1 = \arg\min_{\boldsymbol{\theta}_1 \in \Theta_1} \mathbb{P}_n[\max_{a_2} Q_2(H_{20}, H_{21}, a_2; \hat{\boldsymbol{\theta}}_2) - Q_1(H_{10}, H_{11}; \boldsymbol{\theta}_1)]^2 \tag{5.10}$$

(see Tsitsiklis and Van Roy 1996 for this estimation method in a similar context). The estimated optimal dynamic treatment regime then uses treatments that maximize the estimated Q-functions. In the linear model example, the estimated optimal regime is

$$\hat{d}_2(h_{21}) = \text{sign}(\hat{\psi}_{20} + h_{21}^T \hat{\psi}_{21})$$

$$\text{and} \qquad \hat{d}_1(h_{11}) = \text{sign}(\hat{\psi}_{10} + h_{11}^T \hat{\psi}_{11}). \tag{5.11}$$

Note that sometimes researchers may consider different treatment options for different subgroups of subjects. For example in the addiction management study, different secondary treatment options are offered each of the three subgroups: initial treatment responders (subgroup 1), NTX nonresponders (subgroup 2), and CBI nonresponders (subgroup 3). In this case, we code $A_2 = 1$ for one treatment option and $A_2 = -1$ for the other treatment option in each subgroup. We can use different linear models, say $Q_2^{(j)}, j =$ subgroup 1, 2, 3, for the three subgroups of subjects, respectively. The final model for Q_2^* can be written as $\sum_{j=1}^{3} Q_2^{(j)} 1_{(j)}$, where $1_{(j)}$ is 1 if the subject belongs to the jth subgroup and 0 otherwise. Since each $Q_2^{(j)}$ is a linear model, the model for Q_2^* fits the linear framework (5.9).

In Q-learning we modeled the Q-functions Q_1^* and Q_2^*. However, only part of the Q-function is relevant for the construction of the decision rules. This can be easily seen from the above linear model example where the estimated decision rule at each decision point only depends on the interaction part in the corresponding linear model (e.g., the decision rules (5.11) only depend on ψs not ϕs). In general, each Q_t^* $(t = 1, 2)$ can be written as $Q_t = g_t^* + \max_{a_t} Q_t^*$ (where $g_t^* = Q_t^* - \max_{a_t} Q_t^*$). g_t^* is called the advantage function (Baird 1994) at the tth decision point. It measures the gain in performance obtained by following treatment a_t as compared to following the best treatment at the tth decision point. Since $\max_{a_t} Q_t^*$ does not contain a_t, we only need to model g_t^* instead of modeling Q_t^*. g_t^* may include many fewer variables than the corresponding Q_t^* since it contains only variables in the interaction terms in Q_t^*. Estimation of only the advantage functions was first proposed in Murphy (2003), along with a least squares estimation method. Robins (2004) provided a refined estimating equation to gain efficiency. Chakraborty and Murphy (2009) showed that under appropriate conditions Q-learning with linear models is algebraically equivalent to an inefficient version of Robins' method.

5.5 Other Analyses

5.5.1 *Inference*

It is crucial to attach measures of confidence (e.g., standard errors, confidence intervals, etc.) to the estimated dynamic treatment regimes. Furthermore since collecting patient information in order to apply the decision rules may be relatively expensive in clinical practice, researchers may be interested in assessing whether certain patient variables are necessary for making the decision. For example in the addiction management example, suppose we are to make the initial decision based on a linear decision rule "treat patients with NTX if $\psi_{10}+\psi_{11}\cdot$'addiction severity' $>$ 0, and treat patients with CBI otherwise." It may be possible to simplify the decision rule by removing the variable "addiction severity" if the data do not provide sufficient evidence that this variable is necessary. In this case it would be useful to assess the extent to which the variable "addiction severity" is important in the data set (say via inference on the associated parameter ψ_{11}). To further justify whether the two initial treatments have different effects on the primary outcome, we may also want to test if $\psi_{10} = 0$.

Inferential methods have been discussed in Robins (2004), Moodie and Richardson (2007), and Chakraborty and Murphy (2009). Robins (2004) noted that the treatment effect parameters at any stage prior to the last can be non-regular. As a consequence, it is difficult to provide valid confidence intervals for the optimal dynamic treatment regime estimated from Q-learning or Robins' method. To explain meaning of the non-regularity in this context, we discuss the inference for ψ_{11}. Consider the linear model as described in (5.9) with H_{11} being the variable "addiction severity." Then the Q-learning estimator of ψ_{11} is given by solving (5.10). Note that $\hat{\psi}_{11}$ is a function of $\max_{a_2} Q_2(H_{20}, H_{21}, a_1; \hat{\boldsymbol{\theta}}_2)$. With the linear parameterization, $\max_{a_2} Q_2(H_{20}, H_{21}, a_1; \hat{\boldsymbol{\theta}}_2)$ equals $\hat{\phi}_{20} + H_{20}^T\hat{\phi}_{21} + |\hat{\psi}_{20} + H_{21}^T\hat{\psi}_{21}|$, which is non-differentiable at the point $\hat{\psi}_{20} + H_{21}^T\hat{\psi}_{21} = 0$. Due to the non-differentiability, it can be shown (see Robins 2004 and Moodie and Richardson 2007) that the asymptotic distribution of $\sqrt{n}(\hat{\psi}_{11} - \psi_{11})$ is normal if $P(\psi_{20} + H_{21}^T\psi_{21} = 0) = 0$ and non-normal otherwise ($\psi_{20}+H_{21}^T\psi_{21} = 0$ implies that there is no second-stage treatment effect for patients with past history H_{21}). Thus ψ_{11} is a "non-regular" parameter and $\hat{\psi}_{11}$ is a "non-regular" estimator of ψ_{11} (see Bickel et al. (1993) for a more precise definition of non-regularity). The practical consequence of this is that whenever $\psi_{20} + H_{21}^T\psi_{21}$ is close to zero, both confidence intervals of ψ_{11} based on formulae derived from Taylor series arguments and confidence intervals based on bootstrap will perform poorly (Moodie and Richardson 2007; Chakraborty and Murphy 2009).

Several approaches have been discussed to deal with non-regularity. Robins (2004) constructed a score method that provides a conservative uniform asymptotic confidence interval for ψ_{11}. Other methods reduced bias in the estimation of ψ_{11} due to non-regularity by substituting $|\hat{\psi}_{20} + H_{21}^T\hat{\psi}_{21}|$ in (5.10) with other quantities. Moodie and Richardson (2007) proposed a hard-threshold estimator in which $|\hat{\psi}_{20}+$

$H_{21}^T \hat{\psi}_{21}|$ is replaced by 0 if it is below a threshold. This method reduced the bias of $\hat{\psi}_{11}$ when $|\hat{\psi}_{20} + H_{21}^T \hat{\psi}_{21}|$ is close to zero. However, it is unclear how to select the threshold value. Chakraborty and Murphy (2009) used a soft-threshold estimator, where $|\hat{\psi}_{20} + H_{21}^T \hat{\psi}_{21}|$ in (5.10) is replaced by $|\hat{\psi}_{20} + H_{21}^T \hat{\psi}_{21}|\left(1 - \frac{\lambda}{|\hat{\psi}_{20} + H_{21}^T \hat{\psi}_{21}|}\right)^+$ with tuning parameter λ chosen by an empirical Bayes method. Simulations in Chakraborty and Murphy (2009) provided evidence that, in the non-regular setting in which $P(\psi_{20} + H_{21}^T \psi_{21} = 0) > 0$, the use of bootstrap confidence intervals along with the soft-threshold estimator (and hard-threshold estimator in some cases) reduced bias due to non-regularity and gave correct coverage rate. When $P(\psi_{20} + H_{21}^T \psi_{21} = 0) = 0$, bootstrap confidence intervals based on the original Q-learning estimator performed the best, but the percentile bootstrap with the soft-threshold estimator also performed reasonably well. Theoretical optimality of this method is unclear and worth further investigation. Instead of providing confidence sets for ψ_{10} and ψ_{11}, Lizotte et al. (2009) proposed a "voting method", where the vote for a treatment is an estimation of the probability that the treatment would be selected as the best treatment if the trial were conducted again. The non-regularity problem in this method is addressed using a hard-threshold estimator. However, this approach is relatively new and untested, hence further investigation and refinements are needed.

5.5.2 *Modeling*

The inference problem discussed in the previous section is based on the parametric or semi-parametric modeling of the Q-functions. Note that the approximation for the Q-functions together with the definition of the estimated decision rules as the argmax of the estimated Q-functions places implicit restrictions on the set of regimes that can be considered. More specifically with a given approximation space for the Q-functions the set of regimes under consideration is $\mathscr{D} = \{(d_1, d_2) : d_t \in \arg\max_{a_t} Q_t, Q_t \in \mathscr{Q}_t, t = 1, 2.\}$. Thus at least implicitly the goal becomes the estimation of the best regime in the space $\mathscr{D}$. However, problems occur if the approximation space for Q_1^* and Q_2^* does not contain the true Q-functions. In particular, when the approximation is poor, the mean primary outcome of the estimated regime, $\mu_{(\hat{d}_1, \hat{d}_2)}$, may not be close to $\max_{(d_1,d_2) \in \mathscr{D}} \mu_{(d_1,d_2)}$ even in large samples (Tsitsiklis and Van Roy 1997). That is, $\mu_{(\hat{d}_1, \hat{d}_2)}$ may not be a consistent estimator of $\max_{(d_1,d_2) \in \mathscr{D}} \mu_{(d_1,d_2)}$ when the approximation space does not contain the true Q-functions. The potential bias (i.e., inconsistency) will be eliminated if the approximation space provides a sufficiently good approximation for Q_1^* and Q_2^*, but then the estimated Q-functions will have large variances due to the small sample size. Consequently, $\mu_{(\hat{d}_1, \hat{d}_2)}$ will have a large variance as well. Thus selecting an appropriate approximation space is the key to success. Ormoneit and Sen (2002) used a sequence of kernel-based approximation spaces and made assumptions on the target function to guarantee a sufficiently rich approximation. Another promising avenue is to use model selection techniques in regression/classification with Q-learning to get a good trade off between bias and variance.

5.6 Discussion and Open Questions

Dynamic treatment regimes are a new approach to treatment design. These treatment designs adapt the treatment to patient characteristics and outcomes.

In Sect. 5.3 we reviewed the SMART design. To conduct a SMART trial where there is a plethora of clinical decisions and treatment options, we need to prespecify study components, such as how many critical decisions are of interest, what are the best time-points to make decisions (which can also be viewed as part of the critical decisions), and what treatment options should be investigated at each decision point, etc. A new screening experimental design proposed in Murphy and Bingham (2009) can be used to identify promising components and screening out negligible ones. As discussed in Murphy et al. (2007), another approach is to use the MOST paradigm developed in Collins et al. (2005). This paradigm advocates the use of a series of experimental trials to prospectively determine active components for future investigation. How to effectively integrate the SMART design into the MOST paradigm is an area for future research.

In Sect. 5.4, we reviewed several methods for estimating the optimal dynamic treatment regimes. In fact, many other methods were omitted. For example, Robins (1986) proposed a G-computation formula to estimate the mean primary outcome of a dynamic treatment regime based on modeling the conditional distributions of the data. Lunceford et al. (2002) and Wahed and Tsiatis (2004, 2006) provided semiparametric estimators of the survival function for a given regime in cancer trials. Those estimates can be used to select the best regime among a small set of dynamic treatment regimes. Thall and colleagues (Thall et al. 2000, 2002, 2007) developed several likelihood-based methods, both Bayesian and frequentist, for selecting the best dynamic treatment regime. Instead of estimating the optimal dynamic treatment regimes, van der Laan and his colleagues (van der Laan et al. 2005; Petersen et al. 2007; van der Laan and Petersen 2007) aimed at ascertaining a "statically" optimal dynamic treatment regime. At each stage the "statically" optimal dynamic treatment regime only uses current available information to make all future decisions. The "statically" optimal dynamic treatment regime is not truly optimal in a multistage setting (see Petersen et al. 2007 for a detailed comparison of the optimal dynamic treatment regime and the "statically" optimal dynamic treatment regime).

In Sect. 5.5, we discussed several open questions concerning data analyses including inference and modeling. Still, there many other challenges and open questions in this regard. For example, instead of inference, variable selection techniques can be used to assess whether particular patient variables are necessary for decision making. In fact variable selection is very important here since data collected in clinical trials are often of high dimension while only a few variables are likely to be useful in selecting the best sequence of treatments. Although variable selection techniques developed for prediction can be used here, such methods may miss variables that are useful for decision making. Gunter et al. (2007) developed a variable selection method for decision making in the case of single decision point and empirically showed that their method is better than Lasso (Tibshirani 1996) in decision making. However, theoretical properties of their method have not been developed and the extension to multistage decision making is needed. Another open question is how to construct a dynamic treatment regime when there are multiple primary outcomes (e.g., functionality, side effects, cost, etc). Thall et al. (2008) proposed a Bayesian procedure for finding the best dosage in a single-stage setting involving bivariate (efficacy, toxicity) outcomes. In general, to develop a good dynamic treatment regime and conduct other secondary analyses, the construction of a high-quality composite outcome is very important. Other challenges include feature construction, dealing with missing data, etc. All these issues are worthy of future research.

The current chapter illustrates the benefits of dynamic treatment regimes and the SMART experimental approach for constructing and evaluating dynamic treatment regimes. We also present methods for estimating the optimal dynamic treatment regime and discuss challenges associated with these methods as well as their potentials for intervention scientists aiming to develop dynamic treatment regimes. Overall, this topic deserves continued research attention and a prominent role in intervention research.

Acknowledgment We acknowledge support for this work from NIH grants R01 MH080015 and P50 DA10075.

References

Alcoholics Anonymous (2001) Chapter 5: how it works. Alcoholics Anonymous, 4th edn. Alcoholics Anonymous World Services, New York

Baird L (1994) Reinforcement learning in continuous time: advantage updating. IEEE Internat Conf Neural Networks 4:2448–2453

Banerjee A, Tsiatis A (2006) Adaptive two-stage designs in phase II clinical trials. Stat Med 25(19):3382–3395

Bellman RE (1957) Dynamic programming. Princeton University Press, New Jersey.

Berry D, Mueller P, Grieve A, Smith M, Parke T, Blazek R, Mitchard N, Krams M (2001) Adaptive Bayesian designs for dose-ranging drug trials. In: Gatsonis C, Carlin B, Carriquiry A (eds) Case studies in Bayesian statistics, vol V. Springer, New York, pp 99–181

Berry DA (2002) Adaptive clinical trials and Bayesian statistics (with discussion). Pharmaceutical Report, American Statistical Association, Alexandria, VA.

Berry D (2004) Bayesian statistics and the efficiency and ethics of clinical trials. Stat Sci 19:175–187

Bickel PJ, Klaassen CAJ, Ritov Y, Wellner JA (1993) Efficient and adaptive estimation for semiparametric models. Johns Hopkins University Press, Baltimore, MD

Chakraborty B, Murphy SA (2009) Inference for nonregular parameters in optimal dynamic treatment regimes. Stat Meth Med Res 19(3):317–343. Available online: 16-July-2009 DOI: 10.1177/0962280209105013

Chow SC, Chang M (2008) Adaptive design methods in clinical trials—a review. Orphanet J Rare Diseases 3:11

Cohen J (1988) Statistical power analysis for the behavioral sciences, 2nd edn. Lawrence Erlbaum Associates, Inc., Hillsdale, NJ

Collins LM Murphy SA, Nair V, Strecher V (2005) A strategy for optimizing and evaluating behavioral intervention. Ann Behav Med 30:65–73

COMBINE Study Research Group (2003) Testing combined pharmacotherapies and behavioral interventions in alcohol dependence: rationale and methods. Alcohol Clin Exp Res 27:1107–1122

Dragalin V (2006) Adaptive designs: terminology and classification. Drug Inform J 40:425–435

Ernst D, Geurts P, Wehenkel L (2005) Tree-based batch mode reinforcement learning. J Mach Learn Res 6:503–556

Fava M, Rush AJ, Trivedi MH, Nierenberg AA, Thase ME, Sackeim HA, Quitkin FM, Wisniewski S, Lavori PW, Rosenbaum JF, Kupfer DJ (2003) Background and rationale for the sequenced treatment alternatives to relieve depression (STAR*D) study. Psychiatr Clin North Am 26(2):457–494

Gunter LL, Zhu J, Murphy SA (2007) Variable selection for optimal decision making. Proceedings of the 11th conference on artificial intelligence in medicine, AIME 2007, Lecture notes in computer science/Lecture notes in artificial intelligence, vol 4594. pp 149–154

Hoel P (1984) Introduction to mathematical statistics, 5th edn. John Wiley and Sons, New York

Jennison C, Turnbull B (2000) Group sequential methods with applications to clinical trials. Chapman & Hall, Boca Raton, FL

Lagoudakis MG, Parr R (2003) Least-squares policy iteration, J Mach Learn Res 4:1107–1149

Lavori PW, Dawson R (2000) A design for testing clinical strategies: biased individually tailored within-subject randomization. J Roy Stat Soc A 163:29–38

Lavori PW, Rush AJ, Wisniewski SR, Alpert J, Fava M, Kupfer DJ, Nierenberg A, Quitkin, FM, Sackeim HA, Thase ME, Trivedi M (2001) Strengthening clinical effectiveness trials: equipoise-stratified randomization. Biol Psychiatr 48:605–614

Lavori PW, Dawson R (2004) Dynamic treatment regimes: practical design considerations. Clin Trials 1:9–20

Lizotte DJ, Laber E, Murphy SA (2009) Assessing confidence in policies learned from sequential randomized trials. Technical report, Department of Statistics, University of Michigan, Ann Arbor, Michigan

Lunceford JK, Davidian M, Tsiatis AA (2002) Estimation of survival distributions of treatment policies in two-stage randomization designs in clinical trials. Biometrics 58:48–57

McLellan AT (2002) Have we evaluated addiction treatment correctly? Implications from a chronic care perspective. Addiction 97:249–252

Miller WR (ed) (2004) COMBINE monograph series, combined behavioral intervention manual: a clinical research guide for therapists treating people with alcohol abuse and dependence. DHHS Publication No. (NIH) 04–5288, vol 1. National Institute on Alcohol Abuse and Alcoholism, Bethesda, MD

Moodie EEM, Richardson TS (2007) Bias correction in non-differentiable estimating equations for optimal dynamic regimes. COBRA Preprint Series. Article 17.

Murphy SA, van der Laan MJ, Robins JM, CPPRG (2001) Marginal mean models for dynamic regimes. J Amer Stat Assoc 96:1410–1423

Murphy SA (2003) Optimal Dynamic Treatment Regimes. J Roy Stat Soc B 65(2):331–366

Murphy SA (2005a) An experimental design for the development of adaptive treatment strategies. Stat Med 24:1455–1481

Murphy SA (2005b) A generalization error for Q-learning. J Mach Learn Res 6:1073–1097

Murphy SA, Lynch KG, Oslin D, Mckay JR, TenHave T (2007) Developing adaptive treatment strategies in substance abuse research. Drug Alcohol Depend 88(2):s24–s30

Murphy SA, Bingham D (2009) Screening experiments for developing dynamic treatment regimes. J Am Stat Assoc 104:391–408

Neyman J (1923) On the application of probability theory to agricultural experiments. Stat Sci 5:465–480 (Translated in 1990)

Oetting AI, Levy JA, Weiss RD, Murphy SA (2007) Statistical methodology for a SMART Design in the development of adaptive treatment strategies. In: Shrout PE (ed) Causality and psychopathology: finding the determinants of disorders and their cures. American Psychiatric Publishing, Inc., Arlington VA

Ormoneit D, Sen S (2002) Kernel-based reinforcement learning. Mach Learn 49(2–3):161–178

Oslin DW, Sayers S, Ross J, Kane V, TenHave T, Conigliaro J, Cornelius J (2003) Disease management for depression and at-risk drinking via telephone in an older population for veterans. Psychosom Med 65:931–937

Pampallona S, Tsiatis AA (1994) Group sequential designs for one and two sided hypothesis testing with provision for early stopping in favour of the null hypothesis. J Stat Plann Infer 42:19–35

Petersen ML, Deeks SG, van der Laan MJ (2007) Individualized treatment rules: generating candidate clinical trials. Stat Med 26(25):4578–4601

Robins JM (1986) A new approach to causal inference in mortality studies with sustained exposure periods—application to control of the healthy worker survivor effect. Comput Math App 14:1393–1512

Robins JM (1987) Addendum to "A new approach to causal inference in mortality studies with sustained exposure periods—application to control of the healthy worker survivor effect". Comput Math App 14:923–945

Robins JM, Wasserman L (1997) Estimation of effects of sequential treatments by reparameterizing directed acyclic graphs. In: Geiger D, Shenoy P (eds) Proceedings of the thirteenth conference on uncertainty in artificial intelligence, Morgan Kaufmann, San Francisco

Robins JM (2000) Robust estimation in sequentially ignorable missing data and causal inference models. In: Proceedings of the American statistical association section on Bayesian statistical science 1999, pp 6–10

Robins JM (2004) Optimal structural nested models for optimal sequential decisions. In: Lin DY, Haegerty P (eds) Proceedings of the second Seattle symposium on biostatistics. Lecture notes in stastitics. Springer, New York

Rubin DB (1978) Bayesian inference for causal effects: the role of randomization. Ann Stat 6:34–58

Schneider LS, Tariot PN, Lyketsos CG, Dagerman KS, Davis KL, Davis S, Hsiao JK, Jeste DV, Katz IR, Olin JT, Pollock BG, Rabins PV, Rosenheck RA, Small GW, Lebowitz B, Lieberman JA (2001) National Institute of Mental Health clinical antipsychotic trials of intervention effectiveness (CATIE) alzheimer disease trial methodology. Am J Geriatr Psychiatr 9(4):346–360

Stone RM, Berg DT, George SL, Dodge RK, Paciucci PA, Schulman P, Lee EJ, Moore JO, Powell BL, Schiffer CA (1995) Granulocyte macrophage colony-stimulating factor after initial chemotherapy for elderly patients with primary acute myelogenous leukemia. New Engl J Med 332:1671–1677

Sutton RS, Barto AG (1998) Reinforcement learning: an introduction. MIT, Cambridge, MA

TenHave TR, Coyne J, Salzer M, Katz I (2003) Research to improve the quality of care for depression: alternatives to the simple randomized clinical trial. Gen Hosp Psychiatr 25:115–123

Thall PF, Millikan RE, Sung HG (2000) Evaluating multiple treatment courses in clinical trials. Stat Med 19:1011–1028

Thall PF, Sung HG, Estey EH (2002) Selecting therapeutic strategies based on efficacy and death in multicourse clinical trials. J Amer Stat Assoc 97:29–39

Thall PF, Wathen JK (2005) Covariate-adjusted adaptive randomization in a sarcoma trial with multi-stage treatments. Stat Med 24:1947–1964

Thall PF, Wooten LH, Logothetis CJ, Millikan R, Tannir NM (2007) Bayesian and frequentist two-stage treatment strategies based on sequential failure times subject to interval censoring. Stat Med 26:4687–4702

Thall PF, Nguyen H, Estey EH (2008) Patient-specific dose-finding based on bivariate outcomes and covariates. Biometrics 64(4):1126–1136

Tibshirani R (1996) Regression shrinkage and selection via the Lasso. J Roy Stat Soc B 32:135–166

Tsitsiklis JN, Van Roy B (1996) Feature-based methods for large scale dynamic programming, Mach Learn 22:59–94

Tsitsiklis JN, Van Roy B (1997) An analysis of temporal-difference learning with function approximation. IEEE Trans Automat Contr 42(5):674–690

van der Laan MJ, Petersen ML, Joffe MM (2005) History-adjusted marginal structural models and statically-optimal dynamic treatment regimens. Internat J Biostatistics 1(1):Article 4

van der Laan MJ, Petersen ML (2007) Statistical learning of origin-specific statically optimal individualized treatment rules. Internat J Biostatistics 3(1)

Wahed AS, Tsiatis AA (2004) Optimal estimator for the survival distribution and related quantities for treatment policies in two-stage randomization designs in clinical trials. Biometrics 60:124–133

Wahed AS, Tsiatis AA (2006) Semiparametric efficient estimation of survival distribution for treatment policies in two-stage randomization designs in clinical trials with censored data. Biometrika 93:163–177

Watkins CJCH (1989) Learning from delayed rewards. Ph.D. thesis, Cambridge University

Chapter 6
Statistical Issues with Trial Data and Economic Modeling for Cost-Effectiveness Evaluation

Kathleen A. Boyd, Andrew H. Briggs, Pierre Ducournau, Marlene Gyldmark, Frederic de Reydet, and Jim Cassidy

6.1 Introduction

Economic evaluations are undertaken to help inform decision making, for example, to help determine which health care interventions to fund given limited health care budgets. A systematic approach is taken to compare alternative interventions in terms of their costs and consequences. Cost–effectiveness analysis (CEA) in particular compares the difference in costs and effects between two or more alternatives, reporting the incremental difference as a cost per unit of outcome, known as an incremental cost–effectiveness ratio (ICER). Alternatively a CEA will report the net monetary benefit of an intervention; however, ICERs are the most popular method for presenting CEA results. The larger the value of the ICER, the more it costs per unit of effectiveness and therefore the less cost-effective the intervention is in comparison to the alternative. The ICER value can be compared against a monetary threshold to help aid decisions regarding appropriate resource allocation.

A CEA can be undertaken retrospectively by utilizing a variety of published (secondary) data sources to provide information on costs, resource use, and effectiveness information which can then be synthesized in a decision model.

K.A. Boyd • A.H. Briggs (✉)
Institute of Health and Wellbeing, University of Glasgow, Scotland, UK
e-mail: Kathleen.Boyd@glasgow.ac.uk; Andrew.Briggs@glasgow.ac.uk

P. Ducournau • F. de Reydet
F Hoffmann-La Roche Ltd, Basel, Switzerland

M. Gyldmark
Modelling, Outcomes Research, Statistics & Epidemiology, F Hoffmann-La Roche Ltd, Basel, Switzerland
e-mail: marlene.gyldmark@roche.com

J. Cassidy
Institute of Cancer Sciences, University of Glasgow, Scotland, UK
e-mail: James.Cassidy@glasgow.ac.uk

W. Tang and X. Tu (eds.), *Modern Clinical Trial Analysis*, Applied Bioinformatics and Biostatistics in Cancer Research, DOI 10.1007/978-1-4614-4322-3_6,

Alternatively, the analysis may be conducted prospectively alongside an ongoing clinical trial or study in a primary data analysis. This latter approach has become increasingly common, with cost–effectiveness analyses being undertaken alongside or "piggy-backed" (Kuntz and Weinstein 2001) onto a randomized controlled trial. However, it is also possible to use a mixture of these two approaches, with prospective trial information supporting wider secondary evidence to provide a more robust analysis (Sculpher et al. 2006). Therefore the role of a trial is complementary to the economic model, providing data inputs which can also be synthesized with external secondary data.

An area of health care that would benefit substantially by adopting this complementary approach to decision modeling for economic evaluation is that of oncology. The development of new therapies and technologies for oncology outpaces most other areas of health care, and as there is mounting pressure on health care services to provide fast and effective cancer treatments, there is an increasing desire for large-scale randomized controlled trials (RCTs) to assess the effectiveness of these new therapies. However, there is also a recognized need for CEA due to the added pressure new expensive cancer treatments put on limited health care budgets. Consequently the cost–effectiveness element in a trial will often be considered at a late stage and frequently gets "piggy-backed" onto the trial objectives. In such cases adopting a broad modeling approach, which utilizes the prospective oncology trial as a key input and then synthesizes this with data from a variety of sources to develop the model, leads to a more robust CEA with more meaningful results than those that are restricted solely to the data collected within the trial.

This chapter will consider some statistical issues of importance that arise when prospective trial data is used as a key informant to an economic model. Drawing on examples from oncology, issues with extrapolating beyond the observed data to determine lifetime outcomes, using area under the curve analysis or modeling techniques, incorporation of adverse events which trade off against survival in cancer treatment, and dealing with uncertainty will be illustrated with data from the "Xeloda in adjuvant colon cancer therapy" (X-ACT) study throughout the chapter. A brief summary of the X-ACT colorectal cancer study is now presented.

Xeloda in Adjuvant Colon Cancer Therapy

X-ACT study: Adjuvant Xeloda versus 5-FU/LV for the treatment of Dukes C colon cancer

The X-ACT study (Xeloda in Adjuvant Colon Cancer Therapy) aimed to evaluate the effectiveness and cost–effectiveness of two alternative treatments for Stage III colorectal cancer. Capecitabine (Xeloda) is an oral treatment for adjuvant and metastatic cancer and was compared with the current standard treatment, intravenous 5-fluorouracil plus leucovorin (5-FU/LV) in a randomized controlled trial. The objective of the trial was to establish the efficacy and tolerability of Xeloda in comparison to 5-FU/LV. The primary

(continued)

(continued)
measures of effectiveness were improvements in disease-free survival (DFS) and overall survival (OS), while the incidence and severity of adverse events were also of importance. The trial found Xeloda to be at least equivalent to 5-FU/LV in terms of disease-free survival with fewer adverse events. Six year follow-up data was obtained from the trial and extrapolated to a lifetime horizon, in order to assess the cost per quality adjusted life month (QALM) gained in a Markov model for the cost–effectiveness analysis. The trial results and cost–effectiveness analysis are reported in detail by Twelves et al. (2005) and Cassidy et al. (2006). Data from this study will be used throughout this chapter to illustrate examples of the issues that are discussed.

6.2 Statistical Issues

6.2.1 Survival Analysis

In oncology clinical trials, where the effectiveness of a particular treatment is being assessed, the outcome of interest tends to be mortality (overall survival) and progression of disease (disease-free survival, DFS, or progression-free survival[1]) and therefore survival analysis, also known as time to event analysis, is required. Trial data is used to assess any gains in life expectancy or DFS between the treatment and control arms. The data should be plotted graphically using the Kaplan–Meier (KM) method and the log rank test is often used to assess any difference between the two survival distributions. Traditionally, the effectiveness will be summarized either by assessing the difference in median survival times (the points in time at which only 50% of the treatment and control populations remain alive), or by assessing the difference in survival probabilities between the two arms at a fixed point in time, for example, in adjuvant colorectal cancer studies this tends to be measured at 3 years post treatment (Gill and Sargent 2006). While these two methods are well–established approaches for capturing effectiveness in clinical trials, they each only summarize one dimension of the difference between two treatment options. A more comprehensive understanding of effectiveness in terms of life years gained can be obtained by comparing the full survival curves of the treatment and control groups (Wright and Weinstein 1998), i.e., assessing the mean difference in effectiveness through measuring the area between the two survival curves; this is known as area under the curve (AUC) analysis.

[1]The terminology progression–free survival is used in advanced stages of cancer while disease–free survival is used in cases where surgery has left the patient disease free. The X-ACT study was concerned with adjuvant treatment post-surgery and therefore disease free survival was the outcome of interest. The remainder of this chapter will discuss disease free survival.

In order to undertake an area under the curves analysis, the full survival curves are required and therefore it may be necessary to extrapolate the trial data over the longer term. Ideally all trial patients would be followed up until death in order to have complete time to event data, however, in reality this is impractical, particularly in trials where cure is likely or patients are expected to have a considerable life expectancy beyond the last point of follow-up. Therefore, in order to estimate outcomes for the AUC analysis it is necessary to make lifetime predictions beyond the follow up period to assess the mean difference in survival outcomes. Alternatively the "restricted means" may be used, in which case the longest point of follow-up is used to reflect the restricted mean difference in survival, such that no extrapolation beyond this point is necessary.

6.2.2 Extrapolation

When modeling the survivor function there are a range of probability distributions which can be used, depending on whether a parametric or nonparametric approach is adopted. A parametric approach assumes a skewed distribution for the data which is characterized by a set of parameters, whereas a non-parametric analysis assumes no distribution.[2] Examples of common parametric distributions for modeling survival data include the Exponential, Weibull, Gompertz, gamma, log-normal, and log-logistic distributions. Each of these alternative distributions makes different assumptions about the data and baseline hazard, and therefore each will provide a slightly different survival curve. Careful consideration needs to be given to the choice of distribution; while it is important to ensure the model fits with the actual data, when the aim of the model is to extrapolate beyond the data, it is also critical to consider the logic of the extrapolated model. This can be done by visually checking the estimated survival curves against the Kaplan–Meier curves but also projecting beyond the period of data observation to check the logic of the extrapolated curves. Figure 6.1 below demonstrates this point by using the X-ACT study data to illustrate the observed DFS data in the Kaplan–Meier curves and the extrapolated DFS curves for the treatment (Xeloda) and control (5-FU/LV) arms.

This figure shows that study data meet logical expectations, reflecting a falling probability of DFS over time.[3] The two extrapolated DFS curves appear to fit well with the observed data and they do not cross at any point. The DFS curves indicate that the treatment arm (Xeloda) has slightly greater DFS than the control (5-FU/LV)

[2]Non-parametric approaches will however assume multivariate normality on the log scale for the distribution of coefficients and for this reason the Cox proportional hazards model is often referred to as semi-parametric.

[3]While the survival curve slopes downwards by it very nature indicating a falling probability of survival over time; the corresponding hazard curves will slope upwards indicating an increasing hazard or probability of death over time.

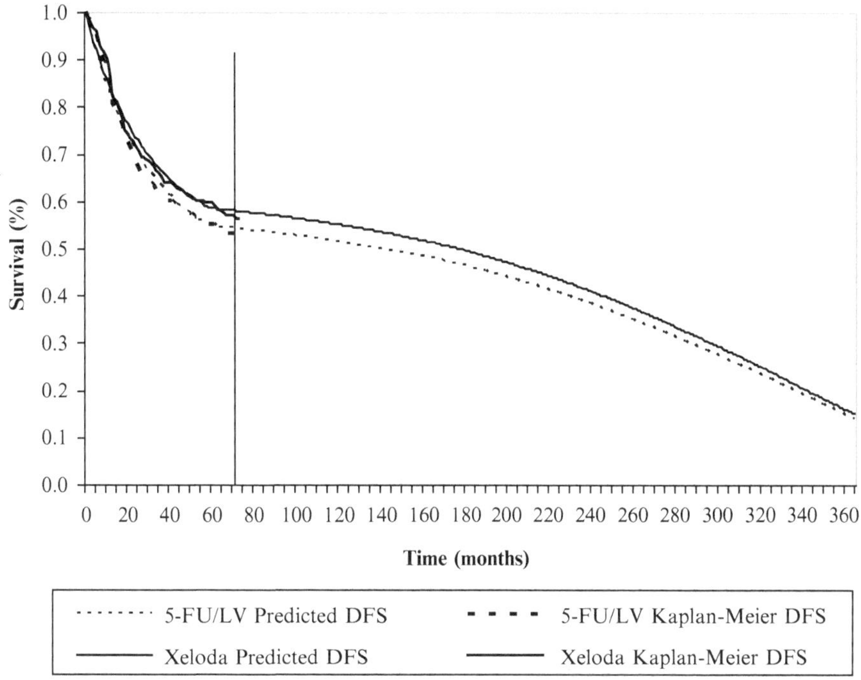

Fig. 6.1 Kaplan–Meier and predicted disease-free survival curves for Xeloda (treatment) and 5-FU/LV (control)

in both the observed and extrapolated curves. The mean difference in DFS between the treatment and control arms is represented by the area between their respective extrapolated DFS curves.

If there is more than one outcome of interest, for example, when both overall survival (OS) and DFS are under study, it is also wise to check the extrapolated curves for both outcomes on the same plot, to ensure that the shape of the curves meet logical expectations. It is a logical impossibility that the DFS curve would cross or overtake the OS curve; however, if an inappropriate distribution is used to extrapolate the data, this situation may arise. A careless fitting of separate models to each of the data from the X-ACT study has been used to demonstrate such as situation in Fig. 6.2.

This diagram illustrates the DFS curve crossing then overtaking the OS curve for both treatment arms, which is a logical impossibility. In such a situation, a better fitting distribution would be required. Therefore, plotting the extrapolated curves for both outcomes of interest, for each alternative distribution, along with assessing the goodness of fit statistics is a means of ensuring that the most appropriate distribution is chosen. In practice it may be necessary to sacrifice model fit during the period of observation for logical consistency of the extrapolated curves.

Further caution is required in undertaking extrapolation if the survivor functions exhibits "fat tails." A fat tail to a survival curve, particularly for overall survival, can

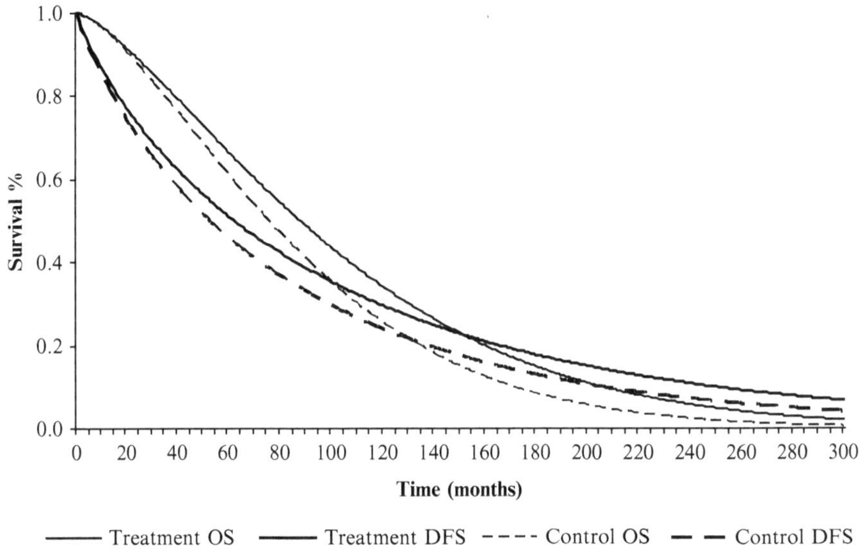

Fig. 6.2 Crossing curves: careless fitting of disease-free survival & overall survival

lead to a greater proportion of survivors than is likely in real life, which in turn can lead to an overestimate of the real survival. However, in a cause specific survival curve, fat tails may be acceptable, in situations that the treatment has provided a "cure."

In the X-ACT study, the follow-up data is truncated at 74 months, at which point the DFS curves for both treatments are exhibiting fat tails; that is they both have a high proportion of survivors. This can be seen in Fig. 6.1 where 56% of the patients in the Xeloda arm and 53% in the 5-FU/LV arm remain disease free and alive at 74 months, the last point of follow-up. Both the treatment and control disease free survivor curves have begun to level off, and it is likely that at this stage the treatments have provided a "cure,". However, this DFS curve is subject to all causes of mortality and therefore caution is required when choosing an appropriate distribution for extrapolating this curve as there is uncertainty around the future causes and risks of mortality.

6.2.3 Markov Modeling

The above discussion illustrates that despite the apparent simplicity of AUC analysis, some analyses are unlikely to be suitable when a high degree of extrapolation is required. In some situations a formal decision model is likely to be preferred.

Decision analytic modeling in health economics refers to a mathematical decision-making tool which structures evidence on clinical and economic outcomes

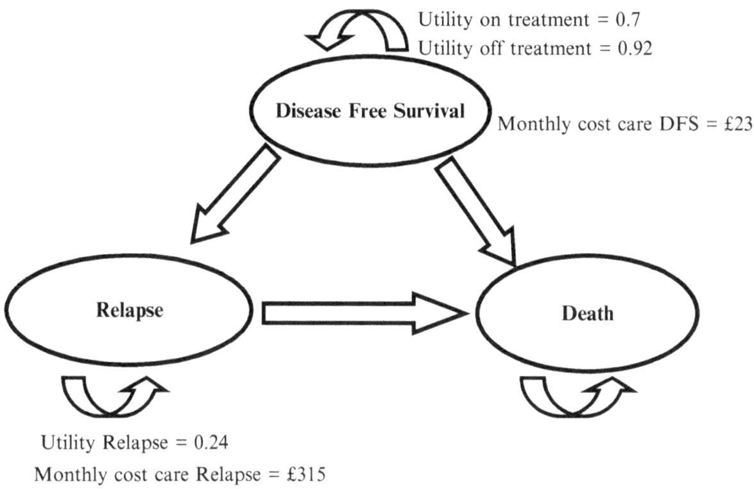

Fig. 6.3 X-ACT study three state Markov model

to inform health care resource use and clinical practices (Weinstein et al. 2003). Models can take many forms but have the common structure of synthesizing evidence on health outcomes and costs from a variety of sources. Models can vary substantially with regard to structure and complexity, but will typically take the form of a decision tree or Markov analysis. A decision tree is ideal for short-term analyses while Markov modeling is more appropriate for longer-term analyses such as those which involve progression of a disease over time (Briggs and Sculpher 1998; Sonnenberg and Beck 1993).

A Markov model assumes a number of specified health states, for which a patient can only be in one at any given time, and the transitions between these states are modeled over time. Where appropriate, each state can be assigned a cost and utility (quality of life) value which is experienced by the patient for the duration of time spent in that state. The time horizon of the model is split into discrete cycles, each of which involves a transition based on assigned probabilities or risks. Therefore Markov modeling is an ideal method for establishing life expectancies. By incorporating an absorbing "death" state to the model and utilizing disease hazard rates and background hazard rates (age- and sex-adjusted population mortality rates) along with appropriate health state risks, the lifetime events experienced by a cohort of patients can be modeled to determine life expectancies along with the costs incurred and utilities experienced for a specific treatment or intervention.

Figure 6.3 illustrates the Markov model developed in the X-ACT study, modeling disease progression over a lifetime horizon. The model has three states; DFS, relapse, and death. At the beginning of the model all patients will be in the DFS state, having undergone surgery followed by adjuvant treatment for the disease. In the following cycles patients can remain in this state, experience a death directly from the DFS state, experience a relapse, or experience death from the relapsed state. The movement between these states is dependent on transition probabilities, which

Table 6.1 X–ACT study baseline model parameters

Model parameters	Mean value	Standard error	Distribution
Deterministic parameters			
Time horizon: lifetime (years)	30		
Mean age	60		
Discount rate: Costs and QALYs (%)	3.5		
Time spent on treatment (months)	6		
Probabilistic parameters			
Utility DFS on treatment	0.70	0.036	Normal
Utility DFS off treatment	0.92	0.05	Normal
Utility relapse	0.24	0.041	Normal
Monthly cost care DFS[a]	£23.00	£4.67	Normal
Monthly cost care Relapse[a]	£315.00	£63.05	Normal
Weibull AFTmodel parameters[b]			
Constant	4.92	0.07	Multivariate normal[c]
Treatment effect[d] (DFS)	−0.16	0.09	Multivariate normal[c]
Scale	1.28	0.04	Multivariate normal[c]
Risk death (from DFS)[e]	21.92	0.30	Normal

[a]Arbitrary standard error set at 20% of the mean value
[b]Accelerated Failure Time Model
[c]Correlations taken into account using Cholesky decomposition
[d]AFT model, treatment effect applied to the control (5-FU/LV), thus taking a negative value
[e]Calculated using exponential model, death rate equivalent to 1/mean value

are applied using estimates such as age- and sex-adjusted background mortality rates, the probability of relapse, the probability of death directly from the DFS state, and probability of death post-relapse. These are ideally obtained from a variety of sources, including a prospective trial.

Survival data directly from a trial can be used to inform the transition probabilities in a Markov model. AUC analysis can be used to independently derive the transition probability from a DFS state to a relapse state and from DFS to a death state; however it does not provide any information regarding transition from the relapse state to death and other non-disease-related risk of death. Markov modeling can be used to formally link these three transitions by incorporating background evidence from other sources.

The X-ACT model derived transition probabilities for the DFS and relapse health states directly from the trial data, using an accelerated failure time (AFT) model[4] (Armitage et al. 2007), while the estimate of overall survival was calculated from the Markov process. A constant mortality rate was used to represent the risk of death from the relapse state; while the risk of death directly from DFS was represented by mortality rates from UK age- and sex-specific life tables. Table 6.1 details the

[4]An accelerated failure time (AFT) model is a parametric alternative to the proportional hazards model. The AFT model assumes that the covariates affect the time scale, either accelerating or decelerating time to failure, as opposed to the proportional hazards model which assumes that the covariates have a multiplicative effect on the hazard function.

parameters used in the baseline model. Each cycle in the model represents 1 month, with a duration of 30 years to incorporate a lifetime time horizon. The mean population age for the analysis was 60 years.

A discount rate of 3.5% was applied to both costs and outcomes in compliance with NICE guidance (NICE 2008). The monthly costs of care for the DFS state and relapse state were £23 and £315, respectively, and therefore the cost of care incurred in the analysis for each treatment arm is dependent upon the time duration patients spent in each of these states.

The integration of quality of life aspects to the analysis can be accomplished in a variety of ways, but in economic evaluations it is most commonly incorporated by using quality adjusted life years (QALYs) as the outcome measure. If a CEA is undertaken based purely on a clinical trial, then utility estimates can be derived directly from the study population if the trial is designed accordingly, enabling the incorporation of both the cost and quality of life attributable to adverse events. Alternatively, the Q-TWiST (quality-adjusted time without symptoms and toxicity) approach (Glasziou et al. 1990), which is an expansion of the area under the curve analysis, can be used. The Q-TWiST approach is a partitioned AUC analysis in which overall survival from the AUC is scaled down by partitioning the survival curve into sections for alternative health states and attributing a low quality of life value to survival during periods of toxicity or symptoms and higher values to the duration of survival without toxicity or symptoms. In this way the quality of life experienced by the patient is incorporated to the life expectancy measure, which is of particular use in oncology treatment economic evaluation studies. Providing that the quality of life part of the Q-TWiST corresponds to a cardinal utility measure with 1 representing full health and 0 representing death then the Q-TWiST approach corresponds to the AUC and quality adjusted survival analysis (Billingham et al. 1999).

In the X-ACT study utilities were applied to each of the model states to represent the quality of life experienced by patients in the analysis. The utility values scale down the survival time of a patient according to the quality of life they experience throughout the model, and in this way QALYs as an outcome measure simultaneously adjust any gains in life expectancy with the quality of life experienced. Ideally, when trial information can be collected prospectively, quality of life values for each of the model states would also be derived, however, the X-ACT trial did not collect any trial-specific information on utilities, relying instead on values derived from published literature sources (NICE 2005; Ramsey et al. 2000) for the Markov health states, as detailed in Fig. 6.3 and Table 6.1. A value of 0.7 was assigned to the DFS state during the treatment period, while a value of 0.92 was assigned to the same state off-treatment, reflecting the difference in quality of life experienced by patients during and after the chemotherapy treatment. Time on treatment was treated deterministically and set at 6 months for both treatments. A value of 0.24 was assigned to reflect the low quality of life in the relapse state and the absorbing state death has a value of 0.

6.2.4 Averse Events in Oncology Trials

Oncology trials typically have numerous outcomes of interest such as overall survival, DFS, and toxicity experienced during treatment. Using separate outcomes for cost–effectiveness analyses may not accurately reflect the best treatment option for the patient due to the trade-off between quality of life and duration of life associated with oncology treatments, and therefore it is important to include the impact of adverse events as well as treatment-related effects on disease free and overall survival.

Information on the types of and probability of experiencing adverse events under both treatment arms were derived directly from the trial and used in the economic evaluation to account for the cost of treating adverse events and the difference in events experienced between the treatment arms. Table 6.2 lists the adverse events, the point estimate probability of experiencing them in each arm of the trial and the mean unit cost of treatment for each event, while also detailing the corresponding mean cost for each adverse event for each treatment arm and the difference between them. The total cost of adverse events for each arm and the corresponding uncertainty intervals are also detailed. The individual costs detailed in the Table 6.2 are deterministic point estimates, derived from the point estimate probability of experiencing a specific adverse event, while the total adverse event costs and surrounding uncertainty intervals for each treatment arm are derived from the probabilistic analysis[5], which is discussed in more detail in the next section.

Table 6.2 shows that there is a mean cost difference of £34 incurred by Xeloda for the adverse events; however, the uncertainty intervals indicate that there is considerable variability surrounding this difference. It would be inappropriate to conclude that Xeloda has a higher mean cost of approximately £34 associated with adverse events in comparison to 5-FU/LV, as the uncertainty intervals indicate that the difference ranges between - £166 and £244, in which case it is highly likely that 5-FU/LV will incur the greater costs when the difference takes a negative value.

Ideally, when trial information can be collected prospectively, quality of life values for each of the adverse events would also be derived; however, as previously discussed the X-ACT trial did not collect any information on utility values for the specific adverse events experienced, and therefore utilized secondary evidence to derive appropriate state-specific utilities for the Markov model. Utilizing key information available from a clinical trial which can then be supported with wider evidence from secondary sources results in a robust cost–effectiveness model which can incorporate all relevant endpoints.

[5]Therefore, the sum of difference of the individual means does not equate to the mean difference between the two treatment arms total costs.

Table 6.2 X-ACT study adverse events(AE)

Treatment related adverse events	Xeloda AE probability	5-FU/LV AE probability	Mean unit costs of treatment	Xeloda AE mean cose	5-FU/LV AE mean cost	Difference
Grade 3 Febrile Neutropenia	0.005	0.017	£ 2,450	£13	£41	−£28
Grade 3 Neutropenia (low)	0.005	0.049	£2,327	£11	£115	−£104
Grade 3 Stomatitis (all)	0.013	0.140	£3,385	£43	£474	−£431
Grade 3 Abdominal pain	0.025	0.016	£3,385	£83	£55	£28
Grade 3 Diarrhoea	0. 116	0.136	£2,237	£271	£317	−£47
Grade 4 Diarrhoea	0.015	0.009	£2,327	£34	£21	£13
Grade 3 Nausea	0.022	0.015	£3,385	£74	£52	£22
Grade 3 Vomiting	0.023	0.018	£3,385	£77	£59	£17
Grade 3 Asthenia (poor)	0.011	0.010	£3,385	£38	£34	£4
Grade 3 Fatigue	0.006	0.011	£3,385	£19	£38	−£18
Grade 3 Hyperbiliru-binaemia	0.020	0.004	£3,385	£67	£13	£54
Grade 3 Palmar-plantar	0.176	0.006	£3,385	£596	£19	£576
			Totals	£1,232	£1,198	£34
			Uncertainty intervals	(1,035,1447)	(988,1448)	(−166,244)

6.2.5 Uncertainty and Probabilistic Sensitivity Analysis

The use of statistical methods for handling uncertainty in patient data has increased alongside the growing prominence of clinical trials used as a framework for economic evaluations; however statistical methods can also be used in dealing with uncertainty when a synthesis and modeling approach has been adopted. (Briggs 2001) considers how to deal with uncertainty for economic evaluation in clinical trials and in modeling approaches. Four key sources of uncertainty are identified: methodological uncertainty (which methods are the most appropriate to use); uncertainty in the study parameters or sampling variation (resource use, health outcomes, and quality of life data); uncertainty surrounding the extrapolation of observed results over time; and uncertainty regarding the generalizability of results.

Methodological uncertainties can be dealt with through the use of a benchmark approach or reference case for undertaking CEAs, while generalizability of results can be dealt with through sensitivity analysis, regardless of whether the CEA is based upon a single trial or a synthesis and modeling approach. However, dealing with uncertainties due to sampling variation (directly from the trial data or from the parameter estimates in a model) and overall model uncertainty (form of modeling approach) should be dealt with differently depending on the evaluation framework adopted.

Evaluations utilizing prospective data from a trial can use standard statistical techniques to obtain point estimates of relevant parameters directly from the trial data and the 95% confidence intervals to characterize the uncertainty surrounding them. To address issues with parameter uncertainty when a data synthesis and modeling approach have been adopted, probabilistic sensitivity analysis should be undertaken. Parameters are random variables which can take a range of values by applying an appropriate distribution to each of the parameter estimates. A Bayesian approach to the decision of appropriate distribution is recommended (Briggs 2001; Ades et al. 2006) whereby the choice of distribution is based on prior beliefs about the parameter. For example, in the X-ACT study, the probabilities for experiencing specific adverse events lie between values of 0 and 1, and therefore based on this knowledge it is appropriate to assume a Beta distribution, which is also bounded on the 0–1 interval. Once distributions have been applied to each of the appropriate parameters, a probabilistic sensitivity analysis can be performed, sampling values from each of these distributions simultaneously to provide different ICERs for each vector of input parameters. Repeated simulations from the distributions of the parameters in the decision model will result in a range of ICERs, from which confidence intervals can be derived. The range of ICERs can also be plotted on a cost–effectiveness plane. Table 6.3 presents the X-ACT study cost and effectiveness outcomes and the probabilistic uncertainty intervals surrounding them.

There are three main cost areas in this analysis comprising of treatment costs, the cost of side-effects or adverse events and the cost of care associated with increases in life expectancy. Table 6.3 shows that while Xeloda has a much higher cost of treatment than 5-FU/LV, the associated costs of administering the treatment are

Table 6.3 X-ACT study cost and effectiveness results

Outcome	Xeloda	5-FU/LV	Incremental difference	Uncertainty intervals around the difference
Mean cost treatment[a]	£2,262	£534	£1,728	(963, 2652)
Mean cost administering treatment[a]	£378	£4,611	-£4,232	(-6261, -2732)
Mean cost adverse events	£1,232	£1,198	£34	(-166, 244)
Mean cost of DFS	£985	£946	£39	(31, 41)
Mean cost relapse	£121	£132	-£12	(-17, -7)
Total cost	£4,977	£7,421	-£2,443	(-4318, -724)
Mean LYG in relapse	0.55	0.61	−0.06	(-0.06, -0.05)
Mean LYG DFS	3.40	3.27	0.13	(0.12, 0.14)
Total LYG	3.95	3.87	0.07	(0.07, 0.08)
Mean QALYs in relapse	0.14	0.16	−0.02	(-0.02, -0.01)
Mean QALYs in DFs	3.02	2.90	0.12	(0.10, 0.13)
Total QALYs gained	3.17	3.06	0.11	(0.08, 0.12)

[a]Arbitrary standard error set at 20% of the mean value

relatively low, while the opposite is the case for 5-FU/LV, resulting in a much lower overall cost of treatment for Xeloda. The mean difference in costs of adverse events experienced is negligible between the two treatment arms, and highly variable as discussed in Table 6.2. The costs of care associated with improvements in life expectancy incur in the DFS and relapse health states of the Markov model and Table 6.3 shows that while Xeloda incurs slightly greater costs in the DFS state, 5-FU/LV incurs slightly greater costs in the relapse state. The overall difference between the two arms shows a cost saving of approximately £2,500 for Xeloda in comparison to 5-FU/LV. This is almost entirely due to cost savings in the administration of treatment. The probabilistic analysis indicated some uncertainty around this total cost saving estimate, ranging between a saving of £4,318 at the lower limit and £724 at the upper limit.

Overall the treatment arm, Xeloda, resulted in an improvement of 0.07 life years in comparison to the 5-FU/LV arm. Once adjusted for the quality of life experienced in each of the model states, the QALY outcomes show a gain of 0.11 QALYs in favor of Xeloda, which is equivalent to an improvement of 1.32 quality-adjusted life months. The uncertainty intervals surrounding the life years and QALY outcomes show a little variability in these estimates.

Overall the cost–effectiveness results from the X-ACT study show that Xeloda is likely to be cost saving in comparison to 5-FU/LV, in the region of £2,500 saved; however there is very little difference between the two treatments in terms of QALY gains, with an improvement of 0.11 QALYs. Figure 6.4 illustrates the X-ACT study cost–effectiveness plane which plots the range of values for the cost–effectiveness outcomes generated in the probabilistic analysis.

The range of outcomes generated in the simulation all fall within the southeastern quadrant, indicating certainty in the conclusion that Xeloda is less costly and slightly more effective than the control, 5-FU/LV. The uncertainty surrounding the mean

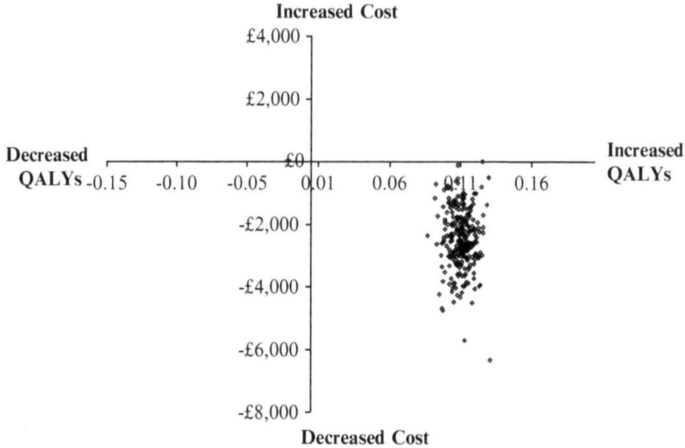

Fig. 6.4 Cost–effectiveness plane for X-ACT probabilistic analysis results

cost-saving estimate of £2,500 can be seen quite clearly, in comparison to the very small uncertainty surrounding the mean QALY gain estimate. If the ICER outcomes generated in the simulation cross over any of the quadrant boundaries, this would be an indication of uncertainty as to whether the treatment is cost-effective. Although the difference in QALY gains is very slight and close to zero, and the difference in cost-saving ranges between £4,000 and £700, none of the simulated values fall on or cross the boundaries and therefore Xeloda was found to be the dominant strategy, saving money in comparison to 5-FU/LV while offering a small improvement in effectiveness.

The results from a probabilistic analysis can also be used to derive a cost effectiveness acceptability curve (CEAC) which represents the probability that the intervention is cost-effective at different monetary thresholds of willingness to pay (Drummond et al. 2007). CEACs are also a convenient way to present uncertainty when there are more than two alternative interventions under consideration, which is often the case. In the case of the X-ACT study, the CEAC would in effect be a straight line, rather than a curve, indicating a 100% probability of being cost-effective at any monetary thresholds, as it is a cost-saving dominant strategy.

As Sculpher et al. (2006) propose, a mixed evaluation approach utilizing prospective study data and other sources of evidence will provide a robust economic analysis; however, it should also consider all relevant uncertainties surrounding the various components of the analysis, i.e., dealing with sampling variation, uncertainty in the extrapolation process and extrapolated results as well as parameter and model uncertainty. The XACT study design incorporated a mixed approach, while it was based alongside a clinical trial and largely informed by this; literature and other sources of information were synthesized in to the model. The analysis therefore required to deal with a range of uncertainties, particularly those surrounding extrapolation of the trial data (five alternative distributions were examined) and probabilistic sensitivity analysis was undertaken and a Monte Carlo simulation to address uncertainty in the parameters, the model structure, and the reported cost–effectiveness results.

6.2.6 Heterogeneity Analysis

Assessing heterogeneity for a CEA utilizing trial-based and secondary data sources requires consideration of various study population subgroups to assess whether other factors inherent to these groups influence the study outcomes by confounding or overriding the actual treatment effects. The UK decision-making body for NHS support and reimbursement of health technologies, the National Institute for Health and Clinical Excellence (NICE), require cost–effectiveness evidence for their decision-making process and formally recognized the importance of heterogeneity and subgroup analysis in their 2004 Health Technology Appraisal guidance. More recently this guidance was revised and now specifically recommends subgroup analysis in all technology appraisals submitted to themselves for consideration (NICE 2008).

There are three main areas of heterogeneity identified by the NICE guide to methods of technology appraisal (NICE 2008) which should be considered in a cost–effectiveness evaluation: differences that arise due to variations in baseline risk, predictive factors which affect the effectiveness outcomes, and variations in post-treatment response.

Where the relative treatment effect is constant (e.g., a hazard ratio or relative risk) then the absolute treatment effect of an intervention is dependent on the underlying baseline risk. When extrapolating observed trial data to estimate overall survival outcomes there are two factors contributing to the overall risk of death: the disease-specific death rate which will initially be constant and then over time will possibly decrease and the non-disease-related risk of death which will increase, for example, as the population ages. The X-ACT study analysis adjusted for differences in age and sex mortality rates to account for variable background hazard.

Another area of potential heterogeneity which may interfere with the treatment effectiveness is predictive factors in certain patient populations. The capacity to benefit from treatment may differ between patient populations, such as socioeconomic groups, and in the case of oncology treatments patients who suffer from other comorbidities may respond differently to treatment.

It is critical that these two different areas or subgroups of potential heterogeneity should be identified and assessed in economic evaluations of oncology treatments, as they cannot only influence the clinical effectiveness of the treatments under study but also impact on the cost of treatment and subsequent cost of care. Therefore they potentially have a large impact on the overall cost–effectiveness estimates of specific treatments and can lead to misleading results if they are not identified and appropriately dealt with.

A third important heterogeneity issue is the post-treatment response of different patients groups and how they are dealt with. Clinical and cost–effectiveness results can be manipulated if the trial exploits early exclusion of patients from the analysis due to poor post-treatment response. Therefore the trial and cost–effectiveness study should use effectiveness data based on an intention to teat analysis, to provide rigor to the analysis and conservative results. Avoiding manipulation of the clinical and cost–effectiveness outcomes gives a true reflection of cost–effectiveness. Cost–effectiveness outcomes can still be manipulated, however, for example when drug companies renegotiate the price of drugs in response to disappointing cost–effectiveness results. For example, a NICE review in 2008 found the cancer treatment Sutent to be not cost-effective. This decision was followed by media and cancer charity outrage as the drug can improve life expectancy by a few months in terminally ill patients. In response the manufacturers renegotiated a discount on the cost of the drug with the NHS, making it more affordable. In light of this NICE revised their stance in February 2009 and approved the use of Sutent in those kidney cancer patients that are deemed terminally ill.

NICE have explicitly addressed heterogeneity issues with post-treatment response, in their 2008 guide to methods of technology appraisal, specifying that treatment continuation groups should not be considered as a separate subgroup for the baseline CEA. Inappropriate handling of post-treatment response groups can lead to misleading results.

6.3 Summary

This chapter has considered some of the statistical issues of importance that arise when prospective trial data is used to inform economic models utilizing full evidence synthesis in the evaluation of oncology treatments. Adopting an approach whereby the clinical trial is a major informant of the model inputs with supplementary data from secondary sources will promote a more robust analysis; however, it also requires careful consideration of the methods utilized in the trial and appropriate handling of them to synthesize the model. Effectiveness data can be derived directly from trials and is often summarized by presenting differences in median survival times between treatment arms; however, this approach is restrictive and one-dimensional and therefore it is more appropriate to use area under the curve analysis which considers the mean differences in effectiveness based on the entire survival curves. In undertaking an AUC analysis issues with extrapolation become of critical importance and consideration must be given to the appropriate choice of distribution, bearing in mind potential pit falls when the data is subject to fat tails and/or short follow-up durations coupled with low disease risk. Markov modeling approaches were discussed as appropriate methods for the synthesis of the trial data with wider secondary sources. Dealing with adverse events in an economic model is of particular relevance to oncology studies which are typically subject to a trade-off between quality of life and duration of life, due to the toxicity levels associated with cancer treatments. There are various means of dealing with this; although the standard QALY approach is the most commonly utilized technique, other methods such as Q-TWiST hold considerable value. Methods for dealing with uncertainty in a mixed data synthesis and trial-based approach should consider all relevant uncertainties surrounding the various components of the analysis, such as dealing with sampling variation, uncertainty in the extrapolation process and extrapolated results as well as parameter and model uncertainty. With regard to heterogeneity there are three main areas that should be given particular consideration, heterogeneity in varying baseline risks, predictive factors which affect mortality rates, and variations in post-treatment response which can potentially manipulate the data.

References

Ades AE, Sculpher M, Sutton A, Abrams K, Cooper N, Welton N, Lu G (2006) Bayesian methods for evidence synthesis in cost-effectiveness analysis. Pharmacoeconomics 24:1–19

Armitage P, Berry G, Matthews J (2007) Statistical methods in medical research. 4th edn. Blackwell, UK

Billingham L, Abrams K, Jones D (1999) Methods for the analysis of quality of life and survival data in health technology assessment. Health Tech Assess 3(10):1–152

Briggs AH (2001) Handling uncertainty in economic evaluations and presenting the results. In: Drummond M, McGuire A (eds) Economic evaluation in health care: merging theory with practice. Oxford University Press, Oxford, p 172–214

Briggs AH, Sculpher M (1998) An introduction to Markov modeling for economic evaluation. Pharmacoeconomics 13(4):397–409

Cassidy J, Douillard JY, Twelves C, McKendrick JJ, Scheithauer W, Bustova I, Johnston PG, Lesniewski-Kmak K, Jelic S, Fountzilas G, Coxon F, Diaz-Rubio E, Maughan TS, Malzyner A, Bertetto O, Beham A, Figer A, Dufour P, Patel KK, Cowell W, Garrison L.P (2006) Pharmacoeconomic analysis of adjuvant oral capecitabine vs intravenous 5-FU/LV in Dukes' C colon cancer: the X-ACT trial. Br J Canc 94:1122–1129

Drummond M, Sculpher M, Torrance G, O'Brien B, Stoddart G (2007) Methods for the economic evaluation of health care programmes. 3rd edn. Oxford University Press, Oxford

Gill S, Sargent D (2006) End points for adjuvant therapy trials: has the time come to accept disease-free survival as a surrogate end point for overall survival? The Oncologist 11:624–629

Glasziou P, Simes R, Gelber R (1990) Quality adjusted survival analysis. Stat Med 9:1259–1276

Kuntz K, Weinstein M (2001) Modelling in economic evaluation. In: Drummond M, McGuire A (eds) Economic evaluation in health care: merging theory with practice. Oxford University Press, Oxford, p 141–171

National Institute for Health & Clinical Excellence (2005) Technology Assessment Report: The use of oxaliplatin and capecitabine for the adjuvant treatment of colon. UK: National Institute for Health & Clinical Excellence. Available online at: http://www.nice.org.uk/nicemedia/pdf/Assessment_Report_(CiC_removed).pdf

National Institute for Health & Clinical Excellence (2008) Guide to Methods of Technology Appraisal. UK: National Institute for Health & Clinical Excellence. Available online at: http://www.nice.org.uk/media/B52/A7/TAMethodsGuideUpdatedJune2008.pdf

Ramsey S, Andreson MR, ETzioni R, Moinpour C, Peacock S, Potosky A, Urban N (2000) Quality of life in survivors of colorectal carcinoma. Cancer 88:1294–1303

Sculpher M, Claxton K, Drummond M, McCabe C (2006) Whither trial-based economic evaluation for health care decision making? Health Econ 15:677–687

Sonnenberg F, Beck RJ (1993) Markov models in medical decision making: a practical guide. Med Decis Making 13:332–338

Twelves C, Wong A, Nowacki M, Abt M, Burris H, Carrato A, Cassidy J (2005) Capecitabine as adjuvant treatment for stage III colon cancer. New Engl J Med 352:2696–2704

Weinstein M, O'Brien B, Hornberger J, Jackson J, Johannesson M, McCabe C, Luce B (2003) Principles of good practice for decision analytic modeling in health-care evaluation: report of the ISPOR task force on good research practices-modeling studies. Value Health 6:9–17

Wright J, Weinstein M (1998) Gains in life expectancy from medical interventions: standardizing data on outcomes. New Engl J Med 339:380–386

Chapter 7
Active-Controlled Clinical Trials*

Yi Tsong and Joanne Zhang

7.1 Introduction

Active control treatment plays an important role in the development of pharmaceutical products. In the setting of evaluating a generic drug product, the innovative drug product serves as an active control reference (Schuirmann 1987) with the objective to demonstrate that the test treatment is equivalent in content and rate of absorption to the active control innovative reference. For nonsystemic generic drug products, the clinical trials are often designed with three arms (placebo, active control, and test treatments), with the innovative product serving as the active control for the therapeutic equivalence of the generic product (Tsong et al. 2004). In the setting of assessing the abuse potential of a test drug product, clinical trials are often designed with three or more arms (placebo, multiple dose of positive active control, and test treatments), the test treatment is compared with the positive control treatment and placebo to show that the test treatment is lack of abuse potential. The trial results are validated by showing the positive control has abuse potential in comparison to placebo (Chen and Tsong 2007). In the clinical trial setting of assessing treatment-induced prolongation of QTc interval, the positive active control treatment is included to validate that the clinical trial has enough power to distinguish the effect on QTc prolongation (ICH 2005; Zhang and Machado 2008; Zhang et al. 2008b; Tsong et al. 2008). The active control treatment has been used in other trials including the "gold" clinical trial, in which the active control treatment

*This chapter does not represent the official position of the Food and Drug Administration.

Y. Tsong (✉) • J. Zhang
Division of Biometrics VI, Office of Biostatistics/Office of Translational Sciences, CDER, 10903 New Hampshire Avenue, Silver Spring, MD 20993–0002, USA
e-mail: yi.tsong@fda.hhs.gov; joanne.zhang@fda.hhs.gov

W. Tang and X. Tu (eds.), *Modern Clinical Trial Analysis*, Applied Bioinformatics and Biostatistics in Cancer Research, DOI 10.1007/978-1-4614-4322-3_7,

serves as the reference in evaluating the efficacy of the test treatment (adjusted for placebo effect) (Hauschke et al. 1999; Pigeot et al. 2003). In evaluating the efficacy of a new treatment in a clinical trial, treating patients with placebo is sometimes considered unethical (Ellenberg and Temple 2000; Temple and Ellenberg 2000; CHMP 2005), and the clinical trial is often designed with the test treatment and a carefully selected active control treatment, but without placebo arm. Although it is not included in the two-treatment active-controlled clinical trial, placebo plays an important role in trial design, hypothesis setting, data analysis, and even in the interpretation of the test results (Tsong et al. 1999; Holmgren 1999; Snapinn 2004).

Depending on the objectives of the application, the study design, the hypothesis to test, the method of analysis, the assumptions, and the requirements of the active control treatment, the decision rule and the result interpretation are different. In this chapter, we will briefly discuss the objective, design, analysis, and interpretation of the active control clinical trials in the various areas of application including bioequivalence trial for generic drug evaluation, clinical trials with three arms (test treatment, active control treatment and placebo) and with two arms (without placebo arm) trials for efficacy.

7.2 Active Control Used in Generic Drug Assessment

In order to evaluate a new generic drug product, bioequivalence clinical trial is often designed to demonstrate that the generic product (i.e., the test treatment, T) is equivalent to the innovative product (i.e., the active control treatment, C) in its content and rate of absorption in the patient's blood system. The primary endpoints of the bioequivalence trials are the pharmacokinetic measures such as area under the curve (AUC), maximum blood concentration of the drug ($C\max$), and time to $C\max$ ($T\max$). Drug T is shown to be bioequivalent to the active control drug C if the geometric mean value μ_T of the measures for T is within 80 % to 125 % of the geometric mean μ_C of C, i.e., $80\,\% < \mu_T/\mu_C < 125\%$. With the assumption that the PK measurement of each treatment group is distributed as lognormal, this requirement is $\ln(80) < \ln(\mu_T) - \ln(\mu_C) < \ln(1.25)$. In another word, it is required to test the null hypothesis $\mathrm{H}_0 : \ln(\mu_T) - \ln(\mu_C) \le -0.223$ or $\ln(\mu_T) - \ln(\mu_C) \ge 0.223$ against $\mathrm{H}_a : -0.223 < \ln(\mu_T) - \ln(\mu_C) < 0.223$. In vivo bioequivalence establishes if H_0 is rejected for all three measures. Rejecting H_0 can be shown if the lower limit of the 90 % confidence interval of $\ln(\mu_T) - \ln(\mu_C) > -0.223$ and the upper limit <0.223.

These clinical trials are designed by randomizing healthy subjects to receive one of the two crossover sequences of the treatments in a crossover trial setting of two treatments/two periods. Subjects receiving the first sequence would receive the new test generic drug T in the first period and then followed by washout period before receiving the active control innovative drug C. Subjects receiving the second treatment sequence would receive T and C in an reversed order.

Schuirmann (1987) proved that by showing the 90 % confidence interval of $\ln(\mu_T)-\ln(\mu_C)$ lies between −0.223 and 0.223; we actually performed two significance tests and rejected the both null hypotheses $H_0 : \ln(\mu_T)-\ln(\mu_C) \leq -0.223$ and $H_0 : \ln(\mu_T)-\ln(\mu_C) \geq 0.223$ at a family-wise 5 % type I error rate. It is often called two one-sided 5 % tests (TOST).

Note that a generic drug product T is composed with the same active chemical ingredients as the innovative drug product C, the only difference is in its formulation. As long as by showing that the generic drug product T has about the same absorption rate and content in blood within the bioequivalence limits, it provides about the same deliverance and efficacy as the innovative drug product C. There is no additional requirement of showing treatment efficacy. For this reason, the type I error rate requirement for the one-sided test is 5 % which differs from new drug applications as described in the later sections.

For a nonsystemic drug product, clinical trial with therapeutic endpoint is required to demonstrate its equivalence to the innovative product. It is typically carried out with a double-blind randomized trial of three parallel arms. After the test generic drug T is shown to be more effective than placebo, i.e., $\mu_T > \mu_C$, the generic drug T is compared with the innovative drug C in order to show that they are equivalent in efficacy within the predetermined equivalence limits. Typically, when the outcome is continuous, μ_T and μ_C represent the population means of the populations of T and C. The test product T is shown to be equivalent to the innovative product C by rejecting the following null hypothesis:

$$H_0 : \mu_T/\mu_C \leq \gamma \text{ or } \mu_T/\mu_C \geq 1/\gamma$$

where γ is the a equivalence ratio value used for generic drug products. By rejecting the null hypothesis, the generic product T is shown to be $\gamma \cdot \mu_C < \mu_T < \mu_C/\gamma$. For most products, $\gamma = 80\%$ is a fixed value. The equivalence test can be performed indirectly using a confidence interval of μ_T/μ_C. The estimation method was described by Hauschke et al. (1999). Bioequivalence establishes when the lower confidence limit of the 90 % confidence interval of μ_T/μ_C is greater than 80 % and the upper confidence limit is smaller than 125 %. The significance test for the ratio of means can be difficult to derive. One may rewrite the null hypothesis as

$$H_0 : \mu_T-\gamma\mu_C \leq 0 \text{ or } \mu_T-(1/\gamma)\mu_C \geq 0$$

Hypothesis testing can be carried out with two one-sided tests, using a simple standardized test statistic in t-test or normal approximation z-test with proper estimation error (Temple and Ellenberg 2000).

Whereas, if the outcome is binary, let μ_T and μ_C represent the proportions of responders of the two drug products, δ be the equivalence limit, T is equivalent to C if it rejects the null hypothesis

$$H_0 : \mu_T-\mu_C \leq -\delta \text{ or } \mu_T-\mu_C \geq \delta$$

By rejecting H_0, it implies that $-\delta < \mu_T - \mu_C < \delta$. For most drug products, δ is defined either as 15 % or 20 %.

The active control innovative drug product C is also used to validate the trial by showing that it is more effective than placebo with the trial population. By rejecting all three null hypotheses, one demonstrates that the test generic product is effective, the trial is properly planned to validate that efficacy of the innovative product, and the test generic product is equivalent to the innovative product (Tsong et al. 2004). Based on the fact that validation of a known drug usually requires a smaller sample size and the concept of minimizing the number of patients receiving non-active treatment, it is desirable to have equivalence trial-designed complete validation before assessing bioequivalence in a hierarchical manner. Tsong et al. (2004) proposed an adaptive group sequential design to validate the trial at interim stage before testing for bioequivalence. The investigator will not spend any type I error rate on bioequivalence test at interim stage before the trial is validated. On the other hand, the adaptive design may reduce the number of patients receiving placebo because once the trial is validated, there is no need to recruit and treat patients with placebo.

7.3 Active Control Used in Risk Assessment with More than Two Treatments

Many risk assessment clinical trials with the objective of equivalence or non-inferiority are designed with more than two treatments. In this section, we describe two types of trials with specific objectives.

Sponsors of drug products composed of chemical substances that may be harmful to the central nervous system and with the potential to be abused are often required to conduct a clinical trial to assess the drug abuse potential. Such trials often consist of placebo P, test drug with low and high doses (LT and HT), and a selected scheduled positive control with low and high doses (LC and HC) (Chen and Tsong 2007). The trials are typically designed with multiple Latin squares of Williams design (Chen and Tsong 2007). Let μ_p, μ_{LT}, μ_{HT}, μ_{LC}, and μ_{HC} be the mean score of placebo, LT, HT, LC, and HC treatments. Typically a drug abuse potential study is designed with a crossover trial with repeated K-treatment, K-period Latin squares, where K is the number of total treatments. For the primary objective of drug abuse potential assessment, we need to first validate the trial by testing the following hypotheses:

$$H_0 : \mu_{CL} = \mu_P \text{ versus } H_a : \mu_{CL} > \mu_P$$

and

$$H_0 : \mu_{CH} = \mu_P \text{ versus } H_a : \mu_{CH} > \mu_P$$

The trial fails the validation if none of the null hypotheses is rejected. Once any of the positive control treatment, say CH, is validated by rejecting H_0, it is used to demonstrate that TH and TL have significantly less abuse potential than CH by testing

$$H_0 : \mu_{TL} = \mu_{CH} \text{ versus } H_a : \mu_{TL} < \mu_{CH}$$

and

$$H_0 : \mu_{TH} = \mu_{CH} \text{ versus } H_a : \mu_{TH} < \mu_{CH}$$

Test treatment T has less abuse potential than CH if both H_0s are rejected. In order to claim that T has no abuse potential, we need to test and reject both of the following null hypotheses:

$$H_0 : \mu_{TL} = \gamma\mu_P \text{ versus } H_a : \mu_{TL} < \gamma\mu_P$$

and

$$H_0 : \mu_{TH} = \gamma\mu_P \text{ versus } H_a : \mu_{TH} < \gamma\mu_P$$

where γ is a prespecified clinical meaningful margin. Each of the above hypotheses is tested with a type I error rate of 2.5 %. Note that the last hypotheses can be tested by comparing the t-statistics or other nonparametric statistic with the proper critical value or by comparing the lower limit of the confidence interval of μ_{TH}/μ_P as described in Sect. 7.2.

On an ECG tracing, the QT interval is measured from the beginning of Q wave to the end of T wave. QT interval reflects the duration of ventricular depolarization and subsequent repolarization. For some drugs, significant prolongation of the absolute QT interval has been associated with the precipitation of a potentially fatal cardiac arrhythmia called Torsades de Pointes (TdP) and can degenerate into ventricular fibrillation, leading to sudden cardiac death. The regulatory concern of QT prolongation of a test drug can be addressed by conducting a thorough QT clinical trial to address the issue prior to the phase 3 efficacy trials. A typical thorough QT trial is designed with placebo, one or two doses of test treatment and an active control treatment. Let μ_{T_k} and μ_{P_k} be the mean QTc (QT corrected by heart rate) change from baseline for the drug and placebo at time k; the test drug is claimed to be free of prolonging QTc interval if it rejects the following null hypothesis at all relevant time point:

$$H_0 : (\mu_{T_k} - \mu_{P_k}) \geq 10$$

versus

$$H_a : (\mu_{T_k} - \mu_{P_k}) < 10, k = 1 \text{ to } K$$

With the objective of the trial being to confirm that the response of the test treatment differs no more than a prespecified margin from the placebo, one may have the concern with the trial population whether the subjects may respond to an active treatment with a known prolongation effect of QT interval. For this reason, ICH E14 recommends that "The positive control should have an effect on the mean QT/QTc interval of about 5 ms (i.e., an effect that is close to the QT/QTc effect that represents the threshold of regulatory concern, around 5 ms)." Let μ_{C_k} and μ_{P_k} denote the mean QTc interval of active control and placebo at kth time point after adjusted for baseline measurement. The trial validation is accomplished if at least one of the following null hypotheses is rejected

$$\mathrm{H}_0(k) : \mu_{C_k} - \mu_{P_k} \leq 5$$

versus

$$\mathrm{H_a}(k) : \mu_{C_k} - \mu_{P_k} > 5 \text{ for } k = 1 \text{ to } K$$

For details of the design and analysis of a thorough QT/QTc trials, please refer to the chapter on thorough QT/QTc clinical trials. Both the prolongation and validation tests can be performed by comparing the t-statistic with the critical value at each time point or comparing the limit of the confidence interval of mean difference with the margin value at each time point with proper adjustment of type I error rate of 5 %.

7.4 Active Control Used in "Gold" Clinical Trials

For efficacy evaluation of a new drug product in certain treatments, e.g., Psychopharm products, CPMP and U.S. FDA recommended using the three parallel arm trials similar to the bioequivalence trials of generic nonsystemic drug products. Such design is also called "gold" clinical trial (Pigeot et al. 2003). For example, for psychopharm drug products, the active control treatment is often used only in comparison with the placebo arm to validate the trial. Test treatment is then compared with placebo in order to establish efficacy. In addition, CPMP recommends that the test treatment is approvable for marketing if it is not only superior to placebo, but shows also that its efficacy (adjusted for placebo) is clinically relevant. (Pigeot et al. 2003) The clinical relevants of the efficacy of the test treatment is demonstrated by comparing with the active control treatment. For example, it needs to show that the test treatment is no less than a given proportion γ (say, $\gamma = 80\%$) of the adjusted efficacy of the active control treatment. It is done by rejecting the following null hypothesis:

$$\mathrm{H}_0 : \mu_T - \mu_P \leq \gamma(\mu_C - \mu_P) \text{ versus } \mathrm{H_a} : \mu_T - \mu_P > \gamma(\mu_C - \mu_P), 0 \leq \gamma \leq 1$$

$$\text{i.e., } \mathrm{H}_0 : \mu_T - \gamma\mu_C - (1-\gamma)\mu_P \leq 0 \text{ versus } \mathrm{H_a} : \mu_T - \gamma\mu_C - (1-\gamma)\mu_P > 0$$

where γ is treated as a constant. These hypotheses involve three parameters, μ_T, μ_C, and μ_P and can be tested by comparing the conventional normalized statistic in the form of t statistic for normally distributed outcomes with the critical value of a t-distribution or compare a z-statistic for approximation test with the critical value of normal distribution. Note that by rejecting the above null hypothesis of non-inferiority, the products are shown $\mu_T - \gamma\mu_C - (1-\gamma)\mu_P > 0$. Since it is shown that $\mu_C > \mu_P$ and $\mu_T - \mu_P - \gamma(\mu_C - \mu_P) > 0$, it implies that $\mu_T - \mu_P > 0$ without specifically requiring that the test treatment is superior to placebo. As indicated in the Sects.7.3 and 7.4, the alternative approach is to estimate the ratio of the test efficacy (adjusted for placebo) and active control efficacy (adjusted for placebo) and compare the lower confidence limit with γ. The null hypothesis is rejected if the lower confidence limit is larger than γ. For details of the estimating method, please refer to Hauschke et al., 1999. Note that showing a test treatment is superior to placebo, although using a two-sided significance test of 5 % type I error rate against $H_0 : \mu_T = \mu_P$, it actually needs to show that $\mu_T > \mu_P$ using a significance test with a 2.5 % type I error rate. Now in order to imply efficacy of the test treatment over placebo (i.e., $\mu_T > \mu_P$) through preservation, one need also use a significance test with a 2.5 % type I error rate.

If it is of interest, data collected in this study can also be used to test if the test treatment is superior to the active control treatment by showing that $\mu_T > \mu_C$. This becomes a two-stage testing issue and there is a concern whether it will affect the overall type I error rate. We will have the issue discussed in later sections.

7.5 Active Control Used in Efficacy Trials Without Placebo Arm

The primary regulatory requirement for a test drug to be approved for marketing is to be able to demonstrate that the test drug is both effective and safe. Hence, the minimum requirement for efficacy is that the test treatment is superior to placebo in treating the disease or symptom. A significant non-inferiority/equivalence clinical trial designed without a placebo arm may sometimes be difficult to be interpreted to satisfy the minimum efficacy requirement. The objectives of non-inferiority hypotheses can be summarized into three types in general (Tsong et al. 2003). The first and most basic objective is to establish efficacy of the test treatment T (over placebo P). It means that it needs to show that the test treatment T is superior to placebo treatment P. But this is to be demonstrated in the clinical trials without P arm. In order to make such a conclusion, one needs to bridge the test-active control comparison to the historical information on the efficacy size (size of superiority over placebo P) of the active control treatment C in the current clinical trial. There are two ways to quantify the efficacy size of the active control treatment. These two approaches in fact represent the other two objectives used in non-inferiority efficacy trials.

The first approach is derived from the concept of historical control study. With this approach, we assume that the efficacy size is no less than δ, a margin that is considered to be medically meaningful and statistically valid. So if the test treatment T is not worse than the active control treatment C by a size equals to δ then it is almost certain that the test treatment T is superior to placebo P. This is often called generalized historical control approach (Tsong et al. 2003) or δ-margin approach

(Hung et al. 2003). The value δ is called the non-inferiority margin. This is easily illustrated with the following three inequalities on the means of the response:

$$\mu_T > \mu_C - \delta \text{ given } \mu_C - \mu_P > \delta \text{ implies that } \mu_T > \mu_C - \delta > \mu_P \quad (7.1)$$

Depending on the disease and symptom to be treated and the active control treatment C chosen, the margin is also determined with the medical judgment. There are situations that the non-inferiority margin is defined in order to be non-inferior to the active control treatment. In order to make sure that the above relationship holds, the value of the non-inferiority margin δ is specifically required in ICH (E9) Guidance so that "[The] margin is the largest difference that can be judged as being clinically acceptable and should be smaller than differences observed in superiority trials of the active comparator." It is further emphasized in ICH (E10) that "The margin chosen for a non-inferiority trial can not be greater than the smallest effect size that the active control drug would be reliably expected to have compared with placebo in the setting of the planned trial." "The determination of the margin in a non-inferiority trial is based on both statistical reasoning & clinical judgment, and should reflect uncertainties in the evidence on which the choice is based, and should be suitably conservative." However, such a margin may not exist for active control C. For example, for community-acquired pneumonia (CAP), most of the historical CAP trials were not to be considered adequate. First, the designs were not adequate: most of the trials are unblended, observational, or not randomized. Second, patients of the historical studies were hospitalized with pneumococcal/lobar pneumonia. Third, studies were evaluated with all-cause mortality. Fourth, information on the resolution of clinical symptoms or signs as measured in the current CAP studies was limited in historical trials. Hence, the effect size of active control treatment becomes not reproducible in current trials with CAP patients of less severe disease.

The second approach is to represent the non-inferiority margin as a specific percentage of the efficacy (over placebo) of the active control treatment C so that $\delta = (1-\lambda)(\mu_C - \mu_P)$, $0 \leq \lambda \leq 1$. By this definition, the objective of a non-inferiority trial is to show that $(\mu_T - \mu_P) > \lambda(\mu_C - \mu_P)$, (Holmgren 1999; Hung et al. 2003) i.e., the test treatment T retains at least $100\lambda\%$ of the effect of the active control treatment C. By moving the parameters around, one needs to test the following hypotheses instead:

$$\mathrm{H}_0 : (\mu_T - \mu_C) \leq (\lambda - 1)(\mu_C - \mu_P) \text{ versus } \mathrm{H}_a : (\mu_T - \mu_C) > (\lambda - 1)(\mu_C - \mu_P) \quad (7.2)$$

Design and analysis of this preservation objective is also called λ-margin approach. The preservation statement is actually quite powerful. For example, for any $\lambda \geq 0$, $(\mu_T - \mu_C) > (\lambda - 1)(\mu_C - \mu_P)$ implies $\mu_T > \mu_P$. Which means that efficacy (over placebo) is indeed implied by the preservation statement. Furthermore, when $\lambda = 1$, the statement implies actually that the new test treatment is superior to the active control treatment.

The hypotheses of the two-arm active controlled non-inferiority clinical trial involve parameter $\mu_C - \mu_P$ with no data collected to estimate in the current trial. This parameters may be determined with medical understanding of what a medically meaningful difference is. But sometimes, it needs to be determined using the external historical data. For example, when the historical data is rich in number of trials, large in sample sizes, and consistent in the estimates of $(\mu_C - \mu_P)$, it would be quite reasonable to use either the minimum of the lower limits of the confidence intervals of $(\mu_C - \mu_P)$ of the historical studies or the lower limit of the confidence interval of $(\mu_C - \mu_P)$ estimated with data combined through meta-analysis as the starting point to determine a non-inferiority margin δ. When δ is determined exactly as the lower confidence limit of $(\mu_C - \mu_P)$, the non-inferiority test is actually carried out by comparing the lower confidence limit $\mathrm{L}(\mu_T - \mu_C)$ of $(\mu_T - \mu_C)$ with the negative value of the lower confidence limit $\mathrm{L}(\mu_C - \mu_P)$ of $(\mu_C - \mu_P)$ of the historical data. When $\mathrm{L}(\mu_T - \mu_C)$ is greater than $-L(\mu_C - \mu_P)$, one claims that the test treatment T is non-inferior to the active control treatment C. This is often called the two confidence intervals approach (Hauck and Anderson 1999).

When the objective is to test if the effect of the test treatment T (adjusted for placebo) preserves at least $100\lambda\% (0 \leq \lambda \leq 1)$ of the effect of the active control treatment C, it is often a common practice to discount the lower confidence limit $\mathrm{L}(\mu_C - \mu_P)$ of $(\mu_C - \mu_P)$ by multiplying it with $(1-\lambda)$. This leads to the comparison of $\mathrm{L}(\mu_T - \mu_C)$ with $-(1-\lambda)\mathrm{L}(\mu_C - \mu_P)$. Snappin (2004) described this approach as double-discounting. For practical purpose, Holmgren (1999) proposed to test

$$\mathrm{H}_0 : (\mu_T - \mu_C) - (\lambda - 1)(\mu_{C|H} - \mu_{P|H}) \leq 0$$

versus

$$\mathrm{H}_\mathrm{a} : (\mu_T - \mu_C) - (\lambda - 1)(\mu_{C|H} - \mu_{P|H}) > 0 \tag{7.3}$$

instead of (7.2), where the subscription H indicates the parameter is of the study population of the historical studies. With this setting, data of the two historical arms (the historical active control arm and the historical placebo arm) of the historical trials are treated as if collected along with the data of the two arms (the active control arm and test arm) of the current non-inferiority trial. The combination of the parameters is then estimated with the corresponding data and the estimation variances are pooled in the combination form for the estimation error of $(\mu_T - \mu_C)$ and $(\lambda - 1)(\mu_{C|H} - \mu_{P|H})$. As long as $(\lambda - 1)(\mu_{C|H} - \mu_{P|H}) \geq (\mu_P - \mu_C)$, showing $(\mu_T - \mu_C) > (\lambda - 1)(\mu_{C|H} - \mu_{P|H})$ would imply that $\mu_T > \mu_P$, had a placebo arm been in the current clinical trial.

It is to be noted that the choice of the active control treatment is extremely important. When the trial is designed to show that the test treatment is superior to the active control treatment, any approved treatment can be used as the active control. But in the design of a two-arm active controlled clinical trial, the active control treatment chosen needs to satisfy the two conditions and assay sensitivity. ICH10 has specifically described assay sensitivity as "A suitable active comparator… could be a widely used therapy whose efficacy in the relevant indication has been clearly established & quantified in well-designed & well documented superiority trials & which can be reliably expected to have similar efficacy in the contemplated AC (active controlled) trial." On the other hand, a non-inferiority trial is valid only if the trial is able to detect differences between treatments if the differences exist. The assay sensitivity required for the active control treatment means that the consistency of the expected efficacy of the active control treatment, $(\mu_{Ck} - \mu_{Pk}) > 0$, exists in any valid historical trial. The factors that may reduce assay sensitivity may include inconsistency of the recruitment criteria, poor compliance with therapy, poor responsiveness of the study population, concomitant medications, a population that improves spontaneously, poor diagnostic criteria, inappropriate measures of drug effect, excessive variability of measurements and biased assessment of end-point, ... etc.

Once an appropriate active control treatment is chosen, one needs to examine the validities of the conditions and assumptions on δ, λ, and the design that provides the bridge to draw a conclusion on the comparison between the test treatment and placebo had it been included in the current trial. First, the sample population of the active control trial needs to be similar to that of the external historical placebo controlled trials of the active control treatment. Second, the study duration, follow-up, and the endpoint of the active controlled trail are the same as in the historical trials of which the non-inferiority margin is derived. Third, the size of efficacy of active control treatment, $(\mu_C - \mu_P)$, had the placebo treatment exist in the current trial, be no less than $(\mu_{C|H} - \mu_{Pk})$.

With the generalized historical control (i.e., δ-margin) method, we often assume that δ is a value with no variability. It may be supported with

1. The sample size of the historical data used is very large
2. The effect size of the active control treatment after adjusted for placebo is expected in any well-designed trial to be much larger than the margin used (i.e., $(\mu_{C|H} - \mu_{P|H}) > \delta$ and $(\mu_C - \mu_P) > \delta$ had placebo used in the current trial)
3. It is a medically meaningful margin but not a statistical estimate

The null hypothesis (7.1) is tested by comparing

$$t(\delta) = (\hat{\mu}_T - \hat{\mu}_C + \delta)/[\text{s.e.}(\hat{\mu}_T - \hat{\mu}_C)] \quad (7.4)$$

with the 97.5th percentile of t-distribution with proper degrees of freedom if the response is normally distributed, or with the 97.5th percentile of the normal distribution for large sample approximation test. When the third condition fails, $t(\delta)$is a good approximation test. When one of the first two conditions fails as well as the third condition, the approximation fails too. The exact form of the test statistic is complicated.

With the two confidence-intervals approach, the hypotheses

$$\text{H}_0 : (\mu_T - \mu_C) \leq (\mu_{P|H} - \mu_{C|H})$$

versus

$$\mathrm{H_a}: (\mu_T - \mu_C) > (\mu_{P|H} - \mu_{C|H}) \tag{7.5}$$

may be tested with the decision rule that reject H_0 if

$$\left[(\hat{\mu}_T - \hat{\mu}_C) - c^1_{0.975}e_1\right] > \left[(\hat{\mu}_{P|H} - \hat{\mu}_{C|H}) + c^2_{0.975}e_2\right]$$

where $\hat{\mu}_T - \hat{\mu}_C$ and $\hat{\mu}_{P|H} - \hat{\mu}_{C|H}$ are the differences of sample means, $e_1$and $e_2$are the errors corresponding to the two estimates, $c^1_{0.975}$ and $c^2_{0.975}$ are the 97.5th percentiles of the corresponding t-distributions or normal distribution. The equation can be re-written as

$$[(\hat{\mu}_T - \hat{\mu}_C) - (\hat{\mu}_{P|H} - \hat{\mu}_{C|H})] > (c^1_{0.975}e_1 + c^2_{0.975}e_2) \tag{7.6}$$

When using the cross-trial comparison (i.e., λ-margin) approach for testing the preservation hypothesis, the unbiased estimate of the linear combination of four parameters $(\mu_T - \mu_C) - (\lambda - 1)(\mu_{C|H} - \mu_{P|H})$ is

$$(\hat{\mu}_T - \hat{\mu}_C) - (\lambda - 1)(\hat{\mu}_{C|H} - \hat{\mu}_{P|H})$$

Here the four arms (i.e., test treatment, active control treatment, active control treatment in historical trials, placebo arm in historical trial) are treated equally as if patients were randomized in the same trial with the same protocol. Hence, the error of the estimate is

$$\sqrt{e_1^2 + (\lambda - 1)^2 e_2^2}$$

where e_1 and e_2 are defined previously as in the two confidence interval approach. The test statistic

$$z(\lambda) = [(\hat{\mu}_T - \hat{\mu}_C) - (\lambda - 1)(\hat{\mu}_{C|H} - \hat{\mu}_{P|H})] / \sqrt{e_1^2 + (\lambda - 1)^2 e_2^2} \tag{7.7}$$

is then compared with the 97.5th percentile of t-distribution with proper degrees of freedom for normally distributed endpoint and compared with the 97.5th percentile of the normal distribution for asymptotic test. On the other hand, H_0 of is rejected if

$$[(\hat{\mu}_T - \hat{\mu}_C) - (\lambda - 1)(\hat{\mu}_{C|H} - \hat{\mu}_{P|H})] > c_{0.975}\sqrt{e_1^2 + (\lambda - 1)^2 e_2^2} \tag{7.8}$$

where $c_{0.975}$ is the critical value. When $\lambda = 0$, (7.8) is

$$[(\hat{\mu}_T - \hat{\mu}_C) - (\hat{\mu}_{P|H} - \hat{\mu}_{C|H})] > c_{0.975}\sqrt{e_1^2 + e_2^2} \tag{7.9}$$

Note that when using $c^1_{0.975} = c^2_{0.975} = c_{0.975} = 1.96$, the right-hand side of (7.6) is $1.96(e_1 + e_2)$ instead of $1.96\sqrt{e_1^2 + e_2^2}$ as in (7.9).

7.6 Statistical Issues in Two-Arm Efficacy Non-inferiority Clinical Trials

There are quite a few issues regarding the two-arm active controlled clinical trials brought to the attention of statisticians and medical researchers. A few of the issues and considerations are discussed below.

1. *Sample size determination for δ-margin and λ-margin approaches*

   For δ-margin approach, sample size n can be determined by assuming a standard deviation of response and a target difference of $[\mu_T - (\mu_C - \delta)]$. For example, let $[\mu_T - (\mu_C - \delta)] = \Delta$ for a prespecified $\Delta > 0$, for continuous variable, the ordinary sample size formula can be used such that

   $$N = \sigma^2(Z_{(1-\alpha)} + Z_{(1-\beta)})^2/(\Delta - \delta)^2$$

   where $2\sigma^2$ is the variance of $X_T - X_C$; α and $(1-\beta)$ are the type I error rate and desired power of the test; $Z_{(1-\alpha)}$ and $Z_{(1-\beta)}$ are the $100(1-\alpha)$th and $100(1-\beta)$th percentiles of normal distribution, respectively. Note that if $\Delta > \delta$, the sample size required for testing the non-inferiority hypothesis H_0: $\mu_T \le (\mu_C - \delta)$ is larger than the sample size needed to test against the superiority hypothesis H_0: $\mu_T = \mu_C$ for the same Δ value. On the other hand, given a small Δ such that $(\mu_T - \mu_C) = \Delta$ and a fixed sample size *n*, there is more power to show $[\mu_T - (\mu_C - \delta)] > 0$ than showing $\mu_T > \mu_C$.

   Sample size for λ-margin approach can be more complicated. A simple approach is to use $\delta = (1 - \lambda)(\mu_{C|H} - \mu_{p|H})$ in the above sample size formula. More appropriate approach for λ-margin approach can be found in Hauschke et al. (1999) and Pigeot et al. (2003)

   When the response variables are binary, the conventional normal approximation formula for sample size estimation

   $$N = \sigma^2(Z_{(1-\alpha)} + Z_{(1-\beta)})^2/(\Delta - \delta)^2$$

   where Δ is the target difference between the two rates ($\Delta = [\Psi_T - (\Psi_C - \delta)]$, where Ψ_T and Ψ_C are the success rates of test and active control treatment, respectively). Let $\Psi = (\Psi_T + \Psi_C)/2$) and $\sigma^2 = \Psi(1 - \Psi)$. Instead σ^2 is to be calculated as the maximum value of the variances under the restriction of $\Psi_T - \Psi_C = -\delta$. It leads to a more complicated solution of a system of equations as discussed by Farrington and Manning (1990).
2. *Choice of non-inferiority margin*

   As pointed out earlier, the choice of δ and λ plays an important role in active controlled clinical trial (D'Agostino et al. 2003). The regulatory requirement of a small δ and large λ is based on two reasons.

   - First, choose a small δ or a large λ is to make sure that by showing that $(\mu_T - \mu_C) > -\delta$ or $(\mu_T - \mu_C) > (\lambda - 1)(\mu_{C|H} - \mu_{P|H})$, one can be certain that $\mu_T > \mu_P$ even if $(\mu_C - \mu_P)$ is much smaller than $(\mu_{C|H} - \mu_{P|H})$ had the placebo arm used in the current trial.

- Second, avoid selecting a margin that is an estimate using the data of historical trials of active control treatment. By determining the margin as an estimate of the function of $(\mu_{C|H} - \mu_{P|H})$, the investigator brought the data of historical trials into the current non-inferiority trial. It is most clear in testing hypotheses (7.3) and (7.5). The parameter $(\mu_{C|H} - \mu_{P|H})$ is to be estimated using the historical data of historical trial in the test statistics. It leads to the dependency of the efficacy decision of the active control treatment, $\mu_{C|H} > \mu_{P|H}$ and the efficacy decision of the test treatment, $(\mu_T - \mu_C) > (\lambda - 1)(\mu_{C|H} - \mu_{P|H})$ in (7.3) and $(\mu_T - \mu_C) > (\mu_{P|H} - \mu_{C|H})$ of (7.5). Tsong et al. (2007a) showed that when using the λ-margin approach, for a given sample size and variance, correlation of the statistics of two independently conducted active control clinical trials is an increasing function of $(1 - \lambda)$. The correlation can be large when λ is less than 80 %. Whereas, when the δ-margin is determined as a function of $(\mu_{C|H} - \mu_{P|H})$ through estimation using the data of the historical trials of the active control treatment, the investigator brought in the data of historical trials of active control treatment to the current non-inferiority trial as well. Tsong et al. (2007b) showed that, for a given sample size and variance, the correlation is actually an increasing function of $(1 - \alpha^*)$, where $(1 - \alpha^*)$ is the confidence level used in determining δ. But the correlation does not change much by using either a smaller or larger α^* value. On the other hand, it is important to have sample size of the historical studies much greater than the sample size of the current active control trial in order to have small correlation. For example, assuming variances are all equal in test, active control, or placebo arms, in planning two active control trials with sample size of 100 subject per arm in each trial, if the total sample size of the historical studies is not larger than 400 per arm, the correlation will be greater than 0.15. In a recent research by Kang and Tsong (2009), they derived the type I error rate of non-inferiority testing adjusted for the correlation caused by bring in data from historical trials.

3. *Testing both superiority and non-inferiority*

With a two-arm active controlled clinical trial, the investigator may show that the test treatment is superior to the active control treatment or to show it is non-inferior to the active control treatment. It has been proven with the δ-margin design that one can test both null hypotheses using either a two-stage procedure or using one-step simultaneous test. With a two-stage procedure, the investigator can test for non-inferiority first. Once shown as significant, the investigator can move on to further test for superiority. On the other hand, the investigator can design to test for superiority first. When the superiority claim fails, the investigator can move on to test for non-inferiority. It has been shown that there is no type I error rate inflation if the non-inferiority margin δ is predetermined before any test (Morikawa and Yoshida 1995; Tsong et al. 2007c). Equivalently, it can be carried out by testing both hypotheses simultaneously by comparing the lower limit of the confidence interval of $(\mu_T - \mu_C)$ with both 0 and $-\delta$.

With the cross-trial comparison (λ-margin) approach, Tsong et al. (2005) showed that for testing superiority hypotheses, the conventional analysis is a comparison between means of the test and active control treatments. Whilst for testing the non-inferiority null hypotheses, it compares the linear combination of the test and active control treatment of the current active control trial and the active control treatment and placebo in the study population of the historical trials. It adopted a four-arm clinical trial design. The two test statistics have different degrees of freedom. Hence, the exact type I error rates of either the superiority test or non-inferiority test of the two-stage procedure differ from that of the single test. Tsong and Zhang (2007) further proposed the simultaneous test for both hypotheses by estimating $[1 + (\mu_T - \mu_C)/(\mu_{C|H} - \mu_{P|H})]$ and comparing the lower limit of the confidence interval with 1 and λ for superiority and non-inferiority claims.

7.7 Summary

Application of clinical trials with active control treatment has a long and important history in drug development. They have been used in bioequivalence assessment and in trial validation. In either case a fixed non-inferiority margin for the study objective is used for all test treatment. Conceptually, there is essentially not much difference between non-inferiority, bioequivalence testing, and superiority testing except a fixed location shift. Because of the ethical concern of randomizing patients in clinical trial to placebo arm without treatment, the practice of designing clinical trials without placebo arm became more regular in the last fifteen years. The non-inferiority efficacy comparison between a test and an active control treatment is no more interpreted in the same way as for in vivo bioequivalence trials. One may interpret that the test treatment is superior to the placebo by showing it is non-inferior to the active control treatment only if the active control treatment and the non-inferiority margin satisfy some of the necessary and sufficient conditions. Because of the choice of active control treatment and the condition on the non-inferiority margin, the fixed margin often turned out to be a parameter to be estimated. On the other hand, validation of these conditions is often difficult if not impossible. In general, in order to have a reasonable margin, large number and large sample size of historical trials of active control product is required. On the other hand, in order to make the interpretation for placebo-adjusted efficacy of the test treatment, assay sensitivity of the active control treatment is needed. In case, the investigator has to rely on statistical estimation from historical trials of active control treatment to determine the non-inferiority margin, the strength of the evidence derived may need to be adjusted through conditional type I error rate or p-value. Furthermore, just like the CAP example discussed in Sect.7.5, for many treatment, a proper active control treatment or a proper non-inferiority margin does not exist with the history of clinical trials. In these cases, the result of the non-inferiority test becomes not interpretable for efficacy against placebo.

Acknowledgment The authors want to acknowledge the contributions of many coauthors involved in the many collaborative works and the colleagues provided important information and examples gathered from their review experiences. This article is the fruit of all the collaborative works. The materials presented in this article are also part of the materials that the authors used in the preparation of a short course offered at the Applied Statistics Symposium in 2006, Korea Food and Drug Administration, and joint Statistical Meeting in 2009. The second International Pharmaceutical statistics Workshop sponsored by China East Normal University in 2012. For that we want to thank International Chinese Statistic Association, Korea Food and Drug Administration, China East normal University, and American Statistical Association for sponsoring the short courses. Last but certainly not the least, the contributions of two editors of the book and the referees of the manuscript should be acknowledged. Without their careful reading, comment, and questions, the manuscript would not be as complete and readable.

References

Committee for Medicine Products for Human (CHMP) (2005) Guidance on the choice of the non-inferiority margin. European Medicines Agency CPMP/EWP/2158/99 (http://www.emea.eu.int/pdfs/human/ewp/215899en.pdf)

Chen L, Tsong Y (2007). Design and analysis for drug abuse potential studies: Issues and strategies for implementing a crossover design. Drug Information J. 41:481–489.

D'Agostino RB, Massaro JM, Sullivan LM (2003) Non-inferiority trials: design concepts and issues – the encounters of academic consultants in statistics. Stat Med 22:169–186

Ellenberg SS, Temple R (2000) Placebo-controlled trials and active controlled trials in the evaluation of new treatment. Part 2: practical issues and scientific issues. Ann Intern Med 133:455–463

Farrington CP, Manning G (1990) Test statistics and sample size formulae for comparative binomial trials with null hypothesis of non-zero risk difference or non-unity relative risk. Stat Med 9:1447–1454

Hauck WW, Anderson S (1999) Some issues in the design and analysis of equivalence trials. Drug Info J 33:109–118

Hauschke D, Kieser M, Diletti E, Burke M (1999) Sample size determination for proving equivalence based on the ratio of two means for normally distributed data Stat Med 18:93–105

Holmgren EB (1999) Establishing equivalence by showing that a prespecified percentage of the effect of the active control over placebo is maintained. J Biopharm Stat 9:651–659

Hung H-MJ, Wang S-J, Tsong Y, Lawrence J, O'Neill RT (2003) Some fundamental issues for non-inferiority testing in active controlled trials. Stat Med 22:213–225

ICH (2005) ICH E14 guidance on the clinical evaluation of qt/qtc interval prolongation and proarrythmic potential for n on-antiarrythmic drugs. In: International Conference Harmonization, Geneva, Switzerland, May 2005

Morikawa T, Yoshida M (1995) A useful testing strategy in phase III trials: Combined test of superiority and test of equivalence. J Biopharm Stat 5:297–306

Pigeot I, Schäfer J, Röhmel J Hauschke D (2003) Assessing non-inferiority of a new treatment in a three-arm clinical trial including a placebo. Stat Med 22:883–899

Schuirmann DJ (1987) A comparison of the two one-sided tests procedures and the power approach for assessing the equivalence of average bioavailability. J Pharmacokinet Biopharm 15:657–680

Snapinn SM (2004) Alternatives of discounting in the analysis of non-inferiority trials. J Biopharm Stat 14:263–273

Temple R, Ellenberg SS (2000) Placebo-controlled trials and active-control trials in the evaluation of new treatments - Part 1: Ethical and Scientific Issues, Part 2: ethical and scientific issues. Ann Intern Med 133:455–463

Tsong Y, Zhang JJ (2005) Testing superiority and non-inferiority hypotheses in active controlled clinical trials. Biometrical J 47(1):62–74

Tsong Y, Zhang JJ (2007) Simultaneous test for superiority and non-inferiority hypotheses in active controlled clinical trials. To appear in J. Biopharm. Statist

Tsong Y, Levenson M, Zhang J (2007) Choice of λ and the dependence of non-inferiority active controlled clinical trials. To appear in Statistics in Medicine

Tsong Y, Wang S-J, Hung H-MJ (2003) Statistical issues in objectives, designs and analysis of non-inferiority test active controlled clinical trials. J Biopharm Stat 13:29–41

Tsong Y, Zhang J, Levenson M (2007) Choice of δ and the dependence of non-inferiority active controlled clinical trials. To appear in Biometrical J

Tsong Y, Zhang J, Wang S-J (2004) Group sequential design and analysis of clinical equivalence assessment for generic non-systemic drug products. J Biopharm Stat 14:359–373

Tsong Y, Higgins K, Wang S-J, Hung H-MJ (1999) An overview of equivalence testing - CDER reviewers' perspective. In: Proceedings of the Biophar. Section, Joint Stat. Meetings, Am. Stat. Assoc, pp 214–219

Zhang J, Machado SG (2008) Statistical issues including design and sample size calculation in thorough Qt/Qtc studies. J Biopharm Stat 18(3):451–467

Zhang L, Dmitrienko A, Luta G (2008) Sample size calculation in thorough qt/qtc studies. J Biopharm Stat 18(3):468–493

Chapter 8
Thorough QT/QTc Clinical Trials*

Yi Tsong and Jinglin Zhong

8.1 Introduction

On an ECG tracing, the QT interval is measured from the beginning of Q wave to the end of T wave. QT interval reflects the duration of ventricular depolarization and subsequent repolarization. For some drugs, significant prolongation of the absolute QT interval has been associated with the precipitation of a potentially fatal cardiac arrhythmia called Torsades de Pointes (TdP) and can degenerate into ventricular fibrillation, leading to sudden cardiac death (Moss 1993). Over the past decade, these cardiac adverse events have resulted in a number of patient deaths and was one of the most frequent adverse events (Wysowski et al. 2001) of a drug that led to either its removal from the market (e.g., terfenadine and cisapride) or the placement of restrictions on its use (e.g., ziprasidone and ranolizine.) With theses experiences, the regulatory agencies now request that each pharmaceutical company conducts at least one thorough QT (TQT) study when submitting a new drug application. The final version of the ICH E14 guidance (ICH 2005) for "The Clinical Evaluation of QT/QTc Interval Prolongation and Proarrhythmic Potential for Nonantiarrhythmic Drugs" was released in May 2005. The abbreviation QTc denotes the QT interval corrected for the heart rate.

In the ICH E14 guidance, it is recommended that the TQT study be conducted generally in early clinical development after some information about the pharmacokinetics of the drug has been obtained. The TQT is a randomized, double blinded,

*This article represents the point of views of the authors. It docs not necessarily represent the official position of U.S. FDA

Y. Tsong (✉) • J. Zhong
Division of Biometrics VI, Office of Biostatistics/Office of Translational Sciences, Center for Drug Evaluation and Research, U.S. Food and Drug Administration, Silver Spring, MD 20993, USA
e-mail: yi.tsong@cder.hhs.gov; jinglin.zhong@fda.hhs.gov

W. Tang and X. Tu (eds.), *Modern Clinical Trial Analysis*, Applied Bioinformatics and Biostatistics in Cancer Research, DOI 10.1007/978-1-4614-4322-3_8,

placebo- and active-controlled, crossover or parallel-arm study with single or multiple doses of the drugs. The guidance states that "A negative thorough QT/QTc study" is one in which the upper bound of the 95% one-sided confidence interval for the largest time-matched mean effect of the drug on the QTc interval excludes 10 ms. This definition is chosen to provide reasonable assurance that the mean effect of the study drug on the QT/QTc interval is not greater than around 5 ms, which is the threshold level of regulatory concern. If the data collected provide enough evidence to show that the drug does not prolong the QTc interval when compared with placebo, the TQT study can be termed as a "negative study." A negative TQT study supports a conclusion that the drug does not prolong the QTc interval to a clinically significant degree. When a TQT study is not negative, i.e., if at least one upper bound of the one-sided 95% CI of the time-matched difference exceeds the threshold of 10 ms, the study is termed as positive or nonnegative. In this case, a positive TQT study indicates that more intense evaluation of the potential for QTc prolongation is necessary during subsequent development of the drug. The purpose of a TQT study is to determine whether the drug has a threshold pharmacologic effect on cardiac repolarization, as detected by QTc prolongation. As noted above, the threshold level of regulatory concern is around 5 ms, as evidenced by the upper bound of the 95% confidence interval around the largest mean effect of the drug and placebo after baseline correction on QTc being less than 10 ms.

With the objective of the trial being to confirm that the response of the test treatment differs no more than a prespecified margin from the placebo one may have the concern with the trial population whether the subjects may respond to an active treatment with a known prolongation effect of QT interval. For this reason, E14 recommends that "The positive control should have an effect on the mean QT/QTc interval of about 5 ms (i.e., an effect that is close to the QT/QTc effect that represents the threshold of regulatory concern, around 5 ms)." Often a positive control treatment with a well-known QT interval prolongation and dose–response profile will be used and an assay validation assessment performed often consists of a quantitative assessment test and a qualitative profile identification.

In this article, we describe the approaches and issues in the design and analysis of a thorough QT clinical trial. We organize the rest of the article as follows. Measurements of QT interval and the methods of QT interval correction by pulse rate will be given in Sect. 8.2. The approaches for confirmative non-inferiority test of drug-induced prolongation of QT interval are given in Sect. 8.3. The quantitative approaches for validation of study are given in Sect. 8.4. Regular fixed sample size designs of the trial and some modeling issues are presented in Sect. 8.5. Finally, a summary and discussion is given in Sect. 8.6.

8.2 Measurement and Correction of QT Interval

The electrocardiogram (ECG) records the polarization activities of a human heart. Figure 8.1 (Li et al. 2004) displays an ECG of two heartbeats. QRS complex represents depolarization occurring within the ventricles of the heart. T wave

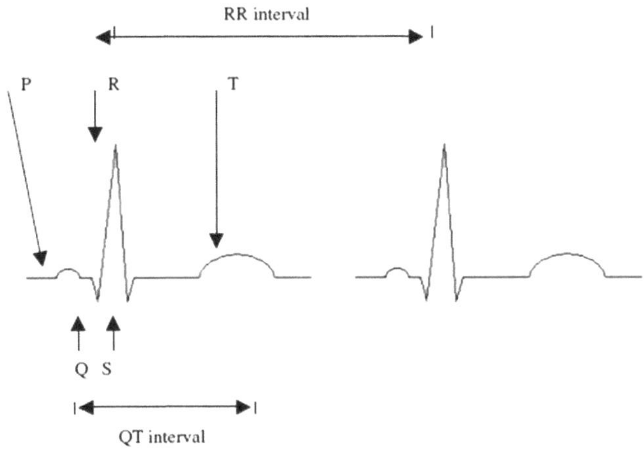

Fig. 8.1 ECG Waves. The P wave represents depolarization occurring within the atria of the heart. The QRS complex represents depolarization occurring within the ventricles of the heart. The T wave represents repolarization occurring within the ventricles of the heart. The QT wave is the period of time it takes to complete both depolarization and repolarization within the ventricles and is measured from the beginning of the Q wave to the end of the T wave.

represents repolarization occurring within the ventricles of the heart. The beginning of the Q wave to the end of the T wave is the QT interval and represents the total time required for both ventricular depolarization and repolarization to occur. The duration of this QT interval is used as a biomarker for the risk of Torsades de Pointes. The time between two consecutive R waves, the RR interval, is inversely proportional to the heart rate. It is known physiologically that QT interval prolonging accompanies the physiological increases in the RR interval. When a drug alters heart rate, it is important to distinguish an effect of drug on the QT interval versus an artifact due to change in the RR interval (inverse of heart rate). Therefore QT interval needs to be corrected before used for the analysis of drug effect on QTc (corrected QT interval). It is desired that the resulting QTc intervals and RR intervals are uncorrelated. Commonly used correction formulas are based on either linear regression or nonlinear regression models for QT–RR relationship (Ahnve 1985; Funk-Brentano and Jaillon 1993; Moss 1993; Malik et al. 2002; Li et al. 2004; Wang et al. 2008).

The individual based or subject-specific correction (IBC) method often fits a regression model to QT–RR data from each individual subject's QT measured at baseline or of placebo against RR data. Under the assumption of constant relationship between QT interval and RR interval, the "slope" of the individual subject estimated using baseline or placebo data will be used on the data collected under treatment to determine the correct QT interval in the assessment of the treatment effect. These are sometimes called off-drug correction method. On the other hand, correction factors may be determined using data collected at off-drug as well as on-drug periods of the same subject. Though it is a one-step modeling, correction factor for off-drug may be different from the factor for on-drug period. These methods are often called off-drug/on-drug (OFFnON) correction method (Li et al. 2004). The population-based correction method (PBC) would assume that subjects under the same treatment (on or off drug) share the same but unknown factor to be estimated. PBC correction method fits a regression model to pooled QT–RR data from all subjects in the current trial. With either PBC or IBC method, there are also variations in assuming that the correction factor may be the same regardless on- or off- drug. In contrast, a fixed correction (FC) method uses a fixed correction formula that is not derived from the current trial data.

There are many proposed fixed QT correction (FC) methods in the medical literature. What is considered the best method of correction among them is debatable. Bazett's formula (1920) is probably the most commonly used by clinicians because it is practical, simple, and convenient. Based on his empirical experience, Bazett assumed the relationship $\mathrm{QT} = \alpha \times \mathrm{RR}^{1/2}$. Hence, $\mathrm{QTc} = \mathrm{QT} \times (1000/\mathrm{RR})^{1/2}$. Fridericia (1920) proposed a different empirical model, $\mathrm{QT} = \alpha \times \mathrm{RR}^{1/3}$, and $\mathrm{QTc} = \mathrm{QT} \times (1000/\mathrm{RR})^{1/3}$. It is often considered as an improvement of Bazett's correction because it gives QTc more consistent from population from population. In addition, Framingham study (1992) used $\mathrm{QTc} = \mathrm{QT} \times (1000/\mathrm{RR})^{0.154}$ and Schlamowitz (1946) proposed to use $\mathrm{QTc} = \mathrm{QT} \times (1000/\mathrm{RR})^{0.205}$ for correction. The FC correction method did not acknowledge the variation of the QT versus RR relationship from population-to-population or subject-to-subject. These limitations were examined by Malik (2001) and Desai et al. (2003).

Let us consider the QT data collected in a crossover design. Let y_{kij} denote the QT or log(QT) measurement during the kth treatment period, for subject i at the time point t_j, where $k = P$ for placebo, T for test, k for baseline of treatment k, $i = 1,\ldots,n$ and $j = 1,\ldots J$, the number of measurements per subject, respectively. Let x_{kij} denote the RR/1000 or log(RR/1000) measurements within the kth treatment periods, respectively. The FC correction methods stated above can be represented by

$$y_{kij} = \alpha_{kij} - \beta x_{kij} \tag{8.1}$$

where β is the fixed correction factor derived from dataset other than the current study and α_{kij} is QTc or log(QTc). It was pointed out that the relationship between the QT interval and the heart rate (HR) is approximately linear by Hodges et al. (1983) and Rautaharju et al. (1993). For this relationship, we let y_{kij} represent QT and x_{kij} represent (HR-60) in (8.1).

A populationbased correction (PBC) method is one that uses the following linear model to determine the correction factor β:

$$y_{kij} = \alpha_{kij} - \beta_k x_{kij} + \varepsilon_{kij} \tag{8.2}$$

where β_k may be estimated using model (8.2) with data of complete study. In thorough QT studies it is often assumed that $\beta_k = \beta$.

The PBC or FC methods can be used even when data are only collected from sparse time points. The PBC method assumes that the underlying true QT–RR (or QT-HR) relationships are the same for all subjects under the same treatment condition (off-drug or on-drug) or different treatment conditions, while the FC method assumes that a specific and completely known QT–RR relationship for all the trial subjects regardless of treatment conditions. Since there may be substantial intersubject variation in QT–RR relationship, the PBC and FC methods may lead to under- or overcorrection of individual QT to a degree that may bias the assessment on QT prolongation in central tendency analysis or categorical analysis.

A thorough QT study is usually powered to detect a small QT interval prolongation (e.g., 0.01 s or 10 ms). A sensitive QT study requires an accurate QT interval correction method to ensure correct attribution of a small QT interval prolongation to potential sources such as treatment or random noise. It has been demonstrated that the QT–RR relationship exhibits a substantial intersubject variability but a low intrasubject variability in human, leading some authors to suggest the use of IBC method whenever data warrant (Batchvarov and Malik 2002; Malik et al. 2002; Batchvarov et al. 2002; Malik et al. 2004). The IBC method assumes different QT–RR relationships for different individual subjects. The IBC method is feasible in thorough QT studies which have QT–RR data from multiple ECGs taken at multiple time points. In practice, the RR range in the QT–RR data from an individual subject may not be sufficiently wide to allow for an accurate estimation of the QT–RR relation.

In order to accommodate the population to population and subject to subject variation of the relation between QT and RR, Malik (2001) proposed the following IBC method:

$$\mathrm{QT}_{ij} = \alpha_i \times \mathrm{RR}_{ij}^{\beta_i}$$

$$\mathrm{QTc}_i = \mathrm{QT}_{ij} \times \frac{1000^{\tilde{\beta}_i}}{\mathrm{RR}_{ij}^{\tilde{\beta}_i}}$$

where $\tilde{\beta}_i$ is the least square estimate of β_i that utilized the baseline or placebo QT and RR data from subject i only. In another word, Malik (2001) proposed to estimate α_i and β_i in the following linear model:

$$y_{pij} = \alpha_i + \beta_i x_{pij} + \varepsilon_{pij} \tag{8.3}$$

using only the baseline or placebo log(QT) and log(RR/1000) or QT and (HR-60) data. The same $\tilde{\beta}_i$ is used to correct QT observed when the subject under treatment. Note that Malik's correction factor is estimated with intrasubject averaged over all time points. It is derived under the assumption that the relationship between QT and RR remains the same over all time points and even under treatment.

Without the assumption of invariant QT/RR relationship and intrasubject averaged over all time points, a general linear model for IC method can be represented as follow. Let y_{kij} denote QT (or log(QT)) measurement during the kth treatment, for subject i at the time point j, k = P for placebo, T for test k_0 baseline of treatment k, $i = 1, \ldots, n_k$ and $j = 1, \ldots J$. J is the number of measurements per subject.

Let x_{kij} denote the RR/1000 (or log(RR/1000)) measurements with the kth treatment, respectively. Most correction methods stated above can be represented by the general fixed effect linear model,

$$y_{kij} = \alpha_{kij} + \beta_{kij} x_{kij} + \varepsilon_{kij} \tag{8.4}$$

α_{kij} is log(QT) or log(QTc)s (intercepts) for the kth treatment of subject i at j time point, respectively, and β_{kij} is the corresponding slopes between QT and RR/100 or between log(QT) and log(RR/1000), and ε_{kij} is i.i.d.N(0, σ_0^2) Note that (8.4) is analyzed assuming QT measurement follows a lognormal distribution when log(QT) and log(RR) are used. Various FC, PBC and IBC can all be represented by (8.4). For example, for a FC, $\beta_{kij} = \beta$ a constant derived from other study For a PBC, $\beta_{kij} = \beta$ or β_k, factors estimated from data of current study. For an IBC intrasubject averaged over all time point, $\beta_{kij} = \beta_{ki}$, estimated for each individual using the data of current study. For a genderspecified PBC, $\beta_{kij} = \beta_k$ Furthermore, when gender difference is of concern, one may analyze the data with gender group in model.

These correction methods have been compared under different models in numerous studies using real data (Ahnve 1985; Funk-Brentano and Jaillon 1993; Moss 1993; Hnatkova and Malik; Malik; Malik et al. 2004). The corrections were aimed at removing the dependency of QTc intervals on RR intervals within subjects. The conclusions of the comparisons are dependent on subject composition and the observed range of RR intervals. Most of the comparisons were focused on the model error which is the intrasubject mean difference between two subsets with each subject. With these constraints, the “error” term would be partially attributed to the data grouping and the outcome of the “error” term may not entirely correspond to the actual biases of the correction methods.

Wang et al. (2008) studied the IBC, PBC, and FC correction methods in order to quantify the biases and variances of the QTc intervals using the IBC, PBC, and FC correction methods conditional on the observed RR intervals. It is easy to see that FC method can be unreliable in comparison to PBC or IBC because it is derived from source of QT–RR data other than the current study data although it has smallest variance conditioning on a given RR value. The biased variance of FC may lead to an inflation of type I error rate when used in treatment comparison. Although Federicia's correction factor has been shown relatively reliable in testing as shown in the numerical example of Wang et al. (2008) as well as in the literature. The use of unvalidated fixed correction factors is discouraged in practice generally.

PBC has smaller variance than IBC conditioning on a given RR, though the difference is usually not large. Wang et al. (2008) showed that the conventional intrasubject averaged PBC is a conditional biased estimate of the true underlying QT interval. But a modified subject-averaged QTc at each time point becomes conditionally unbiased. This leads to the recommendation of using the modified subject-averaged QTc when the data were collected at sparse time points. They also recommended including RR as covariate in the model for the comparison of treatment effect when using either PBC or IBC may have the effect to further reduce the remaining correlation of QTc interval with RR interval.

Note that with PBC or IBC if the correction factor β_{kgij} is estimated with (8.1) using data of the kth treatment of the current study, $\hat{\alpha}_{kgij}$, the corrected QT is a random variable instead of an observation. When treatment comparison is carried out after QT is corrected in a two-step setting, the standard error of $\hat{\alpha}_{kgij}$ needs to be included in the estimation error of the mean treatment difference to avoid standard error bias of comparison.

Li et al. (2004) pointed out that there are three limitations of the approaches. First, the variance of QT interval is usually proportional to its magnitude and the validity of equal variance assumption for the measurement error of QT interval of the models is questionable. Second, subject-specific estimate of $\tilde{\beta}_i$ is often not practical because the number of ECGs is often small for individual subject. Third, in Malik's IBC method, the underline assumption is that the relationship between QT and RR is consistent for each subject regardless the potential changes caused by active treatment. Under these concerns, Li et al. (2004) proposed a linear mixed effects regression model for log-transformed QT and RR,

$$y_{k,gij} = \alpha_{k,g} + a_{gi} + (\beta_{k,g} + b_{gi})x_{k,gij} + \varepsilon_{k,gij} \tag{8.5}$$

where $\alpha_{\mathrm{t,g}}$ and $\alpha_{\mathrm{p,g}}$ are the log(QTc)s (intercepts) for the treatment and placebo periods of gender g, respectively, and $\beta_{\mathrm{t,g}}$ and $\beta_{\mathrm{p,g}}$ are the corresponding slopes between log(QT) and log(RR). Subject random variables a_{gi} and b_{gi} are the same across placebo and test treatment periods. Therefore, Li et al. (2004) assumed that treatment is assumed to shift only the slope and the intercept in the population average level, but not the random effects for each subject. The random effect $\{a_{gi}, b_{gi}\}$ is i.i.d.N(0, d) with $d = \begin{pmatrix} \sigma_a^2, \rho\sigma_a\sigma_b \\ \rho\sigma_a\sigma_b, \sigma_b^2 \end{pmatrix}$ is a 2×2 covariance matrix and $\varepsilon_{k,gij}$ is i.i.d.N(0, σ_0^2). All the α's and β's in the model are estimated jointly by maximizing restrictive maximum likelihood (REML). The true QTc based on the mixed effect model proposed by Li et al. (2004) is hence

$$E(\mathrm{QTc}_{\mathrm{k,male}}) = \mathrm{e}^{\alpha_{k,1}}, \; E(\mathrm{QTc}_{\mathrm{k,female}}) = \mathrm{e}^{\alpha_{k,2}}$$

Model (8.4) leads to the one-step analysis of TQT studies to incorporate QT correction for RR or HR. But on the other hand, assessment of treatment-induced QT prolongation is a non-inferiority test with a margin in original scale. Modeling log transformed data leads to complication in one-step correction-treatment comparison analysis as discussed in Sect. 8.5.

8.3 Assessment of Treatment-Induced Prolongation of QTc Interval

Let $\mu_T(j)$ and $\mu_P(j)$ denote the mean QTc (or baseline-adjusted QTc) interval of test and placebo treatment at time point j. Let $\delta_{TP}(j) = \mu_T(j) - \mu_P(j)$ be the test and placebo difference of mean QTc interval at time point j. As stated in Sect. 8.1, the objective of a thorough QT study is to assess if the test treatment induces medically significant increase in QTc interval on the health subjects. ICH E14 stated that as a reference 10 ms is chosen as the cutoff size of increase. If the test treatment increases QTc interval, the maximum increase is often happened around the time of maximum concentration of dose, T_{max} observed in early bioavailability study. Hence let $j = 1$ to J be time points selected around T_{max}, the ICH E14 maximum-change approach can be presented as testing the following hypotheses:

$$H_0 : \text{Max}_{j=1\text{to}J}\,[\delta_{TP}\,(j)] \geq 10 \text{ against } H_a : \text{Max}_{j=1\text{to}J}\,[\delta_{TP}(j)] < 10 \tag{8.6}$$

In practice, it is represented by testing the following null hypotheses:

$$H_0(j) : \delta_{TP}(j) \geq 10 \text{ against } H_a(j) : \delta_{TP}(j) < 10 \text{ at all } j = 1\text{to}J \tag{8.7}$$

The trial is concluded negative (i.e., "no test treatment induced prolongation of QT interval") when all J null hypotheses in (8.7) are rejected. In practice, TQT trials are mostly carried out in single clinical center. The null hypotheses are rejected by showing that

$$T_j = \frac{\hat{\delta}_{TP}(j) - 10}{e(\hat{\delta}_{TP}(j))} < t(n^*; 0.05) \tag{8.8}$$

where $\hat{\delta}_{TP}(j)$ is the unbiased estimate of $\delta_{TP}(j)$, $e(\hat{\delta}_{TP}(j))$ is the standard error of $\hat{\delta}_{TP}(j)$, and $t(n^*,\ 0.05)$ is the fifth percentile of t distribution of degree freedom n^* (determined by the sample sizes of test and placebo group and design model).

Equivalently, one may reject each of the null hypotheses of (8.7) by showing that the upper confidence limit U of the 90% confidence interval of $\delta_{TP}(j)$,

$$U = \hat{\delta}_{TP}(j) + t(n^*; 0.95)e(\hat{\delta}_{TP}(j)) < 10.$$

at each time point. It has been shown that this intersection-union test approach controls type I error rate α at 0.05 and it is approximately 5% when QT interval prolongation at single time point (Zhang and Machado 2008). Furthermore, it was shown by Berger (1989) to be the uniformly most powerful among all monotone, α-level tests for the linear hypotheses (8.7). However, it is conservative with low power to reject the null hypothesis of (8.6). The power is lower when J is larger.

In fact, it was shown that $\text{Max}_{j=1\text{ to }J}\ \hat{\delta}_{TP}(j)$ is a biased estimate of $\text{Max}_{j=1to J}\delta_{TP}(j)$ and $E[\text{Max}_{j=1\text{ to }J}\hat{\delta}_{TP}(j)] \geq \text{Max}_{j=1\text{ to }J}\delta_{TP}(j)$. The inequality exists when $\hat{\delta}_{\mathrm{TP}}(j) > 0$. Boos et al. (2007) and Cheng et al. (2008) have shown that $\sqrt{n}[\mathrm{Max}_{j=1\,\mathrm{to}\,J}\hat{\delta}_{\mathrm{TP}}(j) - \mathrm{Max}_{j=1\,\mathrm{to}\,J}\delta_{\mathrm{TP}}(j)] \to \mathrm{N}(0,\ 2(\sigma_1^2 + \sigma^2))$, where n is the number of subjects, σ_1^2 and σ^2 are the intrasubject and intersubject variances at each time point in a crossover clinical trial. Let $\delta_{\mathrm{TP}}(j_0) = \mathrm{Max}_{j=1to J}\delta_{\mathrm{TP}}(j)$, Boos et al. (2007) showed that the size of bias is

$$\left[\sum_{j=1,\neq j_0}^{J} \delta_{\mathrm{TP}}(j) \int_{a_j}^{\infty} \varphi_J(z, R_j) - \mathrm{Max}_{j=1\,\mathrm{to}\,J}\delta_{\mathrm{TP}}(j)\right] \int_{-\infty}^{a_{j0}} \varphi_J(z, R_{j_0}) + \sqrt{\frac{2(\sigma_1^2 + \sigma^2)}{n} \sum_{j=1}^{J} \int_{a_j}^{\infty} z_j \varphi_J(z, R_k)} \tag{8.9}$$

where $a_j = \{a_{jj'}\}$ is a vector with $a_{jj'} = \sqrt{n}[\delta_{TP}(j') - \delta_{TP}(j)]/(2\sigma)$ for $j \neq j'$, $j,\ j' = 1$ to J and $a_{jj'} = \infty$; $R_K = \{r_{j,j'j''}\}$ with $r_{j,j'j''} = \frac{\sigma^2}{(\sigma_1^2+\sigma^2)}$ for $j' \neq j''$ and $r_{j,j'j''} = 1$ for $j',\ j,\ j'' = 1$ to J; $\phi_J(z, R_j)$ is the density of J-variate normal distribution with R_j as its covariance matrix.

Boos et al. (2007) provided also the second moment of $\mathrm{Max}_{j=1\,\mathrm{to}\,J}\hat{\delta}_{\mathrm{TP}}(j)$ as follows:

$$E\left\{(\mathrm{Max}_{j=1\,\mathrm{to}\,J}[\hat{\delta}_{\mathrm{TP}}(j)]\right\}^2 = \sum_{j=1}^{J} \delta_{\mathrm{TP}}(j)^2 \int_{a_j}^{\infty} \varphi_J(z.R_j) + \sqrt{\frac{2(\sigma_1^1 + \sigma^2)}{n} \sum_{j=1}^{J} \int_{a_j}^{\infty} z_j \varphi_J(z, R_J) + \frac{2(\sigma_1^2 + \sigma^2)}{n} \sum_{j=1}^{J} \int_{a_j}^{\infty} z_j^2 \varphi_J(z, R_K)} \tag{8.10}$$

and proposed three approaches of bias-corrected confidence interval of $\mathrm{Max}_{j=1\mathrm{to}\ J}\delta_{\mathrm{TP}}(j)$. However, $\mathrm{Var}[\mathrm{Max}_{j=1\,\mathrm{to}\,J}\hat{\delta}_{\mathrm{TP}}(j)]$ is a function of $\delta_{\mathrm{TP}}(j)$. It leads to the difficulties in bias correction of the confidence interval of $\mathrm{Max}_{j=1\,\mathrm{to}\,J}\delta_{\mathrm{TP}}(j)$. Tsong et al. (2009) showed that all three proposed approaches lead to inflation of type I error rate. Formula (8.10) leads also to the problem of unequal variances of $\mathrm{Max}_{j=1\,\mathrm{to}\,J}\hat{\delta}_{\mathrm{TP}}(j)$, $\mathrm{Max}_{j=1\,\mathrm{to}\,J}\hat{\delta}_{\mathrm{TT}}(j)$(the maximum estimate of mean difference between two test groups), and $\mathrm{Max}_{j=1\,\mathrm{to}\,J}\hat{\delta}_{\mathrm{PP}}(j)$ (the maximum estimate of difference between two placebo groups). Hence nonparametric test procedure including permutation or randomization tests may lead to inflation of type I error rate (Tsong et al. 2010a).

Garnett et al. (2008) proposed to assess treatment-induced prolongation of QT interval by comparing the predicted maximum $\Delta\Delta$QTc (treatment QTc difference adjusted by baseline measurement) with 10 ms. It is done by modeling the relationship between concentration (log(concentration)) and $\Delta\Delta$QTc (or log($\Delta\Delta$QTc)) using a linear model. Tsong et al. (2008b) raised the issue on model fitness for the purpose of prediction. The linearity fails when the drug-induced QTc response reaches maximum prior to or after the maximum of drug concentration. Meng (2008) proposed criteria to assess the lag time of QTc response. On the other hand, in order to assess the maximum drug-induced prolongation of QTc interval through concentration-response modeling, one needs to have $\Delta\Delta$QTc (or log(Δ ΔQTc)) at individual subject level. Thorough QT clinical trials are designed either with parallel arms or randomized crossover design. Using the design model, statistical comparison of mean treatment difference will be made with adjustment for other design factors. One may not derive treatment difference at individual subject level based on the analysis with the design model. Concentration-response modeler often took the crossover trial data as if collected in a paired trail and ignored other design factors of the clinical trial. For example, with a crossover clinical trial, period and sequence factors of the design are incorrectly eliminated by the modeler when pairing the treatments of the individual subject. Furthermore, no individual subject $\Delta\Delta$QTc (or log($\Delta\Delta$QTc)) may be obtained when TQT is a parallel arm trial.

8.4 Validation Test

In a thorough QT/QTc trial, positive control treatment is used for the purpose of clinical trial validation (i.e., assay sensitivity) only. They are particularly useful when the outcome of the trial is negative. ICH E14 guidance states that "The positive control should have an effect on the mean QT/QTc interval of about 5 ms (i.e., an effect that is close to the QT/QTc effect that represents the threshold of regulatory concern, around 5 ms)." In current practice, validation is carried out by a statistical hypothesis testing and an informal profile validation procedure. For example, Moxifloxacin, a most frequently used positive control treatment, has a well-document profile (Fig. 8.2). Profile validation is carried out by comparing the profile of TQT with the standard profile.

For the non-profile validation test, let $\mu_{C(j)}$ and $\mu_{P(j)}$ denote the mean QTc interval at jth time point after adjusting for baseline measurement. Let $\delta_{CP}(j) = \mu_{C(j)} - \mu_{P(j)}$. A statistical hypothesis testing is also used for non-profile validation of the TQT. The hypotheses of interest are

$$H_0 : \text{Max}_{j=1 \text{ to } J}\delta_{CP}(j) \leq 5 \text{ ms}$$

versus

$$H_a : \text{Max}_{j=1 \text{ to } J}\delta_{CP}(j) > 5 \text{ ms}, \tag{8.11}$$

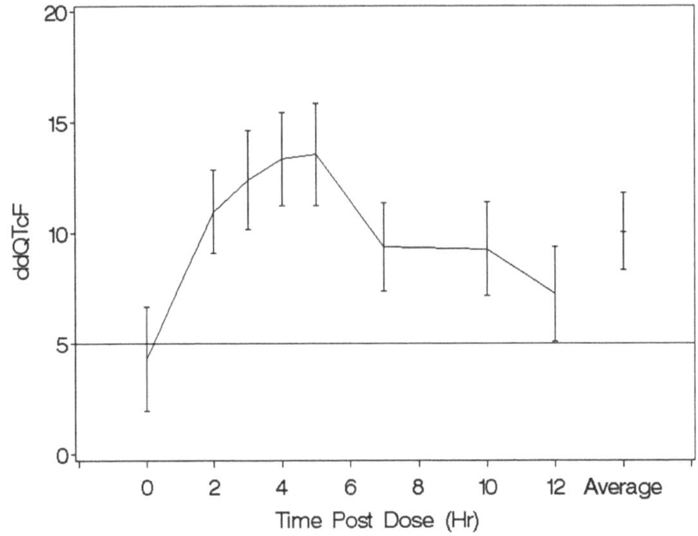

Fig. 8.2 General time course of Moxifloxacin effect of a double-blind randomized study. (Mean baseline adjusted Moxifloxacin - placebo difference and its 90% confidence interval)

In practice, in a setup similar to the non-inferiority test of treatment-induced QT prolongation, the conventional approach is to test the J sets of hypotheses,

$$H_0(j) : \delta_{CP}(j) \leq 5 \text{ ms against } H_a(j) : \delta_{CP}(j) > 5 \text{ ms for } j = 1 \text{ to } J \quad (8.12)$$

Validation holds if at least one of the null hypotheses is rejected. Equivalently, one may do that by showing that the lower limit of at least one confidence interval of $\delta_{CP}(j)$ is greater than 5 ms, i.e.,

$$\text{L} = \hat{\delta}_{CP}(j) + t(n^*; \alpha^*)e(\hat{\delta}_{CP}(j)) > 5. \quad (8.13)$$

for at least one j, $j = 1$ to J with an properly adjusted type I error rate α^*.

Based on the same discussion given in Sect. 8.3, $\text{Max}_{j=1 \text{ to } J}\hat{\delta}_{CP}(j)$ is a biased estimate of $\text{Max}_{j=1 \text{ to } J}\delta_{CP}(j)$. An appropriate bias correction method is yet to be developed for testing hypotheses (8.11). On the other hand, testing (8.12) using $\alpha = 0.05$ for each individual test would lead to the inflation of family type I error rate through multiple comparisons. A simple conservative approach would be using Bonferroni adjustment with $\alpha^* = 0.05/\text{J}$ for each individual test in the conventional intersection-union test for hypotheses (8.12) or estimate each individual confidence interval in (8.13). Power of the validation test can be improved using many Bonferroni modified (stepwise) procedures. For example, Let $p_{(1)} \geq p_{(2)} \geq \ldots \geq p_{(J)}$ be the ordered p-values of the J tests, and $H_{0(1)}$, $H_{0(2)}, \ldots, H_{0(J)}$ be the corresponding ordered null hypotheses. Using Holm's procedure (1979), the study is validated if $p_{(J)}$, the smallest p-value, is less than 0.05/J. Validation is failed if $p_{(J)} \geq 0.05$. On the other hand, using Hochberg's procedure (1988), the study is validated if $p_{(j)} < 0.05/\mathrm{j}$ for any $j = 1, \ldots, J$. Furthermore, when the covariance structure of measurements across time points of the positive control and placebo is known, power can further improved in Holm's procedure. For example, Tsong et al. (2010b) explored the issue with Moxifloxacin as the positive control treatment.

In order to compensate the conservativeness of Bonferroni-type adjustment for multiple comparisons, Zhang (2008) proposed a global test as an alternative to the multiple tests of (8.12). Tsong et al. (2008a) proposed to test (8.11) using both the intersection-union test (8.12) and the global test approach proposed by Zhang (2008) to improve the power of the validation test. This approach was pointed out to be inflating type I error rate in general and may be useful if the QT change profile is in hill or spike shape (Sun and Quan (2009)).

8.5 Designs of Thorough QT Trials

For a typical TQT study, a randomized three or four treatment group design is considered. For demonstration purposes, we focus on a three treatment group case. The three treatment arms are: (1) T—test drug, (2) C—positive control, and (3) P—placebo. A TQT study may have either a crossover design or a parallel design.

In a TQT study, due to large intrasubject variability, the baseline measurements are recommended for OT correction by RR interval. For a parallel study, a whole day of time-matched baseline measurements on the day prior to dosing is customarily collected in order to help adjust for within-subject diurnal variability. For a crossover design, however, it is not essential to have a full day time-matched baseline at the previous day for each period; and baseline measurements at each period, just before treatment, might be enough. This is because in a crossover trial diurnal variability of the QTc is accounted for by the design itself since each subject receives each of the four treatment groups in four post dose periods at exactly the same time points. It is recommended (Patterson et al. 2005; Chow et al. 2008) that replicate ECG at each time point should be collected, and this certainly applies to the baseline ECGs collection. Then within each period, the average of all the measurements before dosing can be used as the baseline value.

A parallel group design may have to be used for drugs with long elimination half-lives, or with metabolites, or a large number of dose groups to be evaluated. With a parallel arm design, subjects will be randomized to each one of the four groups. The choice of time points as well as the number of time points to evaluate the drug's QTc effect should be driven by the pharmacokinetic profile of the drug. The duration of a parallel study might be relatively shorter and the dropout rate might be smaller in compare with a crossover study. The biggest disadvantage of a parallel design is the size of the trial.

A crossover design is often used in TQT trial because it needs much fewer subjects than the parallel arm designs for the same study power, since the precision of the estimated treatment differences with the same subject acts as her/his own control, is greater than that for a parallel design with the same number of subjects. One of the major concerns of a crossover trial is the potential effect of one treatment carried over to the periods follow behind. Hence it is important to have sufficient washout time between two periods.

For a crossover design, a sequence is a prespecified ordering of the treatments to the periods. The randomization of clinical trial is achieved by randomly assign the subjects one of the sequences of treatments. For example, the following sequences may be considered for a three arm trial:

Latin Square 1: Sequence #1: T/P/C;
Sequence #2: C/T/P;
Sequence #3: P/C/T;
Latin Square 2: Sequence #4: T/C/P;
Sequence #5: C/P/T;
Sequence #6: P/T/C;

Here, T, C, and P are the test, positive control and placebo treatment, respectively.

A sequence-balanced complete crossover trial design for more than $K(> 2)$ treatments is often represented by a group of Latin squares. A general Latin square is a design square formed by K distinct sequences of treatment such that the square requires K treatments appear in each row (i.e., sequence) and in each column (i.e., period) exactly once. The first three sequences and the last three sequences given above form two different Latin squares. For a study with 48 subjects, a complete balanced crossover design is to have eight subjects randomized to each of the above six sequences. The most frequently used crossover design in TQT trial is Williams design (Williams 1949). The Williams design consists of one or two specific Latin squares so that two treatments appear in two consecutive periods and in reversed order exactly once. A Williams design can be achieved by a single particular Latin square for an even number of treatments or two particular Latin squares for an odd number of treatments. A Williams design for the three treatments can be formed by the two Latin squares (of six sequences) given previously in this Section. The Williams design is balanced for the first-order carryover effects. It is a variance-balanced design, which means that the variance of difference between any two estimated direct treatment effects is the same so that each treatment will be equally precisely compared with every other treatment (Jones and Kenward 2003). The direct-by-carryover interaction is significant if the carryover effect of a treatment depended on the treatment applied in the immediately subsequent period. If the study model includes period, sequence, treatment, first-order carryover, and direct-by-carryover interaction as fixed effects, the design based on one Williams square or two Williams squares might not have sufficient degrees of freedom to assess direct-by-carryover interaction (Williams 1949). If this is the case, one might consider using a design with repeated multiple Latin squares, Williams designs or complete orthogonal sets of Latin squares. For example, with 48 subjects and three treatments, it can be accomplished by eight sets of the two Latin squares given above. The two Latin squares presented above are orthogonal, which means that when they are superimposed, any combination of the two treatment arms from the two different Latin squares occurs exactly once.

It is also important to note that when repeated using the same Williams square in the trial (to have P precedes C (first order) exactly the same number of times of any two treatments in exactly the same ordering in each square), one may loss the needed degrees of freedom for the analysis of some higher level carryover effect. This problem can be avoided by using the completely orthogonal set of Latin squares. A completely orthogonal set of Latin squares consists of $K - 1$ squares. For example, for the three treatments, the two Latin squares given above form a completely orthogonal set of Latin squares. Again, for the three treatments, with 48 subjects, the Williams design is also a completely orthogonal set of Latin squares (Chen and Tsong 2007; Zhang and Machado 2008).

Let us consider the linear models for the data analysis of the designs. Let us first assume Z_{kij} denote the difference between QTc of the subject i of treatment group k (k = T, C, and P) and baseline QTc value, at time point j. For the parallel arm trial, depending on the covariance structure of Z_{kij}, many linear models may be used to model the relationship of Z_{kij} and the factors. Let k_0 denote the baseline of treatment k and Z_{k_0ij}denote the baseline QTc value. In most situations, one needs to assume that there is interaction between treatment and time and unequal variances across time, and the data need to be modeled at each time point separately as follows:

$$Z_{kij} = R_{kj} + \beta Z_{k_0ij} + \varepsilon_{kij} \tag{8.14}$$

where $R_{kj} + \beta Z_{k_0ij}$ is the fixed effects part of the model, treatment + baseline QTc; ε_{kij} is the random error with variance σ_{kj}^2 for the k-the treatment group. Note that Z_{k_0ij}, the time-matched baseline QTc, value is included as a covariate to prevent conditional bias as described in Sect. 8.2. When the baseline QT measurement is a single pre-dose value, $Z_{k_0ij} = Z_{k_0i}$. The coventional approach for testing hypotheses (8.7) and (8.12) does not include the covariate Z_{k_0ij} in the model.

Under the assumption of equal variance at each time point, the following linear model may be used:

$$Z_{kij} = R_k + T_j + (\mathrm{RT})_{ij} + \beta Z_{k_0ij} + \varepsilon_i(k) + \varepsilon_{kij} \tag{8.15}$$

where $R_k + T_j + (\mathrm{RT})_{ij} + \beta Z_{k_0ij}$ is the fixed effects part of the model, treatment + time + treatment by time + baseline QTc; $\varepsilon_i(k)$ with variance σ_i^2 is a random effect for the ith subject in the k-the treatment group, and ε_{kij} is the random error with with variance σ^2. Note that the random effect $\varepsilon_i(k) + \varepsilon_{kij}$ in (8.15) is the same as the random error ε_{kij} in model (8.14) for a given time j under equal variance in treatment assumption.

Now let us consider applying various correction methods stated in Sect. 8.2 to treatment comparison in (8.14) and (8.15).

When an FC method is used, i.e., QTc $\alpha_{kij} = y_{kij} + \beta x_{kij}$, for a given β, $Z_{kij} = \exp(\alpha_{kij} - \alpha_{k_0ij})$ for time-matched baseline, or $Z_{kij} = \exp(\alpha_{kij} - \alpha_{k_0i})$ for single pre-dose baseline.

When a PBC method is used, $\alpha_{kij} = y_{kij} - \beta x_{kij} - \varepsilon_{kij}$. If the model fits with log-transformed QT values, $Z_{kij} = \exp(\alpha_{kij} - \alpha_{k_0ij}) = \exp[(y_{kij} - \beta x_{kij} - \varepsilon_{kij}) - (y_{k_0ij} - \beta x_{k_0ij} - \varepsilon_{k_0ij})]$, one-step modeling (8.14) or (8.15) including QT correction directly becomes extremely complicated in the composite form of an additive and a multiplicative model. A conventional two-step approach would use actually $Z_{kij} = \exp(\hat{\alpha}_{kij} - \hat{\alpha}_{k_0ij}) = \exp[(y_{kij} - \hat{\beta} x_{kij}) - (y_{k_0ij} - \hat{\beta} x_{k_0ij})]$ in model (8.14) or (8.15) and improperly left out the estimation error of β. On the other hand, if a linear model of PBC method fits well with QT and RR or HR data,

$$\begin{aligned} Z_{kij} &= (\alpha_{kij} - \alpha_{k_0ij}) = (y_{kij} - \beta x_{kij} - \varepsilon_{kij}) - (y_{k_0ij} - \beta x_{k_0ij} - \varepsilon_{k_0ij}) \\ &= (y_{kij} - y_{k_0ij}) - \beta(x_{kij} - x_{k_0ij}) - (\varepsilon_{kij} - \varepsilon_{k_0ij}). \end{aligned}$$

Substituting Z_{kij} in model (8.14) or (8.15) without the the covariate term βZ_{k_0ij}, the linear model provides a one-step approach for treatment comparison of the thorough QT studies such that

$$(y_{kij} - y_{k_0ij}) = \beta(x_{kij} - x_{k_0ij}) + R_{kj} + \varepsilon_{kij} \tag{8.16$'$}$$

$$(y_{kij} - y_{k_0ij}) = R_k + T_j + (RT)_{ij} + \beta(x_{kij} - x_{k_0ij}) + \varepsilon_i\,(k) + \varepsilon_{kij} \tag{8.17$'$}$$

Similarly, when a linear IBC method is applied, $Z_{kij} = (\alpha_{kij} - \alpha_{k_0ij}) = (y_{kij} - \beta_{kij} x_{kij} - \varepsilon_{kij}) - (y_{k_0ij} - \beta_{kij} x_{k_0ij} - \varepsilon_{k_0ij}) = (y_{kij} - y_{k_0ij}) - \beta_{kij}(x_{kij} - x_{k_0ij}) - (\varepsilon_{kij} - \varepsilon_{k_0ij})$, where α_{kij} and α_{k_0ij} are QTc, x_{kij} and x_{k_0ij} are RR of treatment k and baseline of treatment k. The one-step linear model is represented by

$$(y_{kij} - y_{k_0ij}) = \beta(x_{kij} - x_{k_0ij}) + R_{kj} + \varepsilon_{kij} \tag{8.18$''$}$$

and

$$(y_{kij} - y_{k_0ij}) = \beta_{kij}(x_{kij} - x_{k_0ij}) + R_k + T_j + (RT)_{ij} + \varepsilon_i(k) + \varepsilon_{kij} \tag{8.19$''$}$$

Let us consider the mixed effects model of the Williams design. Let Z_{kijl} be QTc, the corrected QT measurement adjusted for baseline QT of the k-th treatment of the i-th subject observed at the j–th time point of the l-th period; $i = 1, \ldots, n,\ j = 1, \ldots J$, and $l = 1, \ldots, L$. The model of this crossover trial is

$$Z_{kijl} = \mu + \tau_k + t_j + \eta_l + \tau_k \cdot t_j + \tau_k \cdot \eta_l + b_i + \varepsilon_{ij} \tag{8.20}$$

where μ is the overall mean response, $\tau_k + t_j + \eta_l + \tau_k \cdot t_j + \tau_k \cdot \eta_l$ are the fixed effects of treatment + time + period + treatment-time interaction + treatment-period interaction, $b_i \sim N(0, \sigma_s^2)$ are independent and identically distributed normal random subject effects with mean 0 and variance σ_s^2, and $(\varepsilon_{i1}, \ldots, \varepsilon_{iJ}), i = 1, \ldots,$ n are independent and identically distributed as

$$\begin{pmatrix} \varepsilon_{i1} \\ \vdots \\ \varepsilon_{iJ} \end{pmatrix} \sim \text{iid}\,\mathrm{N}_j \left[0, \begin{bmatrix} \sigma_{11} & \sigma_{12} & \cdot & \sigma_{1J} \\ \sigma_{21} & \sigma_{22} & \cdot & \cdot \\ \cdot & \cdot & \cdot & \cdot \\ \cdot & \cdot & \cdot & \cdot \\ \cdot & \cdot & \cdot & \cdot \\ \sigma_{J1} & \cdot & \cdot & \sigma_{JJ} \end{bmatrix} \right]$$

Note that $\sigma_{jj}^2 = \sigma^2$ when the variances are equal across all time points and treatment. Again, when incorporating the correction method in the analysis, for an FC method, $Z_{kij} = \exp(\alpha_{kij} - \alpha_{k_0 ij})$ using Bazett's or Fridericia's correct method. When a PBC or IBC method is used with a linear model on log(QT) and log(RR), we have the same difficulty in linear modeling of (8.20). When PBC is used with linear model on QT and RR values, the one-step treatment comparison model including PBC becomes

$$(y_{kijl} - y_{k_0 ijl}) = \mu + \beta(x_{kijl} - x_{k_0 ijl}) + \tau_k + t_j + \eta_l + \tau_k \cdot t_j + \tau_k \cdot \eta_l + b_i + \varepsilon_{ij} \quad (8.21')$$

With IBC, the model becomes

$$(y_{kijl} - y_{k_0 ijl}) = \mu + \beta_{kijl}(x_{kijl} - x_{k_0 ijl}) + \tau_k + t_j + \eta_l + \tau_k \cdot t_j + \tau_k \cdot \eta_l + b_i + \varepsilon_{ij} \quad (8.22'')$$

In many crossover design, there is a single pre-dose QT and RR value in each period and $y_{k_0 ijl} = y_{k_0 il}$ and $x_{x_0}^{ijl} = x_{k_0 il}$.

8.6 Discussion on Conclusions

The history of statistical design and analysis of thorough QT (TQT) studies is short. TQT is typically carried out at pre-phase III stage. Before the publication of ICH E14 and FDA drug regulatory requirement, the study is most frequently carried out by clinical pharmacologist without much involvement of statisticians. However, as a part of the NDA requirement for most of new drugs, efficient designs and stringent statistical analysis procedures were proposed in the NDA submission and research publication. This chapter covers some of the basic considerations in design and analysis of the new area of regulatory statistics. There are some more advanced topics such as dealing with missing data, advanced crossover-parallel arm hybrid design and adaptive design and analysis (Zhang 2009; Tsong et al. 2009), and modeling the repeatedly measured data (Li et al. 2009; Ring 2009; Liu et al. 2009) that were proposed and discussed by the researchers. We regret that we would not be able to include them in this chapter.

Acknowledgments This chapter was prepared with many inputs from statistical reviewers of CDER QTc team of the Office of Biostatistics. The authors like to thank Drs. Stella Machado and Joanne Zhang of CDER, FDA for the support and discussion on the development of this chapter.

References

Ahnve S (1985) Correction of the QT interval for heart rate: review of different formulae and the use of Bazett's formula in myocardial infarction. American Heart J 109:568–574

Batchvarov V, Malik M (2002) Indivdual patterns of QT/RR relationship. Card Electrophysiol Rev 6:282–288

Batchvarov VN, Ghuran A, Smetana P, Hnatkova K, Harries M, Dilaveris P, Camm AJ, Malik M (2002) QT-RR relationship in healthy subjects exhibits substantial intersubject variability and high intrasubject stability. Am J Physiol Heart Circ Physiol 282:H2356–H2363

Bazett JC (1920) An analysis of time relations of electrocardiograms Heart 7:353–367

Berger RL (1989) Uniformly most powerful tests for hypotheses concerning linear inequalities and normal means. J Amer Statist Assoc 84 192–199

Boos DD, Hoffman D, Kringle R, Zhang J (2007) New confidence bounds for QT studies. Stat Med 26:3801–3817

Chen L, Tsong Y (2007) Design and analysis for drug abuse potential studies: issues and strategies for implementing a crossover design. Drug Info J 41:481–489

Cheng B, Chow S-C, Burt D, Cosmatos D (2008) Statistical assessment of QT/QTc prolongation based on maximum of correlated normal random variables. J Biopharm Statist 18(3):494–501

Desai M, Li L, Desta Z, Malik M, Flockhart D (2003) Variability of heart rate correction methods for the QT interval: potential for QTc overestimation with a low dose of haloperidol. Br J Clin Pharmacol 55(6):511–517

Fridericia LS (1920) Die Systolendauer im Elekrokardiogramm bei normalen Menschen und bei Herzkranken. Acta Medica Scandinavica 53:469–486

Funk-Brentano C, Jaillon P (1993) Rate-corrected QT intervals: techniques and limitations Am J Cardiol 72:17B–23B

Garnett, C. E., Beasley, N., Bhattaran, V. A., Jadhav, P. R., Madabushi, R., Stockbridge, N., Tornee, C. W., Wang, Y., Gobburu, J. V. (2008). Concentration-QT relationships play a key role in the evaluation of proarrhythmic risk during regulatory review. J. Clin. Pharmacol. 48:13–18.

Hnatkova K, Malik M (2002) Optimum formulae for heart rate correction of the QT interval. Pacing Clin Electrophysiol 22:1683–1687

Hochberg Y (1988) A sharper Bonferroni procedure for multiple tests of significance. Biometrika 75:800–802

Hodges M, Salerna D, Erline D (1983) Bazett's QT correction reviewed: evidence that a linear correction for heart rate is better. J Am Coll Caddiol 1:694 [Abstract]

Holm S (1979) A simple sequentially rejective multiple test procedure. Scan J Statist 6:65–70

International Conference on Harmonisation, ICH (E14) Guidance (2005) The Clinical Evaluation of QT/QTc Interval Prolongation and Proarrhythmic Potential for Non-Antiarrhythmic Drugs. Geneva, Switzerland: International Conference on Harmonisation; May 2005. Available at http://www.ich.org/

Jones B, Kenward MG (2003) Design and analysis of cross-over trials, 2nd edn. Chapman and Hall, London

Li J (2009) Use of PK/PD modeling principle to address QT/QTc interval prolongation. DIA Cardiovascular safety, QT, and arrhythmia in drug development conference. April 30-May 1, 2009 at Bethesda, MD

Li L, Desai M, Desta Z, Flockhart D (2004) QT analysis: a complex answer to a 'simple' problem. Stat in Medicine 23:2625–2643

Liu W, Bretz F, Hayter AJ, Wynn HP (2008) Assessing nonsuperiority, noninferiority, or equivalence when comparing two regression models over a restricted covariate region. Biometrics, DOI:10,1111/j.1541-0420.2008.01192.x

Malik M (2001) Problems of heart rate correction in assessment of drug-induced QT interval prolongation. J Cardiovasc Electrophysiol 12:411–20

Malik M, Farbom P, Batchvarov V, Hnatkova K, Camm AJ (2002) Relation between QT and RR intervals is highly individual among healthy subjects: implications for heart rate correction of the QT interval. Heart 87:220–228

Malik M, Hnatkova K, Batchvarov V (2004) Differences Differences between study-specific and subject-specific heart rate corrections of the QT interval in investigations of drug induced QTc prolongation Pace 27:791–800

Meng Z (2008) Simple direct QTc-exposure modeling in thorough QT studies. 2008 FDA/Industry Workshop, Rockville, MD

Moss AJ (1993) Measurement of the QT interval and the risk associated with QTc interval prolongation: a review. Am J Cardiol 72:23B–25B

Patterson S, Agin M, Anziano R et al (2005) Investigating drug induced QT and QTc prolongation in the clinic: statistical design and analysis considerations. Drug Inform J 39:243–266

Rautaharju PR, Zhou SH, Wong S, Prineas R, Berenson G (1993) Function characteristics of QT prediction formulas. The concept of QT_{max} and QT rate sensitivity. Computer Biomed Res 26:188–204

Ring A (2009) Impact of delayed effects in the exposure-response analysis of clinical QT trials. DIA Cardiovascular safety, QT, and arrhythmia in drug development conference. April 30-May 1, 2009 at Bethesda, MD

Schlamowitz I (1946) An analysis of the time relationship within the cardiac cycle in electrocardiograms of normal man. I. The duration of the QT interval and its relationship to the cycle length (R-R interval). Am Heart J 31:329

Sun G, Hui Q (2010) Letter to the Editor: Revisit the combination validation test of Tsong et al, for thorough QT/QTc clinical trials. J of Biopharm Statist 20(3): 683–687.

Tsong Y, Zhong J, Chen WJ (2008a) Validation testing in thorough QT/QTc clinical trials. J of Biopharm Statist 18(3):529–541

Tsong Y, Shen M, Zhong J, Zhang J (2008b) Statistical issues of QT prolongation assessment based on linear concentration modeling. J of Biopharm Statist 18(3):564–584

Tsong Y, Zhang J, Zhong J (2009) Letter to Editor: Comment on new confidence bounds for QT studies In: Boos DD, Hoffman D, Kringle R, Zhang J. Stat Med, 2007: 26:3801–3817. Statist Medicine 28:2936–2940

Tsong Y, Zhong J (2010a) Multiple testing of repeated measured outcome: Assessment of treatment induced QT interval prolongation. J of Biopharm Statist 20(3):613–623.

Tsong Y, Yan L, Zhong J, Nie L, Zhang J (2010b) Multiple testing of repeated measured outcome: Validation test of thorough QT study using Moxiflozacin. J of Biopharm Statist

Tsong Y, (2013) On the designs of QTc clinical trials. To appear in J Biopharm Statist

Wang Y, Pan G, Balch A (2008) Bias and variance evaluation of QT interval correction methods J of Biopharm Statist 18(3):427–450

Williams EJ (1949) Experimental designs balanced for the estimation of residual effects of treatments. Australian J Sci Res 2:149–168

Wysowski DK, Corken A, Gallo-Torres H, Talarico L, Rodriguez EM (2001) Postmarketing reports of QT prolongation and ventricular arrhythmia in association with cisapride and Food and Drug Administration regulatory actions. Am J Gastroenterol 96(6):1698–1703

Zhang J (2008) Testing for positive control activity in a thorough QTc study. J of Biopharm Statist 18(3):517–528

Zhang J (2009) Moxifloxacin and placebo can be given in a crossover fashion during a parallel-designed thorough QT study. DIA Cardiovascular safety, QT, and arrhythmia in drug development conference. April 30–May 1, 2009 at Bethesda, MD

Zhang J, Machado SG (2008) Statistical issues including design and sample size calculation in thorough QT/QTc studies. J of Biopharm Statist 18(3):451–467

Chapter 9
Causal Inference in Cancer Clinical Trials

Babette A. Brumback and Wendy B. London

9.1 Introduction

It is generally accepted that results of a randomized clinical trial (RCT) can be interpreted causally, despite what is known as the Fundamental Problem of Causal Inference (FPCI, Holland 1986). Briefly, the FPCI states that because we can observe only one of a patient's potential outcomes to a new or a standard treatment, we can never determine which treatment would have caused a better outcome relative to the other. The potential outcomes framework for causal inference was originally introduced by Neyman (1923), Holland (1986) reviews it in great detail and explains how the "statistical solution" to the FPCI involves estimating an average difference in potential outcomes using the observed difference in mean outcomes across the two treatment groups, assuming the data resulted from a simple RCT. The statistical solution allows for causal inference to be made (subject to the caveat that a statistically significant difference may be due to chance) about the population from which the patients in the RCT are "randomly" accrued; e.g., the new treatment improves survival relative to the standard treatment. In practice, accrual is typically not random, and the population is not always clearly defined. Under these circumstances, all one can readily conclude from an RCT showing superiority of the new treatment is that there exists a subgroup of individuals who would benefit; however, one cannot readily identify the subgroup, and there may even be individuals who would be harmed.

B.A. Brumback (✉)
Department of Biostatistics, University of Florida, Gainesville, FL 32611, USA
e-mail: brumback@ufl.edu

W.B. London
Dana-Farber Cancer Institute, Boston, MA, 02215
e-mail: wendy.london@childrens.harvard.edu

W. Tang and X. Tu (eds.), *Modern Clinical Trial Analysis*, Applied Bioinformatics and Biostatistics in Cancer Research, DOI 10.1007/978-1-4614-4322-3_9,

In recent years, statistical solutions to the FPCI have been developed for many different experimental and observational contexts. Rubin (1974, 1977, and 1978) used the potential outcomes framework to develop statistical methods for causal inference from observational studies based on the concept of ignorable treatment assignment (treatment is independent of the potential outcomes conditional on a sufficient set of confounders). Robins (1989, 1992, 1994, 1997, 1998, 1999a, b, 2000, 2004) and colleagues (e.g., Robins et al. 1992; Robins and Tsiatis 1992; Robins and Greeland 1994; Robins et al. 1999; Robins et al. 1999; Robins et al. 2000; Hernan et al. 2000, 2001, 2002; Murphy et al. 2001; van der Laan and Robins 2003; Brumback et al. 2004, Robins and Rotnitzky 2004) developed methods based on the concept of sequentially ignorability for the analysis of the joint effects of sequential treatments, typically in the context of either a purely observational study or an RCT with secondary nonrandomized treatments or exposures. Several researchers proposed methods for the analysis of clinical trials subject to noncompliance (e.g., Sommer and Zeger 1991, Robins 1994, Angrist et al. 1996, Balke and Pearl 1997, Goetghebeur and Lapp 1997, Robins and Rotnitzky 2004, Cheng and Small 2006). Sommer and Zeger (1991) were among the earliest, and began a tradition of estimating the causal effect of treatment in the so-called "compliers," those who would comply with either the new treatment or the standard treatment if offered. Compliance is a posttreatment variable, and as such one can conceptualize potential compliance outcomes under the new or the standard treatment. Estimating the effect of treatment in the "compliers" amounts to stratifying patients by their potential compliance outcomes. Frangakis and Rubin (2002) extended this approach to general posttreatment variables, including death, and termed the strata defined by the potential posttreatment variable outcomes as "principal strata." Using the principal stratification approach, one can estimate, for example, the effect of a new treatment on a non-survival outcome in the subgroup that would remain alive regardless of which treatments were assigned. In this way, one adeptly avoids the problem of conceptualizing what an outcome would have been had a patient not died. However, instead one must make modeling assumptions to circumvent the problem that the subset of patients randomized to a given treatment who do not die and who also would not have died had they been randomized to the other treatment is an unidentifiable group.

In this book chapter, we will focus on the context of an RCT with a subsequent nonrandomized treatment. It often happens in cancer clinical trials that a surrogate outcome rather than overall survival is used to compare two randomized treatments, for the practical purpose of shortening the duration of the trial and to avoid dilution, by subsequent nonrandomized treatment, of the treatment effect under study in the RCT. However, overall survival is nevertheless often of primary interest rather than a surrogate outcome. The difficulty that ensues is that subsequent to the assessment of the surrogate outcome and hence the end of the RCT, patients enroll in a nonprescribed fashion into subsequent trials or may be given additional treatment as part of routine practice. Thus it becomes exceedingly hard to disentangle the effect of the randomized treatment from the effects of the subsequent, nonrandomized (from the point of view of the original trial) treatments. In what follows, we will compare various methods for attempting to disentangle these effects.

9.1.1 *Motivating Example*

It is usually helpful to consider a specific scenario. The Children's Oncology Group (COG) (Pediatric Oncology Group at the time) study P9462 was a phase II study of relapsed neuroblastoma patients randomized to either Topotecan (TOPO, the standard therapy, $n = 59$) or Topotecan + Cytoxan (TOPO/CTX, the experimental therapy, $n = 58$). The primary outcome was response and was assessed using a two-stage group sequential design. Responders achieved either a complete response, a partial response, or less than a partial response to therapy. Criteria were met to proceed from stage 1 to stage 2. At stage 2, patients who received TOPO achieved a response in 19/59 (32 %) versus 26/58 (45 %) among those who received TOPO/CTX, but the design required a difference of eight or more to conclude superiority of TOPO/CTX. Although statistical significance was not achieved, it appeared there was a trend toward a higher response rate in those patients treated with combination therapy. In addition, there was no statistically significant effect of treatment arm on overall survival ($p = 0.65$ by log rank test).

Although response was the primary endpoint of the P9462 study, an analysis of the proportion of patients surviving at 2 years poses all of the challenges of an RCT with subsequent nonrandomized treatment, and therefore is perfect for the illustrative purposes of this chapter. A factor which complicates the survival analysis is that some patients ($n = 12$) received a bone marrow transplant (BMT) after the response assessment. Nine patients in the TOPO/CTX group had a posttreatment BMT versus only three in the TOPO group. Patients who achieved a response were more likely to have the posttreatment BMT than those who did not, in either treatment group. This raises the interesting question of how to adjust for the second treatment (the posttreatment BMT) in the comparison of survival. For expository purposes in this book chapter, we will focus on the proportion of patients surviving at 2 years.

9.2 Notation and Assumptions

9.2.1 *Observable Data*

The observable data are denoted with four binary variables (T_1, H_2, T_2, Y) measured in consecutive temporal order. T_1 equals 0 for randomization to TOPO and 1 for randomization to TOPO/CTX. H_2 equals 1 for a response to therapy and 0 otherwise. T_2 equals 1 for subsequent BMT and 0 otherwise. Y equals 1 for survival past 2 years and 0 otherwise. Response was measured prior to 2 years, and we assume that patients have been followed up long enough that there is no censoring. We also assume that it is hypothetically possible for each patient to be assigned to a subsequent BMT, even if $H_2 = 0$ due to death. This assumption would be plausible, for example, if the two treatments T_1 and T_2 had been randomly assigned

Table 9.1 The 16 causal types

U	Y_{00}	Y_{01}	Y_{10}	Y_{11}	Description
1	0	0	0	0	Doomed no matter what treatment
2	0	0	0	1	Needs both CTX and BMT to survive
3	0	0	1	0	Needs CTX and cannot have BMT to survive
4	0	0	1	1	Needs CTX to survive
5	0	1	0	0	Needs BMT and cannot have CTX to survive
6	0	1	0	1	Needs BMT to survive
7	0	1	1	0	Needs CTX or BMT but cannot have both to survive
8	0	1	1	1	Needs CTX or BMT to survive
9	1	0	0	0	Cannot have either in order to survive
10	1	0	0	1	Needs neither or both to survive
11	1	0	1	0	Cannot have BMT in order to survive
12	1	0	1	1	Cannot have BMT without CTX in order to survive
13	1	1	0	0	Cannot have CTX in order to survive
14	1	1	0	1	Cannot have CTX without BMT in order to survive
15	1	1	1	0	Cannot have both in order to survive
16	1	1	1	1	Survivor no matter what treatment

at baseline; patients not alive at H_2 would not actually be able to receive $T_2 = 1$ even had they been assigned to it, but as a BMT would not have been able to revive them, the outcome would be the same had they received it or not.

9.2.2 *Potential Outcomes*

Throughout the chapter, we will invoke four binary potential survival outcomes for each patient: Y_{00}, Y_{01}, Y_{10}, and Y_{11}, where Y_{ab} denotes the potential survival outcome to treatment with $T_1 = a$ and $T_2 = b$. Let the sixteen possible combinations of $(Y_{00}, Y_{01}, Y_{10}, Y_{11})$ be indexed by 1,...,16, and let U be a number from 1 to 16 denoting the potential survival outcome combination for each patient (see Table 9.1). When we focus on the principal stratification approach, we will make use of two additional binary potential outcomes: T_2^0 and T_2^1, which contain information on whether a patient would end up receiving a subsequent BMT if assigned to $T_1 = 0$ or assigned to $T_1 = 1$, respectively. When we focus on structural nested models, we will make use of the potential outcomes Y_0, the potential outcome to setting the actual treatments T_1 and T_2 to zero, which always equals Y_{00}, and Y_1, the potential outcome to setting just treatment T_2 to zero, which equals Y_{10} for patients with $T_1 = 1$ and Y_{00} for patients with $T_1 = 0$.

In this book chapter, potential outcomes will, for expository purposes, be assumed deterministic for each individual (though random for a randomly selected individual). It would certainly be possible to extend the discussion to incorporate probabilistic potential outcomes, but this is not often done in the causal inference literature, probably for notational ease.

We also assume that the potential outcomes satisfy consistency axioms necessary for them to be useful and well defined. That is, an individual's potential outcome to $T_1 = a$ and $T_2 = b$ equals the observed outcome when the observed treatments are $T_1 = a$ and $T_2 = b$. Furthermore, an individual's potential outcomes are stable regardless of treatments assigned to other individuals in the population. This latter assumption would typically not be plausible for treatments such as vaccines.

9.2.3 *Causal Types*

Individuals in the population belong to one of 16 possible causal types, based on their four potential outcomes and hence on U. These causal types are displayed in Table 9.1. We cannot identify an individual's causal type; this is an extension of the FPCI.

9.3 Dynamic Treatment Regimes and the Sequential Ignorability Assumption

A dynamic treatment regime is a sequence of treatment assignments that depend upon intermediate variables (variables that are not static and that are measured after the first treatment assignment). In our scenario, patients are following dynamic regimes because T_2 presumably depends on T_1 and H_2. The patients are each following one of eight possible dynamic regimes, displayed in Table 9.2.

Table 9.2 The eight dynamic regimes

Regimen	T_1	T_2
1	TOPO	No BMT regardless of H_2
2	TOPO	No BMT if $H_2 = 0$; BMT if $H_2 = 1$
3	TOPO	BMT if $H_2 = 0$; No BMT if $H_2 = 1$
4	TOPO	BMT regardless of H_2
5	TOPO/CTX	No BMT regardless of H_2
6	TOPO/CTX	No BMT if $H_2 = 0$; BMT if $H_2 = 1$
7	TOPO/CTX	BMT if $H_2 = 0$; No BMT if $H_2 = 1$
8	TOPO/CTX	BMT regardless of H_2

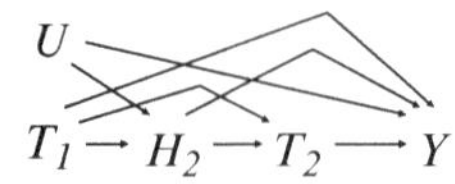

Fig. 9.1 Directed acyclic graph depicting sequential ignorability

9.3.1 Sequential Ignorability

We cannot identify which dynamic regime a given individual is following. However, we will make the assumption that T_1 and T_2 are *sequentially ignorable,* that is,

$$T_1 \coprod \mathrm{U} \tag{9.1}$$

and

$$T_2 \coprod \mathrm{U}|\mathrm{H}_2,\ \mathrm{T}_1, \tag{9.2}$$

where $A \coprod B$ and $A \coprod B|C$ denote that A is independent of B, and that A is independent of B given C, respectively. The first independence holds due to the randomization of T_1, and the second independence is a strong assumption that we make in order to disentangle the effects of T_1 and T_2 on Y. Basically, we pretend that T_2 is more or less randomized for individuals with the same values of H_2 and T_1. This allows us to estimate the effect of T_2 on Y within each of the four strata defined by H_2 and T_1, which in turn enables us to disentangle the effects of T_2 and T_1. This assumption is encapsulated in the directed acyclic graph (DAG) of Fig. 9.1 in terms of the missing arrows from U to T_1 and from U to T_2. See Greenland et al. (1999) and Pearl (1995) for tutorials on how to use DAGs to encode and determine identifiability of causal effects.

One needs to pay special attention to assumption (9.2) in light of the possibility that some patients may have $H_2 = 0$ due to death and thus are predestined, at the time T_1 is assigned, to have $T_2 = 0$ and the potential outcomes $Y_{T_1 0}$ and $Y_{T_1 1}$ both equal to zero. This possibility, by itself, does not render (9.2) untrue. Assumption (9.2) could still hold if these patients are counterbalanced by others with $H_2 = 0$ who are predestined to have $T_2 = 1$ and the potential outcomes $Y_{T_1 0}$ and $Y_{T_1 1}$ both equal to zero. Later in the chapter, when we analyze COG study P9462, we consider the impact of plausible deviations from assumption (9.2) due to the possibility that some patients may have $H_2 = 0$ due to death.

9.3.2 Time-Dependent Confounders

The variable H_2 is sometimes termed a *time-dependent confounder* of the joint causal effects of T_1 and T_2, because it is an effect ofT_1 and a cause of T_2, which complicates analyses of the joint effects of T_1 and T_2.

Table 9.3 Expected and observed observable data for the hypothetical example

T_1	H_2	T_2	n	n with $Y = 1$	Proportion with $Y = 1$
0	0	0	296	72	0.24
0	0	1	74	42	0.57
0	1	0	26	12	0.46
0	1	1	104	72	0.69
1	0	0	248	96	0.39
1	0	1	62	30	0.48
1	1	0	38	6	0.16
1	1	1	152	120	0.79

9.3.2.1 Hypothetical Example

It can be very helpful to investigate different causal target parameters and different approaches to estimating them in the context of a hypothetical example, for which all of the answers are known. We will assume the sampled population consists of just three types of individuals: 30 % are survivors ($U = 16$), 30 % are those in need of BMT in order to survive ($U = 6$), and 40 % are doomed ($U = 1$)–refer to Table 9.1 for their respective potential outcomes. Thus in this hypothetical population, TOPO and TOPO/CTX are equivalent for each and every individual in terms of the causal effect on the proportion of patients surviving at 2 years.

Suppose a sample of 1,000 from this population is randomized with equal probability to TOPO or to TOPO/CTX. Further suppose that, given $T_1 = 0$, the probabilities of $H_2 = 1$ vary by causal type as follows: $P(H_2 = 1|U = 16, T_1 = 0) = 0.4$, $P(H_2 = 1|U = 6, T_1 = 0) = 0.2$, and $P(H_2|U = 1, T_1 = 0) = 0.2$. Thus, in the TOPO group, the survivors are most likely to have responded. Given $T_1 = 1$, we have $P(H_2 = 1|U = 16, T_1 = 1) = 0.2$, $P(H_2 = 1|U = 6, T_1 = 1) = 0.8$, and $P(H_2|U = 1, T_1 = 1) = 0.2$. Thus, in the TOPO/CTX group, those in need of BMT in order to survive are most likely to have responded.

Using the preceding probabilities, one can calculate that $P(H_2|T_1 = 0) = 0.26$, whereas $P(H_2|T_1 = 1) = 0.38$, and thus that TOPO/CTX leads to a greater chance of response, even though it is equivalent to TOPO in terms of the effect on the proportion of patients surviving at 2 years. To complete the hypothetical example, we "randomize" T_2 according to the probabilities $P(T_2 = 1|H_2 = 0, T_1) = 0.2$ and $P(T_2 = 1|H_2 = 1, T_1) = 0.8$. Thus, patients who respond are much more likely to have a subsequent BMT, irrespective of the initial treatment.

We have now specified all of the necessary quantities to determine the expected number of people and the expected proportion of patients surviving at 2 years for each of the strata determined by T_1, H_2, and T_2; the expected data are presented in Table 9.3. To complete the hypothetical example, we will suppose that the observed data actually equal the expected data in Table 9.3, so that we can use it to demonstrate the success of various causal estimators and the lack of success of some standard non-causal estimators.

Most causal analyses have known answers for the hypothetical example. Briefly, TOPO/CTX improves response, which leads to increased chance of BMT, which in turn increases the proportion of patients surviving at 2 years. Thus TOPO/CTX increases the proportion of patients surviving at 2 years relative to TOPO. However, TOPO is equivalent to TOPO/CTX when both are followed up with a BMT (or when both are followed up with no BMT). Suppose that TOPO/CTX leads to greater toxicity due to being a combination rather than single-agent regime; then the best choice would be to administer TOPO followed by BMT.

9.3.3 Scientific Questions of Interest and Answers for the Hypothetical Example

We next make the foregoing causal analyses more precise by considering seven different specific scientific questions of plausible interest, for which we also provide the known answers for the hypothetical example.

1. Q: What is the average difference in potential survival outcomes in the study population following therapy with TOPO versus TOPO/CTX?

   A: From Table 9.3 we calculate that the proportion of patients surviving at 2 years in the TOPO group is $198/500 = 0.396$ versus $252/500 = 0.504$ in the TOPO/CTX group. Thus, TOPO/CTX improves the proportion of patients surviving at 2 years.
2. Q: What is the prediction equation for predicting the proportion of patients surviving at 2 years in the study population from T_1, H_2, and T_2?

   A: This equation can be obtained directly from Table 9.3, using the column labeled "p with $Y = 1$" (proportion with $Y = 1$). For example, $P(Y|T_1 = 0,\ H_2 = 0,\ T_2 = 0) = 0.24$
3. Q: What are the controlled direct effects of T_1 in the study population, that is (a) the effect of T_1 when T_2 is set to 0 for everyone, and (b) the effect of T_1 when T_2 is set to 1 for everyone? Furthermore, what is (c), the causal effect of T_1 in the subgroup randomized to $T_1 = 1$, when T_2 is set to 0 (or when T_2 is set to 1)?

   A: The answers to (a) and (b) are both "no controlled direct effect." This follows from the knowledge that the study population contains only people with $U = 1$, 6, or 16. The answer to (a) is $E(Y_{10} - Y_{00})$ and the answer to (b) is $E(Y_{11} - Y_{01})$, which both equal zero for the study population since for each of the three causal types, $Y_{10} - Y_{00} = 0$ and $Y_{11} - Y_{01} = 0$. The answer to the first part of (c) is $E(Y_1 - Y_0|T_1 = 1)$, which is equal to $E(Y_{10} - Y_{00}) = 0$ when both sets of potential outcomes are presumed to exist, as they are in the hypothetical example. For answering the second part of (c), we need additional potential outcomes. Or, more simply, we can switch the identity of $T_2 = 1$ and $T_2 = 0$; hence the meanings of Y_1 and Y_0 change as well.
4. Q: What is the effect of T_1 in the principal stratum of the study population consisting of people who are "destined" to receive $T_2 = 0$ regardless of whether

they are randomized to $T_1 = 0$ or $T_1 = 1$ (that is, for these people T_2^0 and T_2^1 are both equal to zero)? This might be considered an "uncontrolled direct effect" of T_1.

A: Because $Y_{10} - Y_{00} = 0$ for everyone in the population, it also equals zero for everyone in the principal stratum of interest. Thus there is no effect of T_1 in that principal stratum.

5. Q: What is the causal effect of T_2 on the proportion of patients surviving at 2 years in the study population, as a function of H_2 and T_1?

   A: The answers to this question can be obtained from Table 9.3, because T_2 is assumed independent of U given H_2 and T_1. For example, the effect of T_2 in the subgroup with $T_1 = 0$ and $H_2 = 0$ is 0.57–0.24 = 0.33.
6. Q: What would be the effect on the proportion of patients surviving at 2 years of setting $T_1 = 0$ then $T_2 = 1$ versus $T_1 = 0$ then $T_2 = 0$ in the study population? Of setting $T_1 = 1$ then $T_2 = 1$ versus $T_1 = 1$ then $T_2 = 0$?

   A: The first answer is given by $E(Y_{01} - Y_{00}) = 0.3$, and the second answer is given by $E(Y_{11} - Y_{10}) = 0.3$.
7. Q: What is the optimal dynamic regime for maximizing the effect on the proportion of patients surviving at 2 years in the study population?

   A: The answer to this question is typically determined recursively, although it is more readily determined for this particular hypothetical example given our knowledge of the causal types, which informs us that we cannot do better than to set T_1 to either 0 or 1, and then T_2 to 1 regardless of H_2 and T_1. But for expository purposes, we also determine the answer recursively. First we need to determine the optimal T_2 as a function of H_2 and T_1. Using the expected data in Table 9.3, we can readily see that $T_2 = 1$ is optimal regardless of H_2 and T_1. Second we need to determine the optimal T_1, given that patients will next receive the optimal T_2. Using the expected data in Table 9.3, we calculate that when $T_1 = 0$ and then the optimal T_2 is given, the expected proportion of patients surviving at 2 years is 0.6. And when $T_1 = 1$ and then the optimal T_2 is given, the expected proportion of patients surviving at 2 years is also 0.6. Thus for the hypothetical example, the optimal dynamic regime is to give either $T_1 = 0$ or $T_1 = 1$ and then follow up with $T_2 = 1$ regardless of H_2. Refer to the section on estimating optimal dynamic regimes for more details.

9.4 Standard Approaches to Analysis

We next describe three standard approaches to analysis, and we relate them to the scientific questions that they either correctly answer or are often mistaken to correctly answer.

9.4.1 Intent-To-Treat Analysis

This analysis is based on the obvious estimator of $E(Y|T_1 = 1) - E(Y|T_1 = 0)$, which equals 0.504–0.396 = 0.108 for the hypothetical observed data of Table 9.3. The analysis directly addresses scientific question #1. However, it is important to know that this analysis fails to address questions #3,4, and 7, the answers to which are also very important for deciding upon medical guidelines.

9.4.2 Predictors Analysis

This analysis is based on the regression $E(Y|T_1, H_2, T_2)$. The analysis directly addresses scientific question #2. However, it is important to know that T_1 may not be a "predictor" in this analysis (i.e., its coefficients could be zero) even while it is a cause, because H_2 and T_2 are also included in the analysis and these may be on the causal pathway from T_1 to Y, as in Fig. 9.1. Furthermore, consider the quantity $E(Y|T_1 = 1, H_2 = 0, T_2 = 1) - E(Y|T_1 = 0, H_2 = 0, T_2 = 0)$ in light of the expected hypothetical observable data in Table 9.3: it equals -0.09, meaning that, conditional on $H_2 = 0$ and $T_2 = 0$, TOPO/CTX is associated with a decrease in the proportion of patients surviving at 2 years. But this association cannot be interpreted causally—we know that TOPO/CTX leads to an increase in the proportion of patients surviving at 2 years in the population by way of increased chance of response hence increased chance of BMT, and that it does not effect the proportion of patients surviving at 2 years in the population if BMT is intervened on and held constant (i.e., such that either everyone in the population will have a BMT or everyone in the population will not have a BMT). The predictors analysis can be useful for prognosis, but not generally for causal analysis.

However, it is sometimes useful for causal analysis; for example, we can use this approach to answer causal question #5, because T_2 is assumed to be independent of U given T_1 and H_2 and because T_1 and H_2 are not on the causal pathway from T_2 to Y.

9.4.3 Standard Direct Effects Analysis

This analysis attempts to analyze the effect of T_1 on Y when T_2 is intervened on and held constant (i.e., either equal to one or to zero) for everyone (question #3), and it is based on the regression $E(Y|T_1, T_2)$. The fallibility of this analysis can be discerned from an analysis of the DAG in Fig. 9.1. Suppose that the arrows from T_1 to Y and from H_2 to Y are both missing. Then T_1 has no direct effect on Y unmediated by T_2 (there is no directed path from T_1 to Y that does not pass through T_2). However, the DAG shows that conditional on T_2, there are two "unblocked paths" from T_1

to Y: from $T_1 \rightarrow H_2 \rightarrow T_2 \rightarrow H_2 \rightarrow U \rightarrow Y$ (one can travel from T_2 to H_2 because one has conditioned on T_2) and from $T_1 \rightarrow T_2 \rightarrow H_2 \rightarrow U \rightarrow Y$. One cannot travel directly from T_2 to Y because one has conditioned on T_2. These two unblocked paths will likely lead to an observed conditional association of T_1 with Y, given T_2, and thus to an apparent direct effect of T_1 on Y, controlling for T_2. The fallibility of the standard approach has been noted by Cole and Hernan (2002) for DAGs without H_2 and with an arrow from U to T_2. The foregoing discussion documents fallibility for yet another type of DAG.

We know the answer to question #3 for the hypothetical example: either setting T_2 to 0 or to 1, the direct effect of T_1 on Y is zero. However, when we employ the standard analysis with the data in Table 9.3, we estimate the direct effect of T_1 setting T_2 to 0 to be $(96+6)/(248+38)–(72+12)/(296+26) = 0.36–0.26 = 0.10$. And the direct of T_1 setting T_2 to 1 is estimated to be $(120+30)/(62+152)–(42+72)/(74+104) = 0.70–0.64 = 0.06$. This gives us a concrete example of the fallibility of the standard direct effects analysis.

9.4.4 Summary

In summary, the standard approaches allow us to correctly address questions #1, 2, and 5. To answer questions #3, 4, 6, and 7, alternative approaches are necessary.

9.5 Marginal Structural Model Approach

Robins and colleagues (e.g., Robins 1998, 1999a, b; Robins et al. 2000; Hernan et al. 2000, 2001, 2002) proposed semiparametric modeling of the marginal distributions of the potential outcomes in the study population, possibly conditional on baseline covariates. The resulting model is called an MSM. In our scenario, there is no need to enlist modeling assumptions that impose restrictions on the marginal means. Thus the MSM that we consider is simply $E(Y_{ab}) = \beta_{00}1(a=b=0) + \beta_{01}1(a=0, b=1) + \beta_{10}1(a=1, b=0) + \beta_{11}1(a=b=1)$, where $1(A) = 1$ when A is true and 0 otherwise.

9.5.1 Estimation

Let $\beta = (\beta_{00}, \beta_{01}, \beta_{10}, \beta_{11})$. To estimate β, we need to relate the mean of the potential survival outcomes to functions of the observed data. From the two sequential ignorability assumptions, it follows that

$$E(Y_{ab}) = E(Y_{ab}|T_1 = a) = E_{H_2|T_1=a}E(Y_{ab}|T_1 = a, H_2)$$
$$= E_{H_2|T_1=a}E(Y_{ab}|T_1 = a, H_2, T_2 = b),$$

where the first equality follows from (9.1), the second equality follows from the double expectation theorem, and the third equality follows from (9.2). It then follows from the consistency assumption that

$$E_{H_2|T_1=a}E(Y_{ab}|T_1 = a, H_2, T_2 = b) = E_{H_2|T_1=a}E(Y|T_1 = a, H_2, T_2 = b),$$

and hence that

$$E(Y_{ab}) = E_{H_2|T_1=a}E(Y|T_1 = a, H_2, T_2 = b). \quad (9.3)$$

For high-dimensional H_2, one way to compute (9.3) is with a model for $E(Y|T_1 = a, H_2, T_2 = b)$, e.g., $f(H_2, a, b)$. One fits the model using the observed data, then computes $E(Y_{ab})$ as the average of $f(H_2, a, b)$ in the sample subgroup with $T_1 = a$. A second way to compute (9.3) for high-dimensional H_2 is with a model for $P(T_2 = b|H_2, T_1 = a)$, e.g., $e(H_2, a, b)$. One again fits the model using the observed data, estimates the proportion $P(T_2 = b|T_1 = a)$, and then forms the weight $W = P(T_2 = b|T_1 = a)/e(H_2, a, b)$. This weight is known as an *inverse-probability-of-treatment weight,* because $e(H_2, a, b)$ represents the probability of treatment assignment T_2 as a function of the relevant medical history that is needed for the sequential ignorability assumption, i.e., H_2 and T_1. When the numerator and the denominator of W are known rather than estimated, it happens that

$$E(Y_{ab}) = E(WY|T_1 = a, T_2 = b), \quad (9.4)$$

this relationship then leads to an alternative estimator of $E(Y_{ab})$, known as the inverse-probability-of-treatment-weighted estimator, or IPTW estimator, in which the estimated version of W is substituted for the known version, and the expectation is estimated using simple averaging in the relevant sample subgroups.

Proof of (9.4) is as follows.

$$\int w\, y\, p(h_2, y|T_1 = a, T_2 = b)\mathrm{d}h_2\mathrm{d}y = E_{H_2|T_1=a}E(Y|T_1 = a, H_2, T_2 = b) = E(Y_{ab}),$$

where dh_2 and dy represent either counting measure or Lebesgue measure depending on the nature of H_2 and Y. The first equality follows from the relationship

$$w\, p(h_2, y|T_1 = a, T_2 = b) = p(y|T_1 = a, T_2 = b, H_2 = h_2)p(h_2|T_1 = a),$$

which in turn follows from applications of the multiplication rule $p(A, B) = p(A|B)p(B)$. The second equality follows from the foregoing proof of (9.3).

In our scenario, H_2 is a binary variable, and so no restrictive model is needed for either $f(H_2, a, b)$ or $e(H_2, a, b)$. Therefore, the two methods of estimation are exactly equivalent.

9.5.2 *Inference*

Using either a model for $f(H_2, a, b; \alpha)$ or for $e(H_2, a, b; \theta)$, in which α and θ represent finite-dimensional parameters, the resulting estimating equations for β and either α or θ is of the form $U(\beta, \alpha) = 0$ or $U(\beta, \theta) = 0$. Inferences on β can be made using the asymptotic normality of the estimating equation combined with the delta method, which yields a sandwich estimator. It has been noted (e.g., Robins 1998, 1999a, and Robins et al. 2000) that, when θ is estimated, inferences on β based on treating θ as known are conservative (i.e., the standard error of the estimator for β is larger when θ is assumed known than it is when θ is estimated.)

The nonparametric bootstrap, which resamples the observable data (Y, H_2, T_1, T_2) with replacement, is an easier alternative, which also accounts for finite-sample effects on the standard error of β.

9.5.3 *Answering Questions #3 and 6 for the Hypothetical Example*

We estimated the causal targets of questions #3 and 6 by applying the MSM approach to the observed hypothetical data of Table 9.3. Estimates and standard errors were computed using the IPTW approach and SAS PROC GENMOD; by enlisting the "repeated" option, the conservative sandwich estimator assuming known inverse-probability-of-treatment weights is computed and used to estimate standard errors.

For question #3, the estimate of $E(Y_{11} - Y_{01})$ is 0.0000, with a conservative standard error of 0.0616. The estimate of $E(Y_{10} - Y_{00})$ is also 0.0000, with a conservative standard error of 0.0450. Thus, the MSM approach provides a consistent estimator of the controlled direct effects.

For question #6, the estimate of $E(Y_{01} - Y_{00})$ is 0.3000, with a conservative standard error of 0.0550, and the estimate of $E(Y_{11} - Y_{10})$ is also 0.3000, with a conservative standard error of 0.3000. Again, the MSM approach provides consistent estimators of the target quantities.

9.6 Structural Nested Model Approach

Another approach to handling time-dependent confounding proposed by Robins and colleagues (e.g., Robins 1992, 1994, 1997, 1999a, 2004; Brumback et al. 2003; Robins and Rotnitzky 2004) is based on SNMs. We demonstrate with a structural-nested mean model (SNMM) of the causal effects of T_1 and T_2. Because we have two treatments, the model has two parts to the "nest": $E(Y - Y_1|T_2 = 1, H_2, T_1) = f(H_2, T_1, 1; \alpha)$ (where f(.) is the same function described in the preceding section), and $E(Y_1 - Y_0|T_1 = 1) = \gamma$. These two functions model each model the mean effect of taking away one layer of treatment–first the effect of taking away T_2 and second the effect of taking away T_1.

9.6.1 *Estimation*

Under the assumptions (9.1) and (9.2) of sequential ignorability, it can be shown (e.g., see Brumback et al. (2003) or Robins (1994)) that assumptions (9.1) and (9.2) imply $E(\hat{Y}_1|T_2, H_2, T_1) = E(\hat{Y}_1|H_2, T_1)$ and $E(\hat{Y}_0|T_1) = E(\hat{Y}_0)$, where $\hat{Y}_1 = Y - f(H_2, T_1, 1; \alpha)T_2$ and $\hat{Y}_0 = Y - f(H_2, T_1, 1; \alpha)T_2 - \gamma T_1$. It follows that the parameter α can be consistently estimated using standard regression methods; i.e., one fits the model $E(Y|T_2, H_2, T_1) = f(H_2, T_1, 1; \alpha)T_2 + h(H_2, T_1; \eta)$. The parameter γ can then be consistently estimated as follows. One first forms the random variable $\hat{\hat{Y}}_1 = Y - f(H_2, T_1, 1; \hat{\alpha})T_2$ for each patient (representing the expected outcome had T_2 be taken away if administered). One next fits the regression model $E(\hat{\hat{Y}}_1|T_1) = \gamma T_1$ to estimate γ.

9.6.2 *Inference*

Using models for $f(H_2, a, b; \alpha)$ and for $h(H_2, T_1; \eta)$, in which α and η represent finite-dimensional parameters, the resulting estimating equations for α, γ, and η are of the form $U(\alpha, \gamma, \eta) = 0$. Inferences on α and γ can be made using the asymptotic normality of the estimating equation combined with the delta method, which yields a sandwich estimator.

Inferences on α can also be made conditionally on H_2 and T_1, and hence separately from inferences on γ and η. For this we can use the standard errors obtained from the predictors analysis.

The nonparametric bootstrap, which resamples the observable data (Y, H_2, T_1, T_2) with replacement, is again an easier alternative, which also accounts for finite-sample effects on the standard error of α and γ.

9.6.3 *Answering Questions #3 and 5 for the Hypothetical Example*

We estimated the causal targets of questions #3 and 5 by applying the SNM approach to the observed hypothetical data of Table 9.3. Standard errors were computed using the nonparametric bootstrap approach.

For question #3, the estimate of the controlled direct effect of T_1 in the subgroup of $T_1 = 1$ when no one is given a posttreatment BMT, that is of $E(Y_1 - Y_0|T_1 = 1)$ (the estimate is 0.0000, and the bootstrap standard error of the estimate is 0.0435.) To estimate the controlled direct effect of T_1 in the subgroup of $T_1 = 1$ when everyone is given a posttreatment BMT, we switch the identities of $T_2 = 1$ and $T_2 = 0$ (hence the meaning of Y_1 and Y_0 change as well) and again estimate $E(Y_1 - Y_0|T_1 = 1)$. In this case, the estimate is 0.0000, and the with a bootstrap standard error of the estimate is 0.0612.

For question #5, we need to estimate functions of α. We focus on the causal effect of T_2 in the subgroup with $T_1 = 0$ and $H_2 = 0$, and we estimate standard errors of this effect conditionally on T_1 and H_2. The estimate is 0.3243 with a standard error of 0.0590, which is very close to the true answer of 0.33.

9.7 Principal Stratification

One needs to turn to the principal stratification framework of Frangakis and Rubin (2002) in order to address question #4.

9.7.1 *Naïve Approach to Estimation*

The naïve approach to estimating $E(Y_{10} - Y_{00}|T_2^0 = T_2^1 = 0)$ is to first assume that this quantity equals $E(Y_{10} - Y_{00}|T_2 = 0)$ and to second assume that the latter quantity can be consistently estimated via standard estimators of $E(Y|T_1 = 1, T_2 = 0) - E(Y|T_1 = 0, T_2 = 0)$. Both assumptions are incorrect, and the resulting estimator, which was previously introduced in the section on standard direct effects analysis, is generally inconsistent. In the hypothetical example, the naïve approach would lead to a consistent estimate of 0.1 rather than of zero, the correct answer.

9.7.2 *Principal Stratification*

Whereas Frangakis and Rubin (2002) provide a framework for expressing target quantities such as $E(Y_{10} - Y_{00}|T_2^0 = T_2^1 = 0)$, they do not offer a general approach for estimation. Much work has been contributed (e.g., Roy et al. 2008; Cheng and Small 2006; Barnard et al. 2003) on methods for estimation of such quantities in contexts without H_2 but perhaps with the addition of pre-treatment covariates, X; for example, T_2 might represent compliance, which is a post-randomization variable as is BMT. With the exception of really simple cases, the estimation of principal stratum effects generally proceeds either in terms of a sensitivity analysis in which sensitivity parameters are varied over plausible ranges, or in terms of Bayesian posterior inference with respect to sensitivity priors. Often, strong assumptions are made in addition to ignorability of T_1.

It is an open question, just how one should proceed to identify $E(Y_{10} - Y_{00} | T_2^0 = T_2^1 = 0)$ in our sequential treatment context, with data available on a relevant intermediate variable such as H_2. One would need to incorporate sensitivity parameters and ignorability assumptions concerning all of the potential outcomes, including T_2^0 and T_2^1. The role of ignorability assumption (9.2) would need to be carefully evaluated, because in the principal stratification framework, T_2 would be determined by T_1 rather than "randomly" assigned. Nevertheless, H_2 might still be considered a cause of T_2, and the conditional independence at (9.2) might well approximate relationships in the population. We leave the development of a principal stratification approach as a topic for future research.

9.7.2.1 Estimating Optimal Dynamic Treatment Regimes

Murphy (2003) and Robins (2004) developed methods for inferring optimal dynamic treatment regimes, each using different modeling techniques but both requiring sequential ignorability assumptions (9.1) and (9.2). In our scenario, H_2 is binary; therefore, we do not require restrictive models, and consequently, the methods of Murphy and Robins coincide. Refer to Moodie et al. (2007) for a comparison of the two approaches in more complex scenarios necessitating modeling.

As Moodie et al. (2007) explain, optimal dynamic regimes are defined recursively: first one finds the function $b^{\text{opt}}(T_1 = a, H_2 = h_2)$ which maximizes $E(Y_{ab} | T_1 = a, H_2 = h_2)$, and second one finds the value a^{opt} which maximizes $E(Y_{ab^{\text{opt}}})$. The optimal T_2 is equal to b^{opt}, even if the preceding T_1 was suboptimal. The optimal T_1 is equal to a^{opt}, and it is based on the assumption that the optimal T_2 will follow.

9.7.3 *Estimation*

Estimating the function b^{opt} that maximizes$E(Y_{ab} | T_1 = a, H_2 = h_2)$ is possible because the objective function equals $E(Y | T_1 = a, H_2 = h_2, T_2 = b)$, by (9.2). In the context of our hypothetical example, suppose that $a = 0$ and $h_2 = 0$. Then from Table 9.3, we see that setting $b = 0$ leads to an estimate of 0.24 for the objective function, versus an estimate of 0.57 corresponding to $b = 1$. Thus we

Table 9.4 Actual data from COG study P9462

T_1	H_2	T_2	n	n with $Y = 1$	Proportion with $Y = 1$
0	0	0	41	12	0.29
0	0	1	0	0	NA
0	1	0	16	3	0.19
0	1	1	3	2	0.67
1	0	0	28	3	0.11
1	0	1	2	1	0.50
1	1	0	19	8	0.42
1	1	1	7	2	0.29

estimate $b^{\text{opt}}(0,0)$ to equal 1, and similarly, we estimate $b^{\text{opt}}(0,1) = b^{\text{opt}}(1,0) = b^{\text{opt}}(1,1) = 1$.

To estimate a^{opt}, one can optimize an estimated version of equation (9.3) with b^{opt} substituted for b: $E(Y_{ab^{\text{opt}}}) = E_{H_2|T_1=a}E(Y|T_1 = a, H_2, T_2 = b^{\text{opt}}(a, H_2))$. In the context of our hypothetical example, $a = 0$ leads to an estimate of $((42/74)(74+296)/500)+((72/104)(26+104)/500) = 0.6$ for $E(Y_{ab^{\text{opt}}})$, whereas $a = 1$ corresponds to an estimate of $((30/62)*(62+248)/500)+((120/152)*(38+152)/500) = 0.6$, again. In this case, we estimate that either $T_1 = 0$ or $T_1 = 1$ leads to the optimal overall regime. We will henceforth denote estimates of a^{opt} and b^{opt} as $\hat{a}_{\text{opt}}$ and $\hat{b}_{\text{opt}}$.

9.7.4 *Inference*

One may wish to make hypothesis tests in order to conclude that, say, a^{opt} is equal to one. This can be done by testing whether $E(Y_{0b^{\text{opt}}}) = E(Y_{1b^{\text{opt}}})$ versus a one- or two-sided alternative. One could construct an unbiased estimating equation $U(\delta)$ for estimating $\delta = E(Y_{0b^{\text{opt}}}) - E(Y_{1b^{\text{opt}}})$, and use it to conduct a Wald-type hypothesis test. Alternatively, one could employ the nonparametric bootstrap.

9.8 Analysis of COG Study P9462

Actual data from COG study P9462 are presented in Table 9.4. Note that there are 60 patients in the TOPO group because we include a patient who was randomized after the accrual for the two-stage group sequential design had been reached, and there are 58 patients in the TOPO/CTX group because we excluded two patients with missing survival data. Approaches to accommodate missing and censored data are feasible but beyond the scope of this book chapter. The resulting data are used primarily for illustrative purposes.

Table 9.5 Optimal dynamic regime results for T_2

φ	T_1	H_2	$\hat{b}_{\text{opt}}$	Difference	CI for difference
0	0	0	0	0.293	(0.157, 0.385)
0.15	0	0	0	0.143	(0.007, 0.235)
$12/41 = 0.29$	0	0	0/1	0.000	(-0.092, 0.135)
0.39	0	0	1	0.097	(0.005, 0.233)
$2/3 = 0.67$	0	0	1	0.374	(0.282, 0.509)
$33/41 = 0.80$	0	0	1	0.512	(0.420, 0.648)
Any value	0	1	1	0.479	(0.097, 0.861)
Any value	1	0	1	0.393	(-0.226, 0.885)
Any value	1	1	0	0.135	(-0.523, 0.314)

The first feature of the data to note is that no one with $T_1 = 0$ and $H_2 = 0$ received a BMT. This complicates many of the analyses that we have presented, because $E(Y|T_1 = 0, H_2 = 0, T_2 = 1)$ is not identified. We will represent this quantity with the sensitivity parameter φ.

We will present two analyses. First, the intent-to-treat analysis will be presented because it is the standard method for reporting results of clinical trials. Second, we will estimate the optimal dynamic treatment regimes. We could also employ the MSM or SNM approaches, but we feel that estimating the optimal dynamic regimes yields the most informative analysis in addition to the intent-to-treat analysis. Additionally, the MSM and SNM approaches were clearly illustrated within the context of the hypothetical example, whereas the example was not ideal for illustrating estimation of optimal dynamic regimes. Due to the small sample sizes, we will use the bootstrap for a hypothesis test about the optimal choice for T_1, or a^{opt}, and about the optimal choice for T_2, or b^{opt}.

9.8.1 *Intent-To-Treat Analysis*

The intent-to-treat analysis does not involve the sensitivity parameter φ. We estimate $E(Y|T = 1) - E(Y|T = 0)$ at -0.033, with a 95 % confidence interval of (-0.194, 0.127). The difference in the proportion of patients surviving at 2 years is thus not statistically significant at the 0.05 level.

9.8.2 *Estimating Optimal Dynamic Regimes*

Tables 9.5 and 9.6 present results of estimating the optimal dynamic regimes, and also of comparing the optimal regimes at T_2 and T_1 to the alternatives in terms of estimated differences in expected potential survival outcomes.

Table 9.6 Optimal dynamic regime results for T_1

φ	$\hat{a}_{\text{opt}}$	$\hat{E}(Y_{0b^{\text{opt}}})$	$\hat{E}(Y_{1b^{\text{opt}}})$	CI for difference
0	1	0.41	0.46	(-0.21, 0.35)
0.15	1	0.41	0.46	(-0.21, 0.35)
$12/41 = 0.29$	1	0.41	0.46	(-0.24, 0.35)
0.39	0	0.48	0.46	(-0.28, 0.31)
$2/3 = 0.67$	0	0.67	0.46	(-0.09, 0.49)
$33/41 = 0.80$	0	0.76	0.46	(0.002, 0.59)

Estimates were obtained using the methodology described above, and approximate 95 % confidence intervals for estimated differences were obtained using the nonparametric bootstrap and the percentile method for computing confidence intervals (refer to Efron and Tibshrinani 1993). For the nonparametric bootstrap, we conditioned on the eight sample sizes in the "*n*" column of Table 9.4—thus, we resampled each row of Table 9.4 separately. Conditioning on only the sample sizes in each of the two groups specified by $T_1 = 0$ and $T_1 = 1$ would necessitate additional sensitivity analyses within each bootstrap sample, due to the possibility that zero patients might end up in one of the *other* eight rows of the bootstrap version of Table 9.4, besides row 2. Clearly, we did not do any resampling for row 2; instead, we used the same sensitivity parameter φ within each bootstrap sample, and we treated it as a known rather than estimated quantity.

Table 9.5 focuses on estimating b^{opt} as a function of T_1 and H_2. Also presented (in the "difference" column) is the estimated difference. $E(Y|T_1 = a, H_2 = h_2, T_2 = b^{\text{opt}}) - E(Y|T_1 = a, H_2 = h_2, T_2 = 1 - b^{\text{opt}})$, and the estimated confidence interval (CI) for this difference (in the "CI for difference" column). Only the results for $T_1 = 0$ and $H_2 = 0$ are affected by the value of φ. When φ is less than or equal to 0.15, the confidence interval is greater than zero, indicating that $\hat{b}_{\text{opt}}(0, 0) = 0$. In other words, the proportion of patients surviving at 2 years is statistically significantly higher for patients treated with TOPO who had a response and then did not have BMT, in comparison to not having had BMT. When φ is greater or equal to 0.39, the confidence interval is again greater than zero, indicating that $\hat{b}_{\text{opt}}(0, 0) = 1$, i.e., the proportion of patients surviving at 2 years is statistically significantly higher for patients treated with TOPO who had no response and then had BMT, in comparison to not having had BMT. One might be tempted to set $\varphi = 12/41 = 0.29$ to agree with the results of row 1; this choice would presume that having a BMT and not having a BMT lead to the same expected results, given $T_1 = 0$ and $H_2 = 0$, and, accordingly, it would also imply that there is no statistically significant difference in survival between having a BMT and not having a BMT. However, one might instead think that setting $\varphi = 2/3 = 0.67$ to agree with the results of row 4 is more plausible; this choice would presume that the survival following BMT is the same for the TOPO group regardless of H_2. On this assumption, our analysis would conclude that having a BMT is optimal and that the difference in survival between having one and not having one is statistically significant given $T_1 = 0$ and $H_2 = 0$.

Also shown in Table 9.5 are the results for the other three settings of T_1 and H_2 (last three rows). These results are not affected by the choice of φ. We observe that when $T_1 = 0$ and $H_2 = 1$, it is optimal to have a BMT, and that the difference in survival between having one and not having one is statistically significant. Thus for the TOPO responders, having a BMT appears to be optimal based on this analysis. The differences between having a BMT and not having a BMT are not statistically significant neither for the TOPO/CTX responders nor for the TOPO/CTX nonresponders.

Table 9.6 focuses on estimating a^{opt}. Also presented are $\hat{E}(Y_{0b^{\mathrm{opt}}})$ and $\hat{E}(Y_{1b^{\mathrm{opt}}})$ and the confidence interval for the difference $\hat{E}(Y_{a^{\mathrm{opt}}b^{\mathrm{opt}}}) - \hat{E}(Y_{(1-a^{\mathrm{opt}})b^{\mathrm{opt}}})$, all as functions of the sensitivity parameter φ. One quickly observes that it is impossible to select φ so that TOPO/CTX is statistically significantly optimal at T_1. Setting $\varphi = 2/3 = 0.67$ leads to nearly but not quite statistically significant optimality of TOPO at T_1. If one were prepared to believe that $\varphi = 0.80$, then TOPO would be statistically significantly optimal in terms of the proportion surviving at 2 years.

We next consider establishing the optimal dynamic regime from among the choices in Table 9.2. If $\varphi = 0.80$, then we can conclude that regimen 4 is optimal overall (a statistically significant result). If $\varphi = 0.67$, then we can conclude that regimen 4 is better than regimens 1, 2, or 3 (statistically significant). If φ is 0.15 or less, we can conclude that regimen 2 is better than regimen 1, 3, or 4. For φ less than 0.8, we cannot conclude anything about the relative superiority of regimes 5, 6, 7, and 8 versus one another nor versus any of 1, 2, 3, or 4.

Finally, we consider the impact of plausible departures from assumption (9.2) due to the reality that some patients have $H_2 = 0$ due to death. As we discussed above, assumption (9.2) may still hold in these circumstances. But it is perhaps more plausible that subtle departures would occur, rendering $E(Y|T_1, H_2 = 0, T_2 = 0)$ less than $E(Y_{T_1 0}|T_1, H_2 = 0, T_2 = 0)$ and $E(Y|T_1, H_2 = 0, T_2 = 1)$ higher than $E(Y_{T_1 0}|T_1, H_2 = 0, T_2 = 1)$. Thus the proportions in rows 1 and 4 of Table 9.4 would be biased slightly low and those in rows 2 and 5 would be biased slightly high as estimates of their respective conditional mean potential outcomes. However, in row 2 we do not have an estimated proportion—we only have the sensitivity parameter φ. We could redefine φ to denote the mean conditional potential survival outcome, and thus it would exclude the bias. In this case, for values of φ greater than 0.29, our analyses of the optimal treatment for T_1 would be biased in favor of $T_1 = 1$ (TOPO/CTX); thus the statistical significance of the results concluding optimality of TOPO would remain unchanged.

In any event, the values of φ required for a statistically significantly higher proportion of survivors at 2 years with TOPO were quite high. One can only wonder what the data would have suggested had the trial been larger, or if survival times beyond 2 years had also been considered. But given the data at hand, one would have a very tough time arguing to overturn the null hypothesis that TOPO and TOPO/CTX are equivalent in terms of the proportion of patients surviving at 2 years, even when the estimated optimal dynamic treatment for T_2 is subsequently assigned.

9.9 Discussion

We have described and compared several approaches to causal inference in the setting of a RCT with a subsequent, nonrandomized, treatment. Whereas we focused on the analysis of a phase II study of neuroblastoma patients, there are also other cancer disease areas in which a surrogate endpoint is used but the main interest is overall survival. For example, newly diagnosed patients with follicular non-Hodgkin lymphoma may enroll on a clinical trial with at least two primary outcomes: time to treatment failure and overall survival. Many of these patients survive for a very long time, and if the original treatment fails, they enroll in subsequent clinical trials or pursue treatments in a nonrandomized setting. Thus in this setting, there is often not just one subsequent nonrandomized treatment, but several. Several of the methods discussed in this chapter have been extended to adjust for and estimate the effect of multiple subsequent treatments (for example, see Robins 1999a, 2004 and Murphy 2003). They can also be extended to analyze survival outcomes at multiple points in time; one could use a multiple testing framework with correlated tests, or one could adapt standard methods used in survival analysis, such as the log-rank test, to incorporate adjustment for subsequent treatment.

One can also use extensions of the methods discussed in this chapter to incorporate high-dimensional versions of the intermediate variable H_2. In this case, one needs to make more modeling assumptions than were needed for binary H_2. For example, in the section on marginal structural models, we discussed the need to choose either a model for $f(H_2, a, b)$ or for $e(H_2, a, b)$. Robins (2000) proposes a general class of so-called doubly robust estimators that are consistent if either one or both of these models are correct and semiparametric efficient if both models are correct. Also refer to Bang and Robins (2005).

A practical difficulty that limits successful application of these methods in clinical trial settings is the lack of sufficient statistical power. Clinical trials are often powered for the intent-to-treat analysis, and as such the sample sizes are often too small for detailed secondary analyses, like the ones we have described, to achieve statistically significant results. It may be useful in the future for statistical researchers to develop methods for estimating the sample sizes necessary for these secondary analyses to be sufficiently powered. Alternatively, the underpowered secondary analysis could be used to generate hypotheses that could be tested with subsequent clinical trials.

A final note is that many clinical trials, like COG P9462, utilize group-sequential designs. Our analysis ignored this aspect of the design and relied instead on the sequential ignorability assumptions. It seems very difficult to us to reconcile our methods with the group-sequential design assumptions, but future work could attempt to solve this problem.

References

Angrist J, Imbens GW, Rubin DB (1996) Identification of causal effects using instrumental variables (with discussion). J Am Stat Assoc 91:444–472

Balke A, Pearl J (1997) Bounds on treatment effects from studies with imperfect compliance. J Am Stat Assoc 92:1171–1176

Bang H, Robins JM (2005) Doubly robust estimation in missing data and causal inference models. Biometrics 61:962–972

Barnard J, Frangakis CE, Hill JL, Rubin DB (2003) Principal stratification approach to broken randomized experiments: a case study of school choice vouchers in New York City (with discussion). J Am Stat Assoc 93:299–323

Brumback B, Greenland S, Redman M, Kiviat N, Diehr P (2003) The intensity-score approach to adjusting for confounding. Biometrics 59:274–285

Brumback BA, Hernan MA, Haneuse JPA, Robins JM (2004) Sensitivity analyses for unmeasured confounding assuming a marginal structural model for repeated measures. Stat Med 23:749–767

Cheng J, Small DS (2006) Bounds on causal effects in three-arm trials with non-compliance. J Royal Stat Soc Series B 68:815–836

Cole SR, Hernan MA (2002) Fallibility in estimating direct effects. Int J Epidemiology 31:163–165.

Efron B, Tibshrinani RJ (1993) An Introduction to the Bootstrap. Chapman & Hall., New York

Frangakis CE, Rubin DB (2002) Principal stratification in causal inference. Biometrics 58:21–29

Goetghebeur E, Lapp K (1997) The effect of treatment compliance in a placebo-controlled trial: regression with unpaired data. Appl Stat 46:351–364

Greenland S, Pearl J, Robins JM (1999) Causal diagrams for epidemiologic research. Epidemiology 10:37–48

Hernan MA, Brumback BA, Robins JM (2000) Use of marginal structural models to estimate the effect of prophylaxis therapy for pneumocystis carinii pneumonia on the survival of AIDS patients. Epidemiology 11:561–570

Hernan MA, Brumback BA, Robins JM (2001) Marginal structural models to estimate the joint causal effect of non-randomized treatments. J Am Stat Assoc 96:440–448

Hernan MA, Brumback BA, Robins JM (2002) Estimating the causal effect of zidovudine on CD4 count with a marginal structural model for repeated measures. Stat Med 21:1689–1709

Holland P (1986) Statistics and causal inference. J Am Stat Assoc 81:945–970

Moodie EM, Richardson TS, Stephens DA (2007) Demystifying optimal dynamic treatment regimes. Biometrics 63:447–455

Murphy SA (2003) Optimal dynamic treatment regimes (with discussion). J Royal Stat Soc Series B 65:331–366

Murphy S, van der Laan M, Robins JM (2001) Marginal mean models for dynamic regimes. J Am Stat Assoc 96:1410–1423

Neyman J (1923) On the application of probability theory to agricultural experiments: Essay on principles, Section 9. Translated Stat Sci 5(1990):465–480

Pearl J (1995) Causal diagrams for empirical research. Biometrika 82:669–710

Robins JM (1989) The analysis of randomized and non-randomized AIDS treatment trials using a new approach to causal inference in longitudinal studies. In: Sechrest L, Freeman H, Mulley A (eds) Health Service Research Methodology: A Focus on AIDS. U. S. Public Health Service, National Center for Health Services Research, Washington, DC, pp 113–159

Robins JM (1992) Estimation of the time-dependent accelerated failure time model in the presence of confounding factors. Biometrika 79:321–34

Robins JM, Blevins D, Ritter G, Wulfsohn M (1992) G-estimation of the effect of prophylaxis therapy for pneumocystis carinii pneumonia on the survival of AIDS patients. Epidemiology 3:319–336

Robins JM, Tsiatis AA (1992) Semiparametric estimation of an accelerated failure time model with time-dependent covariates. Biometrika 79:311–319

Robins JM (1994) Correcting for non-compliance in randomized trials using structural nested mean models. Comm Stat Theor Meth 23:2379–2412

Robins JM, Greenland S (1994) Adjusting for differential rates of PCP prophylaxis in high- versus low-dose AZT treatment arms in an AIDS randomized trial. J Am Stat Assoc 89:737–749

Robins JM (1997) Causal inference from complex longitudinal data. In: Berkane M (ed) Latent Variable Modeling and Applications to Causality. Lecture Notes in Statistics, vol 120. Springer, NY, pp 69–117

Robins JM (1998) Marginal structural models. In 1997 Proceedings of the American Statistical Association. Section on Bayesian Statistical Science, pp 1–10

Robins JM (1999a) Marginal structural models versus structural nested models as tools for causal inference. In: Halloran ME, Berry D (eds) Statistical Models in Epidemiology: The Environment and Clinical Trials, IMA, vol 116. Springer, New York, pp 95–134

Robins JM (1999b) Association, causation, and marginal structural models. Synthese 121:151–179

Robins JM, Greenland S, Hu F-C (1999) Estimation of the causal effect of a time-varying exposure on the marginal mean of a repeated binary outcome. J Am Stat Assoc 94:687–700

Robins JM, Rotnitzky A, Scharfstein D (1999) Sensitivity analysis for selection bias and unmeasured confounding in missing data and causal inference models, Sections 6–11. In: Halloran ME, Berry D (eds) Statistical Models in Epidemiology: The Environment and Clinical Trials, IMA, vol 116. Springer, New York, pp 1–92

Robins JM (2000) Robust estimation in sequentially ignorable missing data and causal inference models. In: Proceedings of the American Statistical Association Section on Bayesian Statistical Science, 1999. pp 6–10

Robins JM, Hernan MA, Brumback BA (2000) Marginal structural models and causal inference in epidemiology. Epidemiology 11:550–560

Robins JM (2004) Optimal structural nested models for optimal sequential decisions. . In: Lin DY, Heagerty P (ed) Proceedings of the Second Seattle Symposium on Biostatistics. Springer, New York, pp 189–326

Robins JM, Rotnitzky A (2004) Estimation of treatment effects in randomized trials with non-compliance and a dichotomous outcome using structural mean models. Biometrika 91:763–783

Roy J, Hogan JW, Marcus BH (2008) Principal stratification with predictors of compliance for randomized trials with 2 active treatments. Biostatistics 9:277–289

Rubin DB (1974) Estimating causal effects of treatments in randomized and nonrandomized studies. J Educ Psychol 66:688–701

Rubin DB (1977) Assignment to a treatment group on the basis of a covariate. J Educ Stat 2:1–26

Rubin DB (1978) Bayesian inference for causal effects. Ann Stat 6:34–58

Sommer A, Zeger S (1991) On estimating efficacy from clinical trials. Stat Med 10:45–52

van der Laan MJ, Robins JM (2003) Unified methods for censored longitudinal data and causality. Springer, New York

Chapter 10
Changing Paradigms in Cancer Clinical Trials

Jeanne Kowalski and William Matsui

10.1 Introduction

Genomic technologies have led to a revolution in drug discovery and development; that much is universally agreed upon. The high dimension of data from such technologies has challenged conventional data analytic methods; that much is apparent. Researchers have been hard at work marrying the latest technologies with drug design to more closely examine whether a drug is having any biological impact (clinical trial), what the effect is (measure of endpoint), when it stops working (onset of resistance), and what can be done about it (alternate therapies). Gene therapy, cancer-killing viruses, epigenetic therapy, and cancer stem cell-targeted treatment, highlight a few of the novel approaches to cancer treatment. The emergence of promising new molecular-targeted agents and new technologies for screening and early detection has prompted new opportunities in biostatistics for clinical trial designs and analysis that integrate advances made in molecular and cancer biological process into endpoints.

Cancer has long been recognized to be a genetic disorder and the growing sophistication of genetic resources and tools is contributing greatly to the fight against it. Well before the Human Genome Project, classic cytogenetics revealed gross deletions, amplifications, and rearrangements in cancer cells, changes that can now be analyzed at basepair resolution with technologies including single nucleotide polymorphisms and comparative genomic hybridization-based genomic arrays and next-generation massively parallel sequencing. Expression studies that

J. Kowalski (✉)
Department of Biostatistics and Bioinformatics, Emory University, Atlanta, GA 30322, USA
e-mail: jeanne.kowalski@emory.edu

W. Matsui
Department of Oncology, Division of Hematologic Malignancies, The Sidney Kimmel Comprehensive Cancer Center at Johns Hopkins, Baltimore, MD 21287, USA

W. Tang and X. Tu (eds.), *Modern Clinical Trial Analysis*, Applied Bioinformatics and Biostatistics in Cancer Research, DOI 10.1007/978-1-4614-4322-3_10,

were formerly done one gene at a time are now conducted across the entire transcriptome with 1.5 million feature expression arrays and can be used to evaluate the changes wrought by epigenetic therapeutics such as histone deacetylases and DNA methyltransferase inhibitors. Such genomic advances have resulted in novel therapeutic targets made possible by the discovery of genes whose expression can be changed with treatment. Additionally, they have led to a better understanding of the relationship between the expression of cancer-specific proteins and immune responses to treatment, as in the development of T cell-based cancer vaccine strategies. Along with advances made in the understanding of cancer at the molecular level, there has also been significant progress in understanding specific regulatory processes, most notably in the area of cancer stem cells. Cancer stem cells may help explain in part the dilemma of why so many successful Phase II trials ultimately fail to show improvements in overall survival, and novel anticancer stem cell agents have been developed with the actual potential to rid the patient of cancer cells.

All of these advances create a wealth of data and potential therapeutic targets that will require evaluation through clinical trials which in turn has prompted a reevaluation of set standards and procedures. For example, the optimal use of gene expression data in monitoring patient response to treatment, evaluation of the success of anticancer stem cell agents and cancer vaccines within the confines of short-term follow-up, and evaluation of the success of epigenetic therapies according to yet unknown timing of reexpression of genes. What is needed are not so much ways to improve upon conventional trial design, but rather new approaches to appropriately evaluate whether the treatments' outcomes from new therapies arising from such advances reach the targets they are intended. Moreover, a crucial issue is the accurate evaluation of the clinical impact of these novel treatments. In this chapter, we highlight two basic clinical trial concepts, endpoint definition and monitoring, within the context of the impact of advances made in genomics and cancer biology upon them, using immune-based, epigenetic, and cancer stem cell targeting therapies as illustration, as they represent nascent examples of the translation of such advances into clinical practice.

10.2 Background

In this section, we review key components in clinical trial design and their conventional applications in evaluating therapeutic cancer modalities.

10.2.1 Design

Conventional oncology drug development has evolved to investigate the efficacy of novel drugs within four phases, phase I to determine safety and dose, phase II to evaluate effectiveness and side effects, phase III to verify effectiveness, and phase IV for post-marketing surveillance. In cancer research, traditional Phase II trial designs primarily rely on short-term measurements that are focused on tumor response as a surrogate for long-term patient response.

The design of trials is only as informative as the endpoint chosen to evaluate the intervention. Simply stated, an endpoint is a measure used to evaluate whether a therapy is working or not. Current evaluation methods used in early phase cancer clinical trials rely primarily upon response criteria that reflect changes in tumor bulk. Since these criteria may not accurately reflect targeting of emerging therapies, atypical trail designs have been undertaken, including: (1) continuous administration of a drug followed by close monitoring to detect improved overall survival; (2) discontinuation of a drug followed by "watchful waiting" to evaluate time to relapse; (3) incorporation of new correlative laboratory assays that serially quantify changes in the drug intended target in patients undergoing treatment. The first option lacks feasibility given the large sample size and length of follow-up required. The second option is restricted to conditions where treatment may be stopped without patient harm, though not optimal for patient management, and as a "watch and wait" approach, it is also likely to require a lengthy follow-up. The third option relies on the use of correlative assays that are yet under development, and their operating characteristics and correlation with clinical outcomes have yet to be established and validated.

10.2.2 Endpoint

In general, an endpoint is a measure to determine whether a therapy is working or not. A biomarker (or biological marker) is a characteristic that is objectively measured and evaluated as an indicator of normal biological processes, pathogenic processes, or pharmacologic responses to a therapeutic intervention. A clinical endpoint is a characteristic that reflects a patients' response (e.g., how they feel, function, or survive). These endpoints are distinct measures of disease characteristics that reflect the effect of a therapeutic intervention. A surrogate endpoint, in brief, is a biomarker intended to substitute for a clinical endpoint. While all surrogate endpoints are biomarkers, not all biomarkers are surrogate endpoints, and in fact only a very few biomarkers may be considered for use as surrogate endpoints. For a biomarker to be considered as a surrogate endpoint, it is required to predict clinical benefit based on epidemiologic, therapeutic, pathophysiologic, or other scientific evidence. Additionally, the utility of a biomarker as a surrogate endpoint requires demonstration of its accuracy (the correlation of the measure with the clinical endpoint) and precision (the reproducibility of the measure).

One approach to establish the link between a biomarker and clinical endpoint is to estimate the proportion of treatment effect that is accounted for by the surrogate endpoint for which there are several ways to make this determination. At a very basic, statistical level, this involves two sets of regressions. The first regression is to examine the effect of treatment on outcome and if significant, to then examine the effect of both the treatment and the marker on the outcome. If the treatment effect is no longer significant, then the marker completely mediates the effect of treatment on outcome. If the treatment effect remains significant, then the marker partially mediates the effect of treatment on outcome, in which case, the proportion of variability in the outcome that is explained by the maker may be estimated.

Strictly speaking, for a surrogate endpoint to be valid as a clinical endpoint requires the ability of the biomarker to account for all of the effects of the intervention on the clinical outcome (endpoint). In practice however, it may be too much to ask of a single biomarker to fully capture all of a treatments' effect. To this end, the use of multiple biomarkers representing various components of complex disease pathways may yield surrogate endpoints that are more comprehensive in the ability to assess effects of therapeutic interventions.

10.2.3 Monitoring

In order to control patient's response to therapies and maintain proper doses, monitoring is necessary. The advent of the genomics era has prompted the potential use of molecular endpoints for monitoring patients' response to therapy. The ability of a trial to answer the posed clinical question depends on whether the marker endpoint is indeed a surrogate for the clinical outcome of interest. There is a clear need for guidelines on the use of gene transcript changes as a molecular endpoint to monitor and assess treatment response, as expanded upon below. Molecular techniques will play a large role in monitoring the progress of several cancers and in reassessing therapeutic strategies. Many groups are at work to create standards that can be used to cross validate results from individual labs, with their focus on creating a standardized assay that is universally accepted and used. Alternative strategies have been to create a series of statistical models whose results will be used to devise a monitoring strategy for implementation to achieve standardization of results. The potential use of molecular endpoints for monitoring of patients response to therapy will have tremendous impact, especially with the advent of new efficient drugs and combinations therapies, since results from clinical trials will be obtained faster than with a clinical endpoint, and in turn, clinicians will be able to adapt the new treatment faster.

10.3 Technology Influence: Molecular Targeted Therapies

In this section, we discuss how advances in the molecular understanding of cancer impact the design, monitoring, and analysis of clinical trials by highlighting chronic myeloid leukemia (CML).

10.3.1 Technology in Monitoring

CML is notable as the first human cancer to be associated with a recurrent genetic abnormality, the Philadelphia chromosome (Ph) (Nowell and Hungerford 1960; Rowley 1973). Moreover, the BCR-ABL product generated by the translocation of distinct genes located on chromosomes 9 and 22 was the first oncogenic fusion protein to be described (Ben Neriah et al. 1986). Finally, the development of the tyrosine kinase inhibitor (TKI), imatinib, represents the first and most successful therapy in clinical oncology that is precisely directed against the molecular abnormality responsible for the pathogenesis of the disease (Druker et al. 2001). CML also represents an ideal example to study the integration of genetically based testing strategies into diagnosis, monitoring, and the development of molecular endpoints in clinical trials.

Initial clinical monitoring strategies for CML utilized normalization of peripheral blood counts as this was typically the abnormality associated with the disease. With the development of more reliable tests to quantify cytogenetic abnormalities, the majority of later tests have focused on quantifying the Ph or BCR-ABL fusion gene within the bone marrow or peripheral blood of patients using metaphase cytogenetics or fluorescent in situ hybridization (FISH). More recently, even more sensitive quantification of BCR-ABL transcripts using RT-PCR-based techniques has been incorporated into routine clinical use (Hughes et al. 2006). Therefore, CML provides a clear example of how sophisticated molecularly based assays can be incorporated into disease monitoring strategies.

10.3.2 Technology Innovation: Impact

In the 1980s, the identification of BCR-ABL as a deregulated tyrosine kinase prompted the search for agents that could specifically inhibit its activity. These investigations eventually resulted in the development of small molecule inhibitors and approval of the first TKI, imatinib, as a treatment for CML in 2001 (Druker et al. 2001). Clinical experience with imatinib has suggested that long-term clinical outcomes, namely durability of responses to the drug, can be estimated by achievement of specific clinical milestones at defined points following the initiation of therapy (Hughes et al. 2006). These include the reduction in peripheral blood counts and eradication of the Ph detected by FISH. Data now also exist to support the importance of achieving molecular milestones, based upon quantitative RT-PCR, following the initiation of imatinib. Newly diagnosed chronic phase CML patients who achieved $a > 3$ log reduction in BCR-ABL transcripts at 12 months after initiation of imatinib have been found to experience improved progression-free survival (PFS) compared to patients who did not reach a 3 log decrease (Hughes et al. 2003). The progressive improvement in the sensitivity of disease monitoring has also enhanced the detection of minimal residual disease (MRD) (Hughes et al. 1991).

The majority of newly diagnosed CML patients treated with TKIs demonstrate total eradication of the Ph detected by standard metaphase cytogenetics or FISH. However, most of these Ph negative patients remain persistently positive for BCR-ABL transcripts detected using RT-PCR, and if TKI treatment is discontinued, patients uniformly relapse (Cortes et al. 2004; Hughes et al. 2003). Thus, increased sensitivity may provide meaningful information that can guide clinical decision making.

10.3.3 Technology Innovation: Challenges

With continued improvements in clinical monitoring, several questions regarding the use and interpretation of the results of these molecular assays have emerged. For example, does the magnitude of change in transcript levels measured by RT-PCR impact clinical outcomes (e.g., is a 3 log reduction as informative as a 4 log reduction)? What is more informative, the relative reduction or the absolute level of transcripts? Although specific results may correlate with clinical outcomes utilizing imatinib, will the predictive value of these results hold for newer generations of TKIs or other agents designed to eradicate MRD? What is the true clinical nature of these findings and do they correlate with overall survival in addition to PFS? Given the contemporary nature of molecular testing, the methods utilized to carry out these sensitive assays are continuously changing. Therefore, it is unclear within the context of ongoing modifications and enhancements of a specific assay whether similar results will correlate in an identical fashion with the previous version of the test. It is likely that the operating characteristics of specific assays will require continuous reassessment, especially in validating values obtained near the technical limits of each assays. For example, disease burden quantification at or near the limit of detection of the RT-PCR assay appears to have large sample to sample variability. A problem that is further complicated as the criteria of "undetectable" as an endpoint can be confounded by sample handling, since undetectable values can be due to low quality RNA or poor RNA yield and variations on the limit of detection from sample to sample or run to run.

The introduction of TKIs produced dramatic improvements in the overall response rate, time to response, and toxicity over prior treatments. However, with improvements in CML therapy, the low rate of disease progression has made it impractical to use clinical events that are undeniably important, such as overall survival or transformation to blast crisis, as primary endpoints in clinical trials because of the long periods of monitoring required for their accurate measurement. Therefore, more sensitive measures of disease monitoring, such as FISH or RT-PCR for BCR-ABL, have also become incorporated as endpoints in clinical trials that compare the efficacy of newer generations of TKIs to imatinib. However, despite the improvement in sensitivity of disease detection and incorporation of quantification of BCR-ABL by RT-PCR into almost all contemporary CML trials, the relationship between these results and clinical outcomes remains unclear, and this endpoint alone has only recently been accepted by the FDA as evidence supporting product registration for agents in CML. Although quantitative PCR testing of BCR-ABL in patients with CML has become the predominant molecular monitoring technique for CML therapy, it is regarded as a risky endpoint from a regulatory perspective, largely due to variations in results arising from the both sample processing and the assay itself. Moreover, accepted standards execution and quantification of results are lacking. Therefore, devising a monitoring strategy for CML patients is a challenge ideally suited for a comprehensive statistical approach, requiring the development of models fully informed by current biomedical knowledge, efficient inference that extracts maximal information, and a design that combines the population with individual data with no absolute.

10.4 Technology Influence: Immune-Based Therapies

In this section, we discuss the challenges raised in the design and monitoring of cancer clinical trials by immune-based therapies, such as cancer vaccines. Unlike preventative vaccines that are designed to protect against symptomatic infections, cancer vaccines are administered to treat an existing condition/disease. Such vaccines fall under one of two general types. Cell-based vaccines are created by isolating immune cells, such as T cells or antigen-presenting cells, from the patient, activating them ex vivo in the presence of cancer antigens, then delivering them back to the patient with or without additional factors that facilitate immune activation. In contrast, vector-based vaccines consist of engineered vectors (viruses or plasmids) encoding cancer proteins and other molecules to stimulate the immune system that are directly administered to the patient. In each case, the goal is to induce the patient's immune system into recognizing and eradicating tumor cells. In terms of clinical trial design, cancer vaccine testing fails to "fit" into the conventional four phases of trial design. First, these therapies typically display little to no serious toxicity risks nor follow a linear dose–response relationship, and conventional dose-escalation to establish the maximum tolerable dose is not required. Moreover, the optimal dose and schedule of vaccine administration cannot be determined during early phase testing through toxicity-based escalation schema. Cancer vaccines are also not metabolized, and there is little need for conventional pharmacokinetic correlates. Similarly, many cancer vaccines are designed to address one specific tumor type and trials studying mixed populations of cancer patients for optimal target identification and selection are superfluous. All of these factors may actually facilitate and accelerate early clinical trials. However, the relative benefits afforded by the safety of most immune-based therapies are balanced by the difficulties in actually evaluating and monitoring antitumor efficacy during these same early phases of clinical testing.

10.4.1 Technology in Monitoring

The monitoring of response to immunotherapeutic agents presents new challenges for several reasons. With cancer immunotherapy, the primary goal is similar to all other forms of cancer therapeutics, namely improvement in clinical outcomes. Similar to the clinical testing of conventional antineoplastic agents, surrogate measures of survival endpoints, such as tumor response, have been utilized to evaluate immune-based therapies during early phase testing. However, the effective induction of the host's (patient's) own immune system to react against tumor-specific antigens and ultimately eradicate tumors requires complex, dynamic, and highly coordinated cellular interactions. Therefore, the kinetics of response may be relatively slow, delayed, and highly variable. In fact, therapeutic trials have yet to show definitive evidence that cancer vaccines can actually elicit antitumor immunity that shrink tumors in a measurable or predictable fashion, yet patients who receive these vaccines tend to live longer and respond better to subsequent treatment. Therefore, a major question surrounding cancer vaccine testing remains whether we are appropriately measuring and monitoring immune reactivity using standard response criteria.

In their review, Schlom et al. (2007) examine two cell-based vaccines, Sipuleucel-T (Provenge) and GVAX, in addition to three trials using an engineered pox-virus vector. While their review article focuses on prostate cancer vaccines, the researchers consider these trials as examples of ongoing progress in similar vaccine therapies for lymphoma, melanoma, pancreatic, lung, and other types of cancer. According to their review of five prostate cancer vaccine trials, Schlom et al. (2007) offer evidence that patients who receive vaccines may respond better to subsequent chemotherapy or hormone treatment, leading to improved patient survival. However, the endpoints of these trials were not long-term survival but a reduction in tumor size. With this endpoint, such vaccines may be deemed ineffective and abandoned since the primary endpoint (tumor size reduction) was not achieved, despite their real and more beneficial therapeutic value in prolonging patient survival. These data prompt the rethinking of clinical vaccine trial design and in particular the current approach to measuring cancer vaccine activity. The inability of conventional short-term response criteria (e.g., response criteria in solid tumors, RECIST) to appropriately evaluate cancer vaccines in specific and immunotherapeutic agents in general is well recognized, and recently immune-related response criteria have been proposed (Wolchok et al. 2009; Ribas 2009).

In another review, Wolchok et al. (2009) examine anti-CTLA-4 immunotherapy (ipilimumab) as monotherapy for patients with advanced melanoma in three multi-center Phase II clinical trials. The immunotherapeutic agent, ipilimumab (Yervoy), was designed to target CTLA-4 by preventing its binding to another molecule that inhibits T cell activation, thus allowing for enhanced immune activation by expansion of T cells for antitumor activity. In all three studies four patterns of clinical responses to ipilimumab therapy could be identified. In two of these, response patterns were based on conventional response criteria: (1) response in baseline lesions by week 12, with no new lesions (i.e., a reduction in tumor burden) and (2) stable disease (i.e., inhibition of disease progression). However, the other two response patterns were novel: (1) responses after an initial increase in total tumor burden and (2) reduction in total tumor burden during or after the appearance of a new lesion later than 12 weeks after therapy. Wolchok et al. (2009) provide evidence that responses observed with immunotherapeutic agents may extend beyond those of cytotoxic agents and could include responses after disease progression that are not captured by response evaluation criteria in solid tumors or WHO criteria. Therefore, monitoring for antitumor activity may extend well beyond the time frames required to evaluate standard cytotoxic agents and biomarkers indicative of early vaccine activity are required. However, the ability to accurately define tumor-specific immune reactivity and identify related markers capable of providing evidence of activity during clinical testing of cancer vaccines poses a major challenge.

In addition, the realization that the active antitumor immunity may involve the recognition of multiple antigens generates additional complexity. To this end, innovations in technology have considerably improved the ability to identify and catalogue all proteins recognized following vaccination, but the ability to distinguish critical antigens among the large number identified has raised new difficulties.

10.4.2 Technology Innovation: Impact

The treatment of cancer is increasingly practiced using multiple modalities, and immunotherapeutic approaches are gaining high interest as evidenced by the approvals of both sipuleucel-T and ipilimumab. One major goal of tumor immunotherapy has been to develop more effective methods to identify specific tumor antigens. These defined antigens may be useful for the preparation of vaccines with improved tumor specificity and provide reagents required to effectively monitor the generation of antitumor immunity. For these reasons, research in cancer immunotherapy has recently focused on identifying, isolating, and cloning specific tumor-associated antigens that can be recognized by the host immune system. Many antigens have been identified by using tumor cells as antigens–immunogens to stimulate the production of specific antibodies in laboratory animals. Another approach is to identify antibodies that are formed by the patients themselves against known tumor antigens, such as p53 mutants, HER-2/neu, CEA, and PSA. Their identification, however, is difficult when using conventional methods.

Although there is a long history of analyzing serologically defined tumor antigens, the field was revolutionized by a technique termed *SEREX*, for serological analysis of recombinant cDNA expression libraries of human tumors with autologous serum (Tureci et al. 1997) Here, serum from patients is used to screen cDNA expression libraries, and reactive clones are sequenced to identify antigens recognized by the humoral immune system. During the initial development of this method, tumor antigens, such as MAGE and tyrosinase that had been previously identified as T-cell-recognized epitopes, were detected by autologous antibody that validated this approach. SEREX analysis has since identified a series of provocative cancer antigens that may have relevance to the etiology, diagnosis, and therapy of cancer. Indeed, the method has been so prolific that the antigens identified by SEREX are now catalogued online with the database currently containing some 2743 sequences and 2316 clones

Given the rarity of clear-cut immunemediated tumor responses, a critical limitation of all methods of tumor antigen identification has been the source of the starting material to be used. Most T cell approaches have either used TILs or peripheral blood T cells from patients who may or may not have evidence of an ongoing clinical response. Similarly, the majority of reports using SEREX have simply been a cataloging of antibody specificities found in cancer patients as compared to normal subjects, together with an analysis of the antigen expression pattern in normal versus transformed tissues. While the identification of the targets being recognized is instructive, it is difficult to interpret what role immunity to a given set of antigens plays in the clinical course of the disease. Accordingly, the antibody response has the potential to provide a more comprehensive picture of the targets of recognition, enabling comparisons across patients in settings that can be linked to disease response. Therefore, a main objective in immunotherapy is not only to generate a broad picture of tumor antigens specific to a particular cancer but also to correlate immune recognition of such antigens with clinical response to tumor cell-based immunotherapy.

10.4.3 Technology Innovation: Challenges

The SEREX technology undoubtedly has the potential to identify new tumor-specific antigens, but it presents a number of drawbacks such as the laborious nature of the library screening operations, the high degree of background noise, the large amounts of material necessary, and most importantly, the challenge of being able to develop an analytical approach to distinguishing relevant targets from "bystanders," since adoption into routine use SEREX has had altered the dilemma of having too few identified antigens to too many for use in immunotherapeutic development.

The application of the SEREX technology to identify target antigens, while straightforward, also creates a challenge for analyses of results for a number of reasons. For a given patient, in terms of results, a gene list is obtained whose expression is recorded as either induced or increased. However, unlike other related technologies, such as microarrays, where a same set of genes is tested on all patients, in this case, both the number and actual genes vary according to each patient, some of which may overlap and at the extreme, some of which may be completely distinct among patients. A goal is to identify genes that are characteristic of "responders," as determined by clinical outcomes criteria applied to all patients. To this end, in the simplest case, gene sets among patients may be examined to identify those in common, but this approach is likely to miss important genes, and given that both the number and type of genes vary by patient, a null set is likely.

Additionally, this approach, though simple, does not incorporate patient variability in results. To address these limitations associated with an "intersecting among all patients" approach, one proposal is to instead examine genes in common among patient pairs. While it seems unreasonable to expect overlap in genes among all patients within a clinical response group, it does, however, seem reasonable to expect some overlap in patient pairs within each group. Therefore, one may estimate the proportion of patient pairs with genes in common and compare this proportion between "responders" and "nonresponders" Since the unit of analysis in this case is patient pairs that although identically distributed are not independent, U-statistics are required to address such analyses (Kowalski and Tu 2008).

Alternative to a change observed in a single marker, immunotherapy trials prompt the need for multiple biomarkers in which immune reactivity as a "signal of response" may be defined in terms of clinical response, biologic activity, or immune response. For biologic activity, potential measures may include regulatory T-cell activity or immune response against target cells, or molecular response (MRD). An immune profile may be assessed by sequential samples collected over time points (e.g., baseline, follow-up visits) to assess reproducibility of assay results. As for clinical activity, there is no current mandate to demonstrate clinical activity with conventional oncology endpoints in proof-of-principle trials and thus, typically no end-stage patients and a homogenous population are selected. As for surrogate endpoints in trials with cancer vaccines, molecular response is being considered. Cancer vaccines are expected to work best in MRD population. Molecular markers that allow uniform assessment of MRD and the impact of a vaccine on the target disease may function as a measure of biological and/or clinical activity. Some examples include CML, with a well-defined chromosomal abnormality (BCR-ABL) that is detectable by RT-PCR and acute myeloid leukemia (AML), where multiple heterogeneous chromosomal abnormalities are not present in all patients, requiring an array of markers to determine biological activity.

10.5 Technology Influence: Cancer Stem Cell Therapies

In this section, we discuss issues impacting the development of strategies to monitor clinical trials designed to target cancer stem cells. Cancer stem cells represent an emerging concept in oncology that suggests that tumor growth is primarily driven by phenotypically and functionally distinct cell populations within an individual tumor (Dalerba et al. 2007). Moreover, cancer stem cells have been hypothesized to play a major role in disease relapse and progression, and the development of effective strategies targeting these cells may improve long-term outcomes (Huff et al. 2006b). In contrast to the preceding sections in which novel molecularly targeted therapies and cancer vaccines were discussed, therapies directed against cancer stem cells aim to eliminate specific cell types rather than every cell within an individual tumor. Therefore, the ability to monitor clinical trials targeting cancer stem cells presents a series of challenges that must take into account the specificity of the targeting strategy, the clinical context, and availability of reliable biomarkers strategies.

10.5.1 Technology in Monitoring

It has been long recognized that individual tumors are composed of phenotypically heterogeneous cells despite their clonal origins. Moreover, in many cancer systems, tumorigenic potential does not appear to be a property that is universally shared, but rather limited to a small proportion of cells. Pioneering studies in AML and CML synthesized these data in the 1990s and demonstrated that tumorigenic potential was limited to phenotypically distinct cell populations (Lapidot et al. 1994; Sirard et al. 1996). In these studies, leukemic blasts displaying some aspects of normal myeloid maturation were unable to propagate disease in immunodeficient mice, but tumor cells resembling normal hematopoietic stem cells could engraft and produce disease that was histologically identical to the original tumor. Moreover, the tumors forming in these animals could continue to produce tumors when they were serially transplanted into secondary recipients. Therefore, cancer stem cells are capable of self-renewal, the critical property that maintains the malignant clone over time. These findings broadly impacted tumor biology in several ways. First, they suggested that human tumors may be cellularly organized in a manner paralleling normal tissues with cancer stem cells positioned at the top of a cellular hierarchy and giving rise to more differentiated cells. Moreover, cancer cells could be divided into two broad categories each with distinct clinical implications. Mature tumor cells phenotypically characterize the tumor and are primarily responsible for inducing clinical symptoms. In contrast, relatively primitive cancer stem cells mediate the production of new tumor cells. Finally, the phenotypic similarities between leukemic and normal stem cells suggested that cancers may arise from normal stem cells. Since these initial studies in leukemia, cancer stem cells have been identified in a wide number of human malignancies, including solid tumors (Dalerba et al. 2007). Moreover, cancer stem cells have also been found to be relatively resistant to standard treatment approaches including cytotoxic chemotherapeutic agents and radiation (Bao et al. 2006; Dylla et al. 2008; Matsui et al. 2008) .Therefore, their persistence following treatment coupled with their long-term proliferative potential suggest that cancer stem cells may mediate disease relapse, and therapies that can target and eliminate cancer stem cells may improve long-term survival rates.

10.5.2 Technology Innovation: Impact

The potential that cancer stem cells are responsible for disease relapse and long-term outcomes has generated great interest in developing novel therapeutic strategies that can target and eliminate these drug-resistant and clonogenic cells. For example, several cellular pathways required for proper development of the normal embryo, such as the Hedgehog and Notch signaling pathways, have been hypothesized to regulate stem cells across many human cancers (Reya et al. 2001). Inhibitors of these pathways have been developed and clearly clinical trials testing their toxicity and efficacy have begun (Von Hoff et al. 2009). In most diseases, cancer stem cells are thought to represent a minority of all tumor cells. Therefore, a major challenge in these cancer stem cell targeting clinical trials is the ability to detect efficacy against cells that represent a small proportion of the total tumor burden. This is especially challenging during early phase clinical trials that have traditionally been dependent upon reductions in tumor burden as measures of efficacy. The loss of clonogenic activity by effective cancer stem cell targeting should result in prolonged relapse-free or overall survival. However, these endpoints typically require large cohorts of patients and long periods of follow up. Thus, a major challenge is the development of surrogate endpoints that may be able to detect and quantify the inhibition of cancer stem cells within relatively short time frames.

In many diseases, discrepancies exist between early response markers, such as reduction in tumor burden, and long-term clinical outcomes, especially overall survival (Huff et al. 2006a). The cancer stem cell hypothesis may explain this dilemma, as strategies that target bulk tumor cells would be deemed "effective" and advanced into later stage clinical trials. However, if these treatments fail to impact cancer stem cells that are responsible for disease relapse and progression, then it is likely that they would ultimately improve survival rates. As mentioned above the clinical benefit of therapeutic strategies that actually target cancer stem cells may be missed within early phase clinical trials if cancer stem cells do not contribute to tumor bulk. However, if sufficient time were to pass, the inhibition of new tumor production may be eventually detected as clinical responses. Therefore, effective monitoring methods that are independent of tumor bulk and able to accurately quantify cancer stem cells may provide proof that a particular therapeutic targeting strategy is potentially active and warrants testing in mature clinical trials designed to evaluate survival outcomes.

Cancer stem cells have typically been identified using specific markers that can distinguish them from the remaining cells that make up the tumor bulk. These have included cell surface antigen expression as well as biochemical and biophysical properties that appear to be shared across a number of tumor types as well as with normal self-renewing stem cells (Brennan and Matsui 2009). However, the quantification of tumor cells expressing these stem cell-specific markers may be confounding as no one individual or set of markers has been found to totally encapsulate all cells capable of clonogenic growth in any disease. Another strategy to quantify cancer stem cell frequency is to measure their functional capabilities, rather than their phenotypic properties. These may include their ability to form tumors in immunodeficient mice or their ability to form colonies or tumor spheres in vitro. These functional assays have the advantage of more specifically measuring the clinically relevant property in question, but they are typically more complex than the identification and enumeration of tumor cells displaying a specific phenotype. Few examples serially quantifying cancer stem cells in patients undergoing a defined treatment are available, but examples exist with both hematologic malignancies and solid tumors. In breast cancer, the quantification of phenotypic breast cancer stem cells by flow cytometry or clonogenic growth in patients receiving standard neoadjuvant therapy has provided support that cancer stem cells are resistant in this disease (Li et al. 2008). In multiple myeloma, changes in the relative frequency of cancer stem cells in patients undergoing treatment appear to predict PFS (Huff et al. 2008). Although these studies do not provide definitive evidence that the frequency of cancer stem cells is associated with outcomes, they do suggest that it is possible to carry out relatively complex correlative assays in the clinical setting. Therefore, a clear reduction in overall cancer stem cell burden during early phase clinical trials examining novel targeting strategies may encourage further studies examining survival endpoints regardless of the impact on overall tumor burden.

These biomarker strategies may be especially useful when examining novel therapies primarily designed to target the self-renewal potential of cancer stem cells. In most normal systems, the capacity for self-renewal is inversely related to maturation statues. For example, in normal hematopoiesis, mature blood cells with effector functions that include protection against infectious agents (neutrophils), oxygen-carrying capacity (erythrocytes), and maintenance of vascular integrity (platelets) are incapable of proliferation. In contrast, primitive, self-renewing hematopoietic stem cells that lack these functional capabilities are primarily involved in maintaining blood production over the lifetime of the individual. Many of the strategies that appear to inhibit self-renewal appear to induce the differentiation of cancer stem cells (Matsui et al. 2003; Peacock et al. 2007). Therefore, these strategies may actually increase the production of mature tumor cells following their initiation that would be paradoxically inferred as disease progression, and biomarkers that can characterize effects on cancer stem cells may provide evidence of efficacy.

10.5.3 Technology Innovation: Challenges

Although cancer stem cells have been prospectively identified in a wide variety of human malignancies, the clinical relevance of these findings remains unclear. The ability to produce tumors in the ectopic setting, either in vivo in immunodeficient mice or in vitro in colony or sphere-forming assays, has provided the functional basis to support claims that specific cell populations are clonogenic. However, the extent that these artificial assay systems recapitulate the true biological system within patients is unknown. As described above, the development of biomarkers that can serially quantify cancer stem cells are needed as surrogate endpoints during early clinical trials targeting cancer stem cells. Moreover, these biomarkers may also provide clinical evidence supporting the cancer stem cell hypothesis. If cancer stem cells are truly responsible for disease relapse and progression, then their relative frequency should be associated with long-term outcomes. This relationship has been demonstrated in brain tumors, breast cancer, and pancreatic adenocarcinoma (Ginestier et al. 2007; Rasheed et al. 2010; Zeppernick et al. 2008). Moreover, it is possible that cancer stem cells vary among individual patients with the same diagnosis and that specific cellular features may dictate clinical outcomes. For example, a study in breast cancer has suggested that a specific gene signature identified within breast cancer stem cells by gene expression profiling is associated with long-term outcomes (Liu et al. 2007). Therefore, the development of biomarkers may both provide evidence that the cancer stem cell hypothesis is clinically relevant as well as essential tools for monitoring clinical trials

10.6 Technology Influence: Epigenetic-Based Therapies

In this section, we discuss the impact of genomic advances on agents targeting epigenetic modifications in cancer. Epigenetics is the study of mechanisms that alter a gene's function without requiring a change in its sequence. Like an on and off switch, they change a gene's expression, turning its protein production on and off, without changing the protein product itself, i.e., transcriptional regulation. These mechanisms, which include DNA methylation changes in the chromatin structure, and expression of noncoding RNAs are partially responsible for the differences in gene expression from diverse tissues that share the same DNA sequence. When such mechanisms malfunction, they may give rise to diseases, including cancer, and in as much may be a distinguishing feature of cancer itself.

10.6.1 Technology in Monitoring

A signature of epigenetic alterations is gene silencing, turning off production of the gene's messenger RNA, which is the template or code for their proteins. Tumor suppressor genes (TSGs) help cells die at the appropriate time and prevent their uncontrolled growth that is characteristic of cancer. In cancer cells, TSGs are often either deleted or silenced and therefore unavailable to fulfill their function. Since suppressor gene dysfunction permits the survival of cancer cells, if one can restore normal gene function the cancer cells should die off, ultimately rendering one cancer free. This approach to cancer treatment is the foundation of epigenetic therapy, i.e., "waking up" TSGs that are physically present, but have gone silent. Specifically, epigenetic treatment uses enzymes that regulate histone acetylation and DNA methylation, processes required for epigenetic switching. Thus, an epigenetic therapy target is a gene whose expression can be changed by treatment. TSGs make good targets as some that have been turned off by epigenetic silencing can be turned on with treatment. This therapeutic approach offers potential options to patients whose cancers have become unresponsive to other drugs and, since it typically employs very low drug doses (versus conventional chemotherapies' maximum tolerable doses), patients may experience fewer side effects.

Epigenetic therapies are currently being applied successfully in translational research studies of patients with very poor prognosis myeloid malignancies. Remarkably, the biological targets of these effective therapies are largely unknown. The methylation inhibitor, decitabine (DAC), has shown therapeutic efficacy in poor prognosis AML patients and been found to induce gene expression changes in

AML cell lines. With similar pharmacology and perhaps even better characterized therapeutic benefits, azacytidine (AZA) is known to activate expression of once silenced genes in cancer cell lines derived from multiple tumor types.

10.6.2 Technology Innovation: Impact

While TSGs are attractive therapeutic targets, many are as yet undiscovered. A noteworthy characteristic of TSGs is that they express very low amounts of message, which makes these genes difficult to discover, and makes their response to epigenetic therapy hard to define, let alone monitor based solely on microarray expression-measuring technology. A clinical trials dilemma with such a therapeutic approach is how to identify genes that are meaningful to response as epigenetic modifying agents influence global changes in gene transcription, prompting questions of whether to focus directly at epigenetic modification of a single gene, group of genes, or across the entire genome, with additional consideration of whether to examine DNA methylation, histone/chromatin modifications, the timing of such modifications, and how to integrate their results with gene expression.

The monitoring of responses to epigenetic therapies, such as HDAC and DNMT inhibitors, motivates not only questions of which genomic change to consider and when to examine them, but equally important, new bioinformatic approaches and tools to address the confluence of molecular changes, their relationship among each other, and to clinical response. As agents designed to "wake up" genes are silenced through methylation, responses to HDAC and DNMT inhibitors may be examined by genes' reexpression and herein lies an issue, i.e., what expression profile defines a TSG or stated another way, how to use microarray gene expression arrays to identify TSGs?

A disconnect exists between biological and biostatistical principles in evaluating genes based on data from biotechnologies, such as microarrays, for candidate TSGs. On the one hand, a key biological feature of cancer genes that have been modified by epigenetic silencing is that they are turned virtually off in cancer, which, in the case of microarray expression data, is indicated by very low expression. On the other hand, such genes are unable to be directly filtered and evaluated, and oftentimes, are even discarded by current analytical methods, due primarily to the difficulty in discerning very low expression signal levels from that of noise. Consider three genes whose expression shows the same large fold-change between pharmacologically treated and untreated sample cancer tissues, but which differ greatly in their absolute expression levels. A fold-change (treated/untreated) of four may be obtained from arbitrary expression level ratios of 40/10 (gene 1), as well as from 16/4 (gene 2) or 4/1 (gene 3).

Expression differences such as that of genes 1 and 2, though readily detectable and favored by most analytical tools, are from a biologic standpoint, not relevant as a TSG. By comparison, the expression pattern of gene 3 illustrates the following defining characteristics: (1) low expression in control relative to other genes; (2) low expression in treated relative to other genes; and (3) large fold-change between treated and untreated cells relative to other genes. The characteristics of low expression, large-fold change are most likely to represent a TSG, since as a subset of cancer genes, TSGs tend to regulate cell cycles or have other functions that call for low level expression in the cell.

10.6.3 Technology Innovation: Challenges

Expression levels, as mRNA concentrations, are often measured by gene expression microarrays, but this and other technologies are subject to certain limitations. It would be difficult, because of the fine line between low expression and noise, to filter gene expression microarray data for such genes. The difficulty of this task is in obtaining an accurate and precise estimate of noise, generated by many documented biological and technological sources some of which are unobservable, and in defining a threshold for low level expression, thus prompting the need for new bioinformatic tools that filter expression data based on a TSGs' biological profile. As an alternative screening strategy, Kowalski et al. (2009) propose filtration of such expression data according to a defined empirical expression profile that is an exemplar of a TSG and as such, it inherently represents a TSGs' biological characteristic. With this strategy, the challenging task of having to directly address low level expression in the presence of noise is circumvented; a task would otherwise be required based on current approaches.

Gene expression aside, there is yet another issue on methylation changes, in terms of the timing of when to conduct microarray gene expression and/or tiling and related arrays on samples in order to capture the occurrence of the event, "reversal of methylation." The event, "reversal of methylation," may be identified in a number of ways, with as yet, no consistent criteria or guidelines. In terms of microarray gene expression arrays, the effect of methylation reversal is gene reexpression, the profile of which has been previously discussed, along with the limitation in both technology and bioinformatic tools with which to address. While tiling arrays may be used to examine genome-wide methylation changes, it remains a costly technology and its use requires some differential methylation assay prior to the tiling portion. The possibility of combining expression alterations with methylation data is appealing and quite logical, but to do so effectively requires the use of validated methylation assays, some of which remain under development (e.g., Help), and the approach used to filter low level, high-fold expression changes characteristic of previously silenced genes. Moreover, there is the possibility of genes whose expression may be delayed in response to a gene being "turned on," thus requiring a longer follow-up time for the microarray gene expression array component to be considered as part of epigenetic therapy clinical trial design.

While success with most epigenetic therapies is defined by achieving reversal of methylation, an event which has yet to be defined consistently within and among various biotechnologies, there remains the correlation of this event with clinical response. Berg et al. (2007) reported reversal of p15 methylation associated with reexpression of p15 protein, in a subset of patients treated with DAC, a DNMT inhibitor. Those patients in whom methylation reversal was documented also manifested a response to therapy; however, disease responses were also seen in a group of patients who did not manifest hypermethylation of p15 and in whom no methylation reversal was documented, suggesting that another mechanism might also be implicated. Another study by Issa et al. (2004) also examined reversal of p15 methylation following treatment with low dose DAC, but in this case no evidence of significant methylation reversal was seen. These differing reports highlight that the mechanisms underlying the activity of DNA methyltransferase inhibitors remain unclear. Additionally, it underscores the need for well-designed clinical studies that include correlative laboratory data at key time points throughout a patient's therapy.

10.7 Future Directions

We discuss the future direction of cancer trials by considering the combined roles of biostatistics and bioinformatics in cancer research, using cancer stem cell and vaccine trials as premier examples of issues that arise in translating knowledge into clinical practice. Gene expression microarrays and gene expression databases provide new opportunities for the discovery of drug targets and for determining a drug's mode of action. Bioinformatics provides the computational tools by which to extract this information. Biostatistics provides the analytical tools to incorporate this information into clinical trial design and analysis.

The demand for new and improved biomarkers is a reflection of the emerging drug development due in part to genomic advances leading to a better understanding of the disease processes. There are many new ways to discover drugs in light of gene target discoveries. The use of molecular targets to design new chemical compounds has provided many new candidates for testing, resulting in a pressing need for more efficient ways to design trials. The identification of highly precise and accurate biomarkers could enable the testing of more candidate drugs, reduce the number of patients required to conduct trials, expand on our capacity to predict adverse events, and potentially improve regulatory decision making. In oncology in particular, we are entering an era of sophistication in making more precise diagnoses and more informed choices about therapy.

Acknowledgments This work was supported by a Leukemia and Lymphoma Society grant LLS 6094–10 and a Georgia Cancer Coalition distinguished scholar award.

References

Bao S, Wu Q, McLendon RE, Hao Y, Shi Q, Hjelmeland AB, Dewhirst MW, Bigner DD, Rich JN (2006) Glioma stem cells promote radioresistance by preferential activation of the DNA damage response. Nature 444:756–760

Ben Neriah Y, Daley GQ, Mes-Masson AM, Witte ON, Baltimore D (1986) The chronic myelogenous leukemia

Berg T, Guo Y, Abderlkarim M, Fliegauf M, Lubbert M (2007) Reversal of p15/INK4b hypermethylation in AML1/ETO-positive and negative myeloid leukemia cell lines. Leuk Res 31(4):497–506

Brennan S, Matsui W (2009) Cancer stem cells: Controversies in multiple myeloma. J Mol Med 87(11):1079–85

Cortes J, O'Brien S, Kantarjian H (2004) Discontinuation of imatinib therapy after achieving a molecular response. Blood 104:2204–2205

Dalerba P, Cho RW, Clarke MF (2007) Cancer stem cells: models and concepts. Ann Rev Med 58:267–284

Dakalakis M, Nguyen TT, Nguyen C, Guldberg P, Kohler G, Wijermans P, et al. (2002) Demethylation of a hypermethylated P15/INKI4B gene in patients with myelodysplastic syndrome by 5-aza-2' deoxycytidine (decitabine) treatment. Blood 100(8):2957–2964

Druker BJ, Talpaz M, Resta DJ, Peng B, Buchdunger E, Ford JM, Lydon NB, Kantarjian H, Capdeville R, Ohno-Jones S, Sawyers CL (2001) Efficacy and safety of a specific inhibitor of the BCR-ABL tyrosine kinase in chronic myeloid leukemia. New Engl J Med 344:1031–1037

Dylla SJ, Beviglia L, Park IK, Chartier C, Raval J, Ngan L, Pickell K, Aguilar J, Lazetic S, Smith-Berdan S et al. (2008) Colorectal cancer stem cells are enriched in xenogeneic tumors following chemotherapy. PLoSONE 3:e2428

Ginestier C, Hur MH, Charafe-Jauffret E, Monville F, Dutcher J, Brown M, Jacquemier J, Viens P, Kleer CG, Liu S et al. (2007) ALDH1 is a marker of normal and malignant human mammary stem cells and a predictor of poor clinical outcome. Cell Stem Cell 1:555–567

Huff CA, Matsui W, Smith BD, Jones RJ (2006a) The paradox of response and survival in cancer therapeutics. Blood 107:431–434

Huff CA, Matsui WH, Douglas SB, Jones RJ (2006b) Strategies to eliminate cancer stem cells: clinical implications. Eur J Cancer 42:1293–1297

Huff C, Wang Q, Rogers K, Jung M, Bolanos-Meade J, Borrello I, Jones R, Matsui W (2008) Correlation of clonogenic cancer stem cell (CSC) growth with clinical outcomes in multiple myeloma (MM) patients undergoing treatment with high dose cyclophosphamide (Cy) and rituximab. AACR Meeting Abstracts 2008, LB–87

Hughes T, Morgan G, Martiat P, Goldman J (1991) Detection of residual leukemia after bone marrow transplant for chronic myeloid leukemia: role of polymerase chain reaction in predicting relapse. Blood 77:874–878

Hughes T, Deininger M, Hochhaus A, Branford S, Radich J, Kaeda J, Baccarani M, Cortes J, Cross NCP, Druker BJ, et al. (2006) Monitoring CML patients responding to treatment with tyrosine kinase inhibitors: review and recommendations for harmonizing current methodology for detecting BCR-ABL transcripts and kinase domain mutations and for expressing results. Blood 108:28–37

Hughes TP, Kaeda J, Branford S, Rudzki Z, Hochhaus A, Hensley ML, Gathmann I, Bolton AE, van Hoomissen IC, Goldman JM, et al. (2003) Frequency of major molecular responses to imatinib or interferon alfa plus cytarabine in newly diagnosed chronic myeloid leukemia. New Engl J Med 349:1423–1432

Issa JP, Garcia-Manero G, Giles FJ, Mannari R, Thomas D, Faderl S, et al. (2004) Phase I Study of Low-Dose prolonged exposure schedules of the Hypomethylating Agen 5-aza-2'-deoxycytidine (decitabine) in hematopoietic malignancies. Blood 103(5):1635–1640

Kowalski J, Tu X.M (2008) Modern Applied U-Statistics. Wiley, New York

Kowalski J, Talbot C, Tsai HL, Prasad N, Umbricht C, Zeiger M (2009) From ambiguities to insights in cancer diagnosis via query-based comparisons. Pattern Recogn 42:575–580

Lapidot T, Sirard C, Vormoor J, Murdoch B, Hoang TC-CJ, Minden M, Paterson B, Caligiuri MA, Dick JE (1994) A cell initiating human acute myeloid leukaemia after transplantation into SCID mice. Nature 367:645–648

Li X, Lewis MT, Huang J, Gutierrez C, Osborne CK, Wu MF, Hilsenbeck SG, Pavlick A, Zhang X, Chamness GC, et al. (2008) Intrinsic resistance of tumorigenic breast cancer cells to chemotherapy. JNCI J N Canc Inst 100:672–679

Liu R, Wang X, Chen GY, Dalerba P, Gurney A, Hoey T, Sherlock G, Lewicki J, Shedden K, Clarke MF (2007) The prognostic role of a gene signature from tumorigenic breast-cancer cells. New Engl J Med 356:217–226

Matsui W, Huff CA, Vala M, Barber J, Smith BD, Jones RJ (2003) Anti-tumour activity of interferon-alpha in multiple myeloma: role of interleukin 6 and tumor cell differentiation. Br J Haematol 121:251–258

Matsui W, Wang Q, Barber JP, Brennan S, Smith BD, Borrello I, McNiece I, Lin L, Ambinder RF, Peacock C, et al. (2008) Clonogenic multiple myeloma progenitors, stem cell properties, and drug resistance. Canc Res 68:190–197

Nowell PC, Hungerford DA (1960) Minute chromosome in human chronic granulocytic leukemia. Science 132:1497–1497

Peacock CD, Wang Q, Gesell GS, Corcoran-Schwartz IM, Jones E, Kim J, Devereux WL, Rhodes JT, Huff CA, Beachy PA, et al. (2007) Hedgehog signaling maintains a tumor stem cell compartment in multiple myeloma. Proc Natl Acad Sci 104:4048–4053

Rasheed ZA, Yang J, Wang Q, Kowalski J, Freed I, Murter C, Hong S-M, Koorstra J-B, Rajeshkumar NV, He X, et al. (2010) Prognostic significance of tumorigenic cells with mesenchymal features in pancreatic adenocarcinoma. JNCI 102(5):340–51

Reya T, Morrison SJ, Clarke MF, Weissman IL (2001) Stem cells, cancer, and cancer stem cells. Nature 414:105–111

Ribas A, Chmielowski B, Glaspy JA (2009) Do we need a different set of response assessment criteria for tumor immunotherapy? Clin Cancer Res 15(23): 7116–7118

Rowley JD (1973) Letter: A new consistent chromosomal abnormality in chronic myelogenous leukaemia identified by quinacrine fluorescence and Giemsa staining. Nature 243:290–293

Schlom J, Arlen PM, Gulley JL (2007) Cancer vaccines: moving beyond current paradigms. Clin Cancer Res 13(13):3776–3782

Sirard C, Lapidot T, Vormoor J, Cashman JD, Doedens M, Murdoch B, Jamal N, Messner H, Addey L, Minden M, et al. (1996) Normal and leukemic SCID-repopulating cells (SRC) coexist in the bone marrow and peripheral blood from CML patients in chronic phase, whereas leukemic SRC are detected in blast crisis. Blood 87:1539–1548

Tureci O, Sahin U, Pfreundschuh M (1997) Serological analysis of human tumor antigens: molecular definition and implications. Mol Med Today 3(8):342–349

Von Hoff D, Lorusso P, Rudin C, Reddy J, Yauch R, Tibes R, Weiss G, Borad M, Hann C, Brahmer J, et al. (2009) Inhibition of the hedgehog pathway in advanced basal-cell carcinoma. New Engl J Med 361(12):1164–1172

Wolchok JD, Hoos A, O'Day S, Weber JS, Hamid O, Lebbé C, Maio M, Binder M, Bohnsack O, Nichol G, Humphrey R, Hodi FS (2009) Guidelines for the evaluation of immune therapy activity in solid tumors: Immune-related response criteria. Clin Cancer Res 15(23):7412–7420

Zeppernick F, Ahmadi R, Campos B, Dictus C, Helmke BM, Becker N, Lichter P, Unterberg A, Radlwimmer B, Herold-Mende CC (2008) Stem Cell Marker CD133 Affects Clinical Outcome in Glioma Patients. Clin Canc Res 14:123–129

Index

W. Tang and X. Tu (eds.), *Modern Clinical Trial Analysis*, Applied Bioinformatics and Biostatistics in Cancer Research, DOI 10.1007/978-1-4614-4322-3,

MIX
Papier aus verantwortungsvollen Quellen
Paper from responsible sources
FSC® C105338
FSC
www.fsc.org

If you have any concerns about our products,
you can contact us on
ProductSafety@springernature.com

In case Publisher is established outside the EU,
the EU authorized representative is:
Springer Nature Customer Service Center GmbH
Europaplatz 3, 69115 Heidelberg, Germany

Printed by Libri Plureos GmbH
in Hamburg, Germany